Lecture Notes in Computer Science 1432

Edited by G. Goos, J. Hartmanis and J. van Leeuwen

T0223210

Springer

Berlin
Heidelberg
New York
Barcelona
Budapest
Hong Kong
London
Milan
Paris
Singapore
Tokyo

Stefan Arnborg   Lars Ivansson   (Eds.)

Algorithm Theory – SWAT'98

6th Scandinavian Workshop
on Algorithm Theory
Stockholm, Sweden, July 8-10, 1998
Proceedings

Springer

Series Editors

Gerhard Goos, Karlsruhe University, Germany
Juris Hartmanis, Cornell University, NY, USA
Jan van Leeuwen, Utrecht University, The Netherlands

Volume Editors

Stefan Arnborg
Lars Ivansson
Nada, KTH
SE-100 44 Stockholm, Sweden
E-mail: {stefan, ivan}@nada.kth.se

Cataloging-in-Publication data applied for

Die Deutsche Bibliothek - CIP-Einheitsaufnahme

Algorithm theory : proceedings / SWAT '98, 6th Scandinavian
Workshop on Algorithm Theory, Stockholm, Sweden, July 8 - 10,
1998. Stefan Arnborg ; Lars Ivansson (ed.). - Berlin ; Heidelberg ;
New York ; Barcelona ; Budapest ; Hong Kong ; London ; Milan ;
Paris ; Santa Clara ; Singapore ; Tokyo : Springer, 1998
    (Lecture notes in computer science ; Vol. 1432)
    ISBN 3-540-64682-5

CR Subject Classification (1991): F.2, G.2, E.1-2, F.1, G.3, I.3.5

ISSN 0302-9743
ISBN 3-540-64682-5 Springer-Verlag Berlin Heidelberg New York

Typesetting: Camera-ready by author
SPIN 10637621   06/3142 – 5 4 3 2 1 0   Printed on acid-free paper

Preface

The papers in this volume were presented at SWAT '98, the Sixth Scandinavian Workshop on Algorithm Theory. The Workshop continues the tradition of previous SWAT Workshops, and the Workshops on Algorithms and Data Structures, WADS, with which it alternates. The Call for Papers sought contributions presenting original research on algorithms and data structures in all areas, including computational geometry, parallel and distributed computing, graph theory, computational biology, and combinatorics. The program committee chose 28 of the 56 submitted papers for presentation. In addition, invited lectures were presented by Andrew V. Goldberg (NEC Research Institute), Johan Håstad (KTH Stockholm), and David Zuckerman (University of Texas at Austin).

SWAT '98 was held in Stockholm, July 8-10, and was organized in cooperation with Kungliga Tekniska Högskolan (the Royal Institute of Technology) and its department of Numerical Analysis and Computing Science, Nada. The organizing committe consisted of Jens Lagergren (KTH Stockholm) and Lars Ivansson (KTH Stockholm).

The program committee wishes to thank all referees who aided us in evaluating the papers.

Finally, we are very grateful to the Swedish Natural Science Research Council and The Swedish Research Council for Engineering Sciences for sponsoring the workshop, and Janne Carlsson the President of the Royal Institute of Technology.

Stockholm, April 1998   Stefan Arnborg

Program Committee

Stefan Arnborg (KTH Stockholm, chairman)
Rusins Freivalds (University of Latvia)
Rolf Karlsson (Lund University)
Sanjeev Khanna (Bell Labs)
Valerie King (University of Victoria)
Jens Lagergren (KTH Stockholm)
Christos Levcopoulos (Lund University)
Peter Bro Miltersen (University of Aarhus)
Thomas Ottmann (Albert-Ludwigs-Universität Freiburg)
David Peleg (Weizmann Institute of Science)
Martti Penttonen (University of Joensuu)
Alexander C. Russell (University of Texas at Austin)
Aravind Srinivasan (The National University of Singapore)

Referees

Table of Contents

Recent Developments in Maximum Flow Algorithms (Invited Lecture) 1
    A. V. Goldberg

An ϵ - Approximation Algorithm for Weighted Shortest Paths on
Polyhedral Surfaces 11
    L. Aleksandrov, M. Lanthier, A. Maheshwari, J.-R. Sack

Facility Location with Dynamic Distance Functions 23
    R. Bhatia, S. Guha, S. Khuller, Y. J. Sussmann

An Approximation Scheme for Bin Packing with Conflicts 35
    K. Jansen

Approximations for the General Block Distribution of a Matrix 47
    B. Aspvall, M. M. Halldórsson, F. Manne

An Optimal Algorithm for Computing Visible Nearest Foreign Neighbors
Among Colored Line Segments 59
    T. Graf, K. Veezhinathan

Moving an Angle Around a Region 71
    F. Hoffmann, C. Icking, R. Klein, K. Kriegel

Models and Motion Planning 83
    M. de Berg, M. J. Katz, M. Overmars, A. F. van der Stappen,
    J. Vleugels

Constrained Square-Center Problems 95
    M. J. Katz, K. Kedem, M. Segal

Worst-Case Efficient External-Memory Priority Queues 107
    G. S. Brodal, J. Katajainen

Simple Confluently Persistent Catenable Lists 119
    H. Kaplan, C. Okasaki, R. E. Tarjan

Improved Upper Bounds for Time-Space Tradeoffs for Selection with
Limited Storage 131
    V. Raman, S. Ramnath

Probabilistic Data Structures for Priority Queues 143
    R. Sridhar, K. Rajasekar, C. Pandu Rangan

Extractors for Weak Random Sources and Their Applications (Invited
Lecture) 155
    D. Zuckerman

Comparator Networks for Binary Heap Construction 158
    G. S. Brodal, M. C. Pinotti

Two-Variable Linear Programming in Parallel 169
    D. Z. Chen, J. Xu

Optimal Deterministic Protocols for Mobile Robots on a Grid 181
    R. Grossi, A. Pietracaprina, G. Pucci

Concurrent Multicast in Weighted Networks 193
    G. De Marco, L. Gargano, U. Vaccaro

Some Recent Strong Inapproximability Results (Invited Lecture) 205
    J. Håstad

Minimal Elimination of Planar Graphs................................. 210
    E. Dahlhaus

Memory Requirements for Table Computations in Partial k-tree
Algorithms 222
    B. Aspvall, A. Proskurowski, J. A. Telle

Formal Language Constrained Path Problems 234
    C. Barrett, R. Jacob, M. Marathe

Local Search Algorithms for SAT: Worst-Case Analysis 246
    E. A. Hirsch

Speed Is More Powerful than Clairvoyance 255
    P. Berman, C. Coulston

Randomized Online Multi-threaded Paging 264
    S. S. Seiden

Determinant: Old Algorithms, New Insights 276
    M. Mahajan, V Vinay

Solving Fundamental Problems on Sparse-Meshes 288
    J. F. Sibeyn

Output-Sensitive Cell Enumeration in Hyperplane Arrangements......... 300
    N. Sleumer

Fast and Efficient Computation of Additively Weighted Voronoi Cells for
Applications in Molecular Biology 310
    H.-M. Will

On the Number of Regular Vertices of the Union of Jordan Regions 322
  B. Aronov, A. Efrat, D. Halperin, M. Sharir

Distribution-Sensitive Algorithms 335
  S. Sen, N. Gupta

Author Index 347

Recent Developments in Maximum Flow Algorithms (Invited Lecture)

Andrew V. Goldberg

NEC Research Institute, Inc.
4 Independence Way, Princeton, NJ 08540
avg@research.nj.nec.com

1 Introduction

The maximum flow problem is a classical optimization problem with many applications; see *e.g.* [1, 18, 39]. Algorithms for this problem have been studied for over four decades. Recently, significant improvements have been made in theoretical performance of maximum flow algorithms. In this survey we put these results in perspective and provide pointers to the literature. We assume that the reader is familiar with basic flow algorithms, including Dinitz' blocking flow algorithm [13].

2 Preliminaries

The maximum flow problem is to find a flow of the maximum value given a graph G with arc capacities, a source s, and a sink t, Here a flow is a function on arcs that satisfies *capacity constraints* for all arcs and *conservation constraints* for all vertices except the source and the sink. For more details, see [1, 18, 39].

We distinguish between *directed* and *undirected* flow problems. In the standard problem formulation, capacities are reals. Two important special cases of the problem are the integral and the unit capacity cases.

When stating complexity bounds, we denote the number of vertices of the input graph by n and the number of arcs by m. We use the O^* notation to ignore the factors of $\log n$. For the integral case, we assume that capacities are in the interval $[1 \ldots U]$, make the *similarity assumption* [19] $\log U = O(\log n)$, and use the O^{**} notation to ignore the factors of $\log n$ and $\log U$.

The directed and undirected flow problems are closely related. A well-known reduction of Ford and Fulkerson [18] reduces the undirected problem to the directed problem with comparable size and capacity values. A more surprising reduction in the other direction, due to Picard and Ratliff [45], does not significantly increase the problem size but may increase capacities. When applied to a unit capacity capacity problem, the reduction produces a problem with capacities which are polynomial in n (equivalently, an undirected multigraph with unit capacities and the number of edges polynomial in n and m). No efficient reduction is known from the undirected unit capacity problem to the directed unit capacity problem.

The *residual capacity* of an arc tells how much the flow on the arc can be increased without violating capacity constraints. A *residual arc* is an arc with a positive residual capacity. Residual arcs induce the *residual graph*. An *augmenting path* is a path from the source to the sink in the residual graph. The *value* of a flow is the total flow into the sink. The *residual flow value* is the difference between the maximum and the current flow values.

3 History

The first maximum flow algorithms, the network simplex method [11] and the augmenting path method [17], were developed in the 50's. These algorithms are pseudopolynomial. In the early 70's, Dinitz [13] and Edmonds and Karp [15] showed that the shortest augmenting path algorithm is polynomial. The main idea of this algorithm is to assign unit lengths to residual arcs and augment flow along a shortest augmenting path. One can show that the source to sink distance is nondecreasing and must increase after polynomially many augmentations.

During the next decade, maximum flow algorithms became faster due to the discovery of powerful new techniques. Dinitz [13] developed the blocking flow method in the context of the shortest augmenting path algorithm; Karzanov [36] stated blocking flows as a separate problem and suggested the use of preflows to solve it. Edmonds and Karp [15] and independently Dinitz [14] developed capacity scaling. Dinitz [14] also shows how to combine blocking flows and capacity scaling to obtain an $O^{**}(nm)$ maximum flow algorithm. (See also [19].) Galil and Naamad [20] developed data structures to speed up flow computations and obtained an $O^{*}(nm)$ algorithm. We use E instead of O when stating expected time bounds.

For the unit capacity case, Karzanov [35] and independently Even and Tarjan [16] show that Dinitz' blocking flow algorithm solves the maximum flow problem in $O(m^{3/2})$ time on multigraphs and in $O(\min(n^{2/3}, m^{1/2})m)$ time on simple graphs.

Table 1 shows the state of the maximum flow time bounds in 1980. Note the gap between the unit capacity and the other cases.

The push-relabel method, first used in [22] and fully developed by Goldberg and Tarjan [26], leads to better theoretical understanding of the problem and substantial improvements in computational performance of the maximum flow codes (see e.g. [5, 10, 12, 23, 42]). However, for a long time all bound improvements were limited to logarithmic factors.

Recently, polynomial improvements have been made for two classes of problems: undirected unit capacity problems and integer capacity problems. We discuss these cases below, starting with the latter.

capacities	directed	undirected
unit	$O(\min(n^{2/3}, m^{1/2})m)$ Karzanov [35] Even & Tarjan [16]	$\leftarrow$
integral	$O^{**}(nm)$ Dinitz [14] Gabow [19]	$\leftrightarrow$
real	$O^{*}(nm)$ Galil & Naamad [20]	$\leftrightarrow$

Table 1. The best bounds known in 1980. Arrows show problem reductions: left arrow means the problem reduces to the problem on the left; double arrow means the problem is equivalent to that on the left.

4   Integer Capacity Flows

Table 2 shows the history of time bound improvements for the capacitated maximum flow problem. Dinitz' 1973 algorithm was the first one to achieve an $O^{**}(nm)$ bound. For a long time, the $O(nm)$ bound was the goal of much of the theoretical research in the area. Recently, however, Goldberg and Rao [24] developed an $O^{**}(\min(n^{2/3}, m^{1/2})m)$ algorithm, bridging the gap between the unit and integer capacity cases (see Table 1). This result suggests that the $O(nm)$ goal may be too modest.

The Goldberg-Rao algorithm is a generalization of the blocking flow method that uses a non-unit length function. Recall that the main loop of the blocking flow method finds a blocking flow in an acyclic graph. Such a flow can be found in $O(m \log n)$ time [27]. Recall that the original blocking flow method assigns unit lengths to all residual arcs. The use of non-unit length functions for maximum flow computations has been first suggested by Edmonds and Karp [15]. In the context of minimum-cost flows, the idea has been studied in [47, 48]. However, prior to the Goldberg-Rao work, non-unit length functions never gave asymptotic time bound improvements.

Goldberg and Rao use an *adaptive binary length function.*[1] The length of a residual arc depends on the estimated value F of the residual flow. Arcs whose residual capacity is small compared to F have zero lengths and arcs whose residual capacity is large have unit lengths. Zero-length arcs, which seem essential for the algorithm, introduce technical complications, resulting from the fact that the graph G' induced by the residual arcs on shortest paths to the sink may not be acyclic. To deal with this complication, the algorithm contracts strongly connected components of G'. The algorithm limits the amount of flow through each contracted node so that it is easy to route the flow inside the nodes. Because the zero-length arcs have large capacity, the limit is large, and the algorithm makes significant progress even if the limit comes into play.

---

[1] One can use other length functions.

year	discoverer(s)	bound
1951	Dantzig [11]	$O(n^2 mU)$
1956	Ford & Fulkerson [17]	$O(nmU)$
1970	Dinitz [13] Edmonds & Karp [15]	$O(nm^2)$
1970	Dinitz [13]	$O(n^2 m)$
1972	Edmonds & Karp [15] Dinitz [14]	$O(m^2 \log U)$
1973	Dinitz [14] Gabow [19]	$O(nm \log U)$
1974	Karzanov [36]	$O(n^3)$
1977	Cherkassky [9]	$O(n^2 m^{1/2})$
1980	Galil & Naamad [20]	$O(nm \log^2 n)$
1983	Sleator & Tarjan [46]	$O(nm \log n)$
1986	Goldberg & Tarjan [26]	$O(nm \log(n^2/m))$
1987	Ahuja & Orlin [2]	$O(nm + n^2 \log U)$
1987	Ahuja et al. [3]	$O(nm \log(n\sqrt{\log U}/m))$
1989	Cheriyan & Hagerup [7]	$E(nm + n^2 \log^2 n)$
1990	Cheriyan et al. [8]	$O(n^3/\log n)$
1990	Alon [4]	$O(nm + n^{8/3} \log n)$
1992	King et al. [37]	$O(nm + n^{2+\epsilon})$
1993	Phillips & Westbrook [44]	$O(nm(\log_{m/n} n + \log^{2+\epsilon} n))$
1994	King et al. [38]	$O(nm \log_{m/(n \log n)} n)$
1997	Goldberg & Rao [24]	$O(\min(n^{2/3}, m^{1/2})m \log(n^2/m) \log U)$

Table 2. History of the capacitated bound improvements. The dates correspond to the first publication of the result; citations are to the most recent versions. Bounds containing U apply to the integral capacity case only.

Here is a brief outline of the algorithm; see [24] for details. We denote $\min(n^{2/3}, m^{1/2})$ by Λ. The threshold between length one and length zero arcs is $\Theta(F/\Lambda)$.

The algorithm maintains a flow and a cut. The residual capacity of this cut gives an upper bound on the residual flow value. The algorithm also maintains a parameter F, which is an upper bound on the residual flow value. When the algorithm finds a cut with residual capacity of at most $F/2$, it sets F to this residual capacity, thereby decreasing F by at least a factor of two. Distances to the sink are monotone while F does not change. At each iteration, dominated by a blocking flow computation on the contracted graph G', either the flow value increases by at least F/Λ or the source to sink distance increases by at least one. One can show that when the distance exceeds $c\Lambda$, for a fixed constant c, F must decrease. This after $O(\Lambda)$ iterations, F must decrease, either because the flow value increased by at least $F/2$ or because the distance between the source and the sink exceeded $c\Lambda$. This leads to the desired time bound.

5 Undirected Unit Capacity Flows

year	discoverer(s)	bound
1951	Dantzig [11]	$O(n^2 m)$
1956	Ford & Fulkerson [17]	$O(nm)$
1973	Karzanov [35] Even & Tarjan [16]	$O(\min(n^{2/3}, m^{1/2})m)$
1997	Karger [30]	$E^*(n^{4/3} m^{2/3})$
1997	Goldberg & Rao [25]	$O(n^{3/2} m^{1/2})$
1998	Karger & Levine [32]	$O(n^{7/6} m^{2/3})$ $E^*(n^{20/9})$

Table 3. History of the bound improvements for the undirected unit capacity problem. The dates correspond to the first publication of the result; citations are to the most recent versions.

In the undirected case, the unit capacity flow bound of $O(\min(n^{2/3}, m^{1/2})m)$ can be improved using *sparsification techniques*. Two different classes of sparcification techniques, originally developed for the minimum cut problem, proved useful for the maximum flow problem: sparse certificates and random sampling. The former was developed by Nagamochi and Ibaraki [41, 40]. The latter was developed by Karger [28]. (See also [29, 33].)

In this section we assume that the input graph is undirected and all edge capacities are one and all distances are for the unit length function on the residual graph. For simplicity we assume that the input graph is simple and state the bounds in terms of the graph size. For the bounds depending on the flow value and the results for multigraphs, see [30–32, 34].

The $O(\min(n^{2/3}, m^{1/2})m)$ bound on Dinitz' blocking flow algorithm for the unit capacity case is based on the following two theorems.

Theorem 1. [13] *Each iteration of the blocking flow algorithm increases the distance between the source and the sink.*

Theorem 2. [16, 35] *If the distance between the source and the sink is d, then the residual flow value is $O(\min(m/d, (n/d)^2))$.*

5.1 Sparse Certificates

Next we discuss how to use sparse certificates to get a better bound. Let T_i, $i \geq 1$, be the set of edges of G constructed as follows. T_1 is the set of edges in a maximal spanning forest of G. For $i > 1$, T_i is the set of edges in the maximal spanning forest of G with $T_1 \cup \ldots \cup T_{i-1}$ deleted. Let $E_k = \bigcup_{i=1}^{k} T_i$. Nagamochi and Ibaraki show that if we are interested in connectivity of k or less, we can restrict our attention to the graph induced by E_k.

Theorem 3. [41] *If the minimum cut capacity between v and w in G is at most k, then the minimum cut capacity between v and w in (V, E_k) is the same as in G.*

Nagamochi and Ibaraki also give an efficient and elegant algorithm to find E_k.

Theorem 4. [40] *For $k = O(n)$, E_k can be found in $O(m)$ time.*

For weighted graphs and large k, E_k can be found in $O(m + n \log n)$ time [41].

Since G_f is a directed graph even if G is undirected, one cannot apply Theorem 3 directly. Consider a flow f in G and let E^0 and E^1 be the edges of G with flow value zero and one, respectively. Edges in E^0 are not used by the flow and therefore undirected, so we can sparsify the corresponding subgraph. Let E_k^0 be a k-sparse certificate of (V, E^0).

Theorem 5. [25] *If the residual flow value in G is at most k, then the residual flow value in $(V, E^1 \cup E_k^0)$ is the same as in G.*

We know that $|E_k^0| \leq k(n-1)$. If E^1 is small, we can use the above theorem for sparcification during a maximum flow computation. In many cases, E^1 is small. One example is a flow obtained by the shortest path augmentations [16]. Another case is acyclic flows. The following theorem is a stronger version of one in [21].

Theorem 6. [32] *An acyclic flow f in a simple graph uses $O(n\sqrt{|f|})$ edges.*

Karger and Levine [32] use Theorems 5 and 6 to obtain an $O(n^{7/6}m^{2/3})$ algorithm.

The ideas of this section may lead to better computational results. A computational study to determine practicality of these ideas would be interesting.

5.2 Random Sampling

The currently best randomized algorithm, due to Karger and Levine [32], uses the sparse certificates as well as random sampling techniques. This algorithm uses several nontrivial results developed in a sequence of papers [6, 34, 30–32], and the reader should see these papers for details. We do not attempt an accurate description of the Karger-Levine algorithm, but highlight the mail ideas behind the algorithm and give some intuition behind these ideas.

One additional sparcification technique is captured in the following result, which a variant of a result of Benczúr and Karger [6].

Theorem 7. [30] *For any k, one can, in $O^*(m)$ time, find a set of $O(kn \log n)$ edges whose removal partitions G into k-connected components.*

Intuitively, the graph formed by the removed edges is relatively sparse. Karger [30] uses Theorem 7 to find approximate minimum s-t cuts quickly and uses these s-t cuts in a divide-and-conquer algorithm to find maximum flows quickly when the flow value is small.

capacities	directed	undirected
unit	$O(\min(n^{2/3}, m^{1/2})m)$ Karzanov [35] Even & Tarjan [16]	$O(n^{7/6}m^{2/3})$ $E^*(n^{20/9})$ Karger & Levine [32]
integral	$O^{**}(\min(n^{2/3}, m^{1/2})m)$ Goldberg & Rao [24]	$\leftrightarrow$
real	$O^*(nm)$ Galil & Naamad [20]	$\leftrightarrow$

Table 4. The current state of the maximum flow bounds, ignoring logarithmic factors. Arrows show problem reductions.

The next ingredient of the Karger-Levine algorithm is a fast algorithm for finding flows through k-connected graphs. This algorithm is based on random sampling. We partition the edges of the graph into two sets by sampling the edges with probability one half and recursively find maximum flows in the two resulting subgraphs. Then we apply augmentations to convert the union of the two flows into a maximum flow in the original graph. The following theorem can be used to bound the expected number of augmentations.

Theorem 8. [34] *If G is k-connected and edges are sampled with probability p, then with high probability all cuts in the sampled graph are within $(1 \pm \sqrt{8 \ln n/(pk)})$ of their expected values.*

The Karger-Levine randomized algorithm, and especially its subroutines borrowed from the previous papers, are complicated. It would be interesting to simplify this algorithm without losing efficiency. Simplification may lead to improved theoretical bounds or make the algorithm practical.

6 What Next?

Table 4 summarizes the current state of the maximum flow bounds (modulo logarithmic factors). Comparing to Table 1, one can see that significant progress has been made for the integral and the undirected unit capacity cases. The corresponding algorithms use powerful new design and analysis techniques. Better understanding and refinement of these techniques may lead to even faster algorithms. Although logarithmic factor improvements in time bounds are more likely, polynomial improvements may be possible as well, in particular in the undirected unit capacity case.

Table 4 contains several polynomial gaps between "adjacent" cases. One gap is between the integral and the unit capacity cases. A result of Orlin [43] for minimum-cost flows gives some hope that it may be possible to close this gap.

Another gap is between the undirected and the directed unit capacity cases. Extending the undirected case results to the directed case requires sparcification techniques for directed graphs. This is a very challenging open problem. Similarly, closing the gap between the unit and the integral undirected cases seems to require directed graph sparcification because of the reduction to the integral directed case.

What is a natural limit of the current techniques? Suppose we have ideal sparsification: in $O(m)$ time, we reduce the number of arcs in the graph we work on to $O(n)$. Furthermore suppose we can close the gaps between the directed and undirected and the integral and real capacity cases. Then we would be able to solve the maximum flow problem in $O^*(m+n^{3/2})$ time. This ambitious goal may be feasible: Goldberg and Rao [24] show that random sampling of [6] can be used to find a cut of capacity within $(1+\epsilon)$ factor of the minimum in $O^*(m+n^{3/2})$ time, for any constant $\epsilon > 0$. (A related result, due to Karger [31], is that a flow of value within $(1-\epsilon)$ of the maximum can be found in $O^*(m\sqrt{n}/\epsilon)$ time.)

We conclude by stating the following open question. Can the maximum flow problem in sparse unit capacity networks be solved in $o(n^{3/2})$ time? A polynomial improvement to the $O(n^{3/2})$ bound for this problem would be a real breakthrough.

Acknowledgements

The author is greatful to Bill Cunningham, David Karger, Maurice Queyranne, and Cliff Stein for their help with this survey.

References

1. R. K. Ahuja, T. L. Magnanti, and J. B. Orlin. *Network Flows: Theory, Algorithms, and Applications*. Prentice-Hall, 1993.

2. R. K. Ahuja and J. B. Orlin. A Fast and Simple Algorithm for the Maximum Flow Problem. *Oper. Res.*, 37:748–759, 1989.

3. R. K. Ahuja, J. B. Orlin, and R. E. Tarjan. Improved Time Bounds for the Maximum Flow Problem. *SIAM J. Comput.*, 18:939–954, 1989.

4. N. Alon. Generating Pseudo-Random Permutations and Maximum Flow Algorithms. *Information Processing Let.*, 35:201–204, 1990.

5. R. J. Anderson and J. C. Setubal. Goldberg's Algorithm for the Maximum Flow in Perspective: a Computational Study. In D. S. Johnson and C. C. McGeoch, editors, *Network Flows and Matching: First DIMACS Implementation Challenge*, pages 1–18. AMS, 1993.

6. A. A. Benczúr and D. R. Karger. Approximating s-t Minimum Cuts in $\tilde{O}(n^2)$ Time. In *Proc. 28th Annual ACM Symposium on Theory of Computing*, pages 47–56, 1996.

7. J. Cheriyan and T. Hagerup. A randomized maximum flow algorithm. *SIAM Journal on Computing*, 24:203–226, 1995.

8. J. Cheriyan, T. Hagerup, and K. Mehlhorn. An $o(n^3)$-time Maximum Flow Algorithm. *SIAM J. Comput.*, 25:1144–1170, 1996.

9. B. V. Cherkassky. Algorithm for Construction of Maximal Flows in Networks with Complexity of $O(V^2\sqrt{E})$ Operations. *Mathematical Methods of Solution of Economical Problems*, 7:112–125, 1977. (In Russian).

10. B. V. Cherkassky and A. V. Goldberg. On Implementing Push-Relabel Method for the Maximum Flow Problem. *Algorithmica*, 19:390–410, 1997.

11. G. B. Dantzig. Application of the Simplex Method to a Transportation Problem. In T. C. Koopmans, editor, *Activity Analysis and Production and Allocation*, pages 359–373. Wiley, New York, 1951.

12. U. Derigs and W. Meier. Implementing Goldberg's Max-Flow Algorithm — A Computational Investigation. *ZOR — Methods and Models of Operations Research*, 33:383–403, 1989.

13. E. A. Dinic. Algorithm for Solution of a Problem of Maximum Flow in Networks with Power Estimation. *Soviet Math. Dokl.*, 11:1277–1280, 1970.

14. E. A. Dinic. Metod porazryadnogo sokrashcheniya nevyazok i transportnye zadachi. In *Issledovaniya po Diskretnoĭ Matematike*. Nauka, Moskva, 1973. In Russian. Title translation: Excess Scaling and Transportation Problems.

15. J. Edmonds and R. M. Karp. Theoretical Improvements in Algorithmic Efficiency for Network Flow Problems. *J. Assoc. Comput. Mach.*, 19:248–264, 1972.

16. S. Even and R. E. Tarjan. Network Flow and Testing Graph Connectivity. *SIAM J. Comput.*, 4:507–518, 1975.

17. L. R. Ford, Jr. and D. R. Fulkerson. Maximal Flow Through a Network. *Canadian Journal of Math.*, 8:399–404, 1956.

18. L. R. Ford, Jr. and D. R. Fulkerson. *Flows in Networks*. Princeton Univ. Press, Princeton, NJ, 1962.

19. H. N. Gabow. Scaling Algorithms for Network Problems. *J. of Comp. and Sys. Sci.*, 31:148–168, 1985.

20. Z. Galil and A. Naamad. An $O(EV \log^2 V)$ Algorithm for the Maximal Flow Problem. *J. Comput. System Sci.*, 21:203–217, 1980.

21. Z. Galil and X. Yu. Short Length Version of Menger's Theorem. In *Proc. 27th Annual ACM Symposium on Theory of Computing*, pages 499–508, 1995.

22. A. V. Goldberg. A New Max-Flow Algorithm. Technical Report MIT/LCS/TM-291, Laboratory for Computer Science, M.I.T., 1985.

23. A. V. Goldberg. *Efficient Graph Algorithms for Sequential and Parallel Computers*. PhD thesis, M.I.T., January 1987. (Also available as Technical Report TR-374, Lab. for Computer Science, M.I.T., 1987).

24. A. V. Goldberg and S. Rao. Beyond the Flow Decomposition Barrier. In *Proc. 38th IEEE Annual Symposium on Foundations of Computer Science*, pages 2–11, 1997.

25. A. V. Goldberg and S. Rao. Flows in Undirected Unit Capacity Networks. In *Proc. 38th IEEE Annual Symposium on Foundations of Computer Science*, pages 32–35, 1997.

26. A. V. Goldberg and R. E. Tarjan. A New Approach to the Maximum Flow Problem. *J. Assoc. Comput. Mach.*, 35:921–940, 1988.

27. A. V. Goldberg and R. E. Tarjan. Finding Minimum-Cost Circulations by Successive Approximation. *Math. of Oper. Res.*, 15:430–466, 1990.

28. D. R. Karger. Global Min-Cuts in RNC, and Other Ramifications of a Simple Min-Cut Algorithm. In *Proc. 4th ACM-SIAM Symposium on Discrete Algorithms*, 1993.

29. D. R. Karger. Minimum Cuts in Near-Linear Time. In *Proc. 28th Annual ACM Symposium on Theory of Computing*, pages 56–63, 1996.

30. D. R. Karger. Using Random Sampling to Find Maximum Flows in Uncapacitated Undirected Graphs. In *Proc. 29th Annual ACM Symposium on Theory of Computing*, pages 240–249, 1997.

31. D. R. Karger. Better Random Sampling Algorithms for Flows in Undirected Graphs. In *Proc. 9th ACM-SIAM Symposium on Discrete Algorithms*, pages 490–499, 1998.

32. D. R. Karger and M. Levine. Finding Maximum Flows in Undirected Graphs Seems Easier than Bipratire Matching. In *Proc. 30th Annual ACM Symposium on Theory of Computing*, 1997.

33. D. R. Karger and C. Stein. A New Approach to the Minimum Cut Problem. *J. Assoc. Comput. Mach.*, 43, 1996.

34. D.R. Karger. Random Sampling in Cut, Flow, and Network Design Problems. In *Proc. 26th Annual ACM Symposium on Theory of Computing*, pages 648–657, 1994. Submitted to *Math. of Oper. Res.*

35. A. V. Karzanov. O nakhozhdenii maksimal'nogo potoka v setyakh spetsial'nogo vida i nekotorykh prilozheniyakh. In *Matematicheskie Voprosy Upravleniya Proizvodstvom*, volume 5. Moscow State University Press, Moscow, 1973. In Russian; title translation: On Finding Maximum Flows in Networks with Special Structure and Some Applications.

36. A. V. Karzanov. Determining the Maximal Flow in a Network by the Method of Preflows. *Soviet Math. Dok.*, 15:434–437, 1974.

37. V. King, S. Rao, and R. Tarjan. A Faster Deterministic Maximum Flow Algorithm. In *Proc. 3rd ACM-SIAM Symposium on Discrete Algorithms*, pages 157–164, 1992.

38. V. King, S. Rao, and R. Tarjan. A Faster Deterministic Maximum Flow Algorithm. *J. Algorithms*, 17:447–474, 1994.

39. E. L. Lawler. *Combinatorial Optimization: Networks and Matroids*. Holt, Reinhart, and Winston, New York, NY., 1976.

40. H. Nagamochi and T. Ibaraki. A Linear-Time Algorithm for Finding a Sparse k-Connected Spanning Subgraph of a k-Connected Graph. *Algorithmica*, 7:583–596, 1992.

41. H. Nagamochi and T. Ibaraki. Computing Edge-Connectivity in Multigraphs and Capacitated Graphs. *SIAM J. Disc. Math.*, 5:54–66, 1992.

42. Q. C. Nguyen and V. Venkateswaran. Implementations of Goldberg-Tarjan Maximum Flow Algorithm. In D. S. Johnson and C. C. McGeoch, editors, *Network Flows and Matching: First DIMACS Implementation Challenge*, pages 19–42. AMS, 1993.

43. J. B. Orlin. A Faster Strongly Polynomial Minimum Cost Flow Algorithm. *Oper. Res.*, 41:338–350, 1993.

44. S. Phillips and J. Westbrook. Online Load Balancing and Network Flow. In *Proc. 25th Annual ACM Symposium on Theory of Computing*, pages 402–411, 1993.

45. J. C. Picard and H. D. Ratliff. Minimum Cuts and Related Problems. *Networks*, 5:357–370, 1975.

46. D. D. Sleator and R. E. Tarjan. A Data Structure for Dynamic Trees. *J. Comput. System Sci.*, 26:362–391, 1983.

47. C. Wallacher. A Generalization of the Minimum-Mean Cycle Selection Rule in Cycle Canceling Algorithms. Technical report, Preprints in Optimization, Institute für Angewandte Mathematik, Technische Universität Braunschweig, Germany, 1991.

48. C. Wallacher and U. Zimmermann. A Combinatorial Interior Point Method for Network Flow Problems. Technical report, Preprints in Optimization, Institute für Angewandte Mathematik, Technische Universität Braunschweig, Germany, 1991.

An ε - Approximation Algorithm for Weighted Shortest Paths on Polyhedral Surfaces *

Lyudmil Aleksandrov[1], Mark Lanthier[2], Anil Maheshwari[2],
Jörg-R. Sack[2]

[1] Bulgarian Academy of Sciences, CICT,
Acad. G. Bonchev Str. Bl. 25-A, 1113 Sofia, Bulgaria
[2] School of Computer Science, Carleton University,
Ottawa, Ontario K1S5B6, Canada

Abstract. Let $\mathcal{P}$ be a simple polyhedron, possibly non-convex, whose boundary is composed of n triangular faces, and in which each face has an associated positive weight. The cost of travel through each face is the distance traveled multiplied by the face's weight. We present an ε-approximation algorithm for computing a weighted shortest path on $\mathcal{P}$, i.e. the ratio of the length of the computed path with respect to the length of an optimal path is bounded by $(1 + \epsilon)$, for a given $\epsilon > 0$. We give a detailed analysis to determine the exact constants for the approximation factor. The running time of the algorithm is $O(mn \log mn + nm^2)$. The total number of Steiner points, m, added to obtain the approximation depends on various parameters of the given polyhedron such as the length of the longest edge, the minimum angle between any two adjacent edges of $\mathcal{P}$ and the minimum distance from any vertex to the boundary of the union of its incident faces and the ratio of the largest (finite) to the smallest face weights of $\mathcal{P}$. Lastly, we present an approximation algorithm with an improved running time of $O(mn \log mn)$, at the cost of trading off the constants in the path accuracy. Our results present an improvement in the dependency on the number of faces, n, to the recent results of Mata and Mitchell [10] by a multiplicative factor of $n^2/\log n$, and to that of Mitchell and Papadimitriou [11] by a factor of n^7.

1 Introduction

1.1 Problem Definition

Shortest path problems are among the fundamental problems studied in computational geometry and other areas such as graph algorithms, geographical information systems (GIS) and robotics.[5] Let s and t be two vertices on a given possibly non-convex polyhedron $\mathcal{P}$, in $\Re^3$, consisting of n triangular faces on its

---

* Research supported in part by ALMERCO Inc. & NSERC

[5] We encountered several shortest path related problems in our R&D on GIS (see [15]); more specifically, e.g., in emergency response time modeling where emergency units are dispatched to emergency sites based on minimum travel times.

boundary, each face has an associated weight, denoted by a positive real number w_i. A Euclidean *shortest path* $\pi(s,t)$ between s and t is defined to be a path with minimum Euclidean length among all possible paths joining s and t that lie on the surface of $\mathcal{P}$. A *weighted shortest path* $\Pi(s,t)$ between s and t is defined to be a path with minimum cost among all possible paths joining s and t that lie on the surface of $\mathcal{P}$. The *cost* of the path is the sum of lengths of all segments, the path traverses in each face multiplied by the corresponding face weight. A path $\Pi'(s,t)$ between two points s and t is said to be an ϵ-approximation of a (true) shortest path $\Pi(s,t)$ between s and t, if $\frac{\Pi'(s,t)}{\Pi(s,t)} \leq 1 + \epsilon$, for some $\epsilon > 0$. The problem addressed in this paper is to determine an ϵ-approximate shortest path between two vertices on a weighted polyhedron.

1.2 Related Work

Shortest path problems in computational geometry can be categorized by various factors which include the dimensionality of space, the type and number of objects or obstacles (e.g., polygonal obstacles, convex or non-convex polyhedra, ...), and the distance measure used (e.g., Euclidean, number of links, or weighted distances). Several research articles, including surveys (see [5, 13]), have been written presenting the state-of-the-art in this active field. Due to the lack of space, here we discuss those contributions which relate more directly to our work; these are in particular 3-dimensional weighted scenarios.

Mitchell and Papadimitriou [11] introduced the weighted region problem in which each face has an associated positive weight, denoted by a real number $w_i > 0$. They presented an algorithm that computes a path between two points in a weighted planar subdivision which is at most $(1 + \epsilon)$ times the shortest weighted path cost. Their algorithm requires $O(n^8 L)$ time in the worst case, where $L = \log(nNW/w\epsilon)$ is a factor representing the bit complexity of the problem instance. Here N is the largest integer coordinate of any vertex of the triangulation and W (w) is the maximum (minimum) weight of any face of the triangulation. Johannson discussed a weighted distance model for injection molding [6]. Lanthier et al. [8] presented several practical algorithms for approximating shortest paths in weighted domains. In addition, to their experimental verification and time analysis, they provided theoretically derived bounds on the quality of approximation. More specifically, the cost of the approximation is no more than the shortest path cost plus an (additive) factor of $W|L|$, where L is the longest edge, W is the largest weight among all faces. They also used graph spanners to get at most β times the shortest path cost plus $\beta W|L|$, where $\beta > 1$ is an adjustable constant. Mata and Mitchell [10] presented an algorithm that constructs a graph (pathnet) which can be searched to obtain an approximate path; their path accuracy is $(1 + \frac{W}{kw\theta_{min}})$, where θ_{min} is the minimum angle of any face of $\mathcal{P}$, W/w is the largest to smallest weight ratio and k is a constant that depends upon ϵ.

Table 1 compares the running times for the ϵ-approximation algorithms developed by [11], [10], and the one presented in this paper in the case where all vertices are given as integer coordinates. From the table we can clearly see that our algorithm improves substantially the dependence on n, but the dependence on the geometric parameters is somewhat worse. Since in many applications n is quite large (larger than 10^5) the objective for this work has been to find an ϵ-approximation algorithm, where the dependence on n is considerably smaller.

Algorithm	Running Time ($K = O(\frac{N^2 W}{\epsilon w})$)
Mitchell and Papadimitriou [11]	$O(n^8 \log(\frac{nK}{N}))$
Mata and Mitchell [10]	$O(n^3 K)$
Our Results	$O(n(K \log K) \log(nK \log K))$

Table 1. Comparison of weighted shortest path algorithms that rely on geometric precision parameters. N represents the largest integer coordinate.

Although the objective of [8] was different, the schemes are ϵ-approximations in which the dependence on n becomes comparable to [10] (see [9]).

1.3 Our Approach

Our approach to solving the problem is to discretize the polyhedron in a natural way, by placing Steiner points along the edges of the polyhedron (as in our earlier subdivision approach [8]). We construct a graph G containing the Steiner points as vertices and edges as those interconnections between Steiner points that correspond to segments which lie completely in the triangular faces of the polyhedron. The geometric shortest path problem on polyhedra is thus stated as a graph problem so that the existing efficient algorithms (and their implementations) for shortest paths in graphs can be used. One of the differences to [8] and to other somewhat related work (e.g., [8, 3, 7]) lies in the placement of Steiner points.

We introduce a logarithmic number of Steiner points along each edge of $\mathcal{P}$, and these points are placed in a geometric progression along an edge. They are chosen w.r.t. the vertex joining two edges of a face such that the distance between any two adjacent points on an edge is at most ϵ times the shortest possible path segment that can cross that face between those two points.

Our discretization method falls into the class of edge subdivision algorithms. Grid-based methods as introduced e.g., by Papadimitriou [12], are instances of this class. As concluded by Choi, Sellen and Yap [2]: *"... grids are a familiar practical technique in all of computational sciences. From a complexity theoretic viewpoint, such methods have been shunned in the past as trivial or uninteresting. This need not be so, as Papadimitriou's work has demonstrated. In fact, the grid methods may be the most practical recourse for solving some intractable problems. It would be interesting to derive some general theorems about these approaches"* Lanthier et al. [8] and Mata and Mitchell [10] are proofs of such practical methods based on edge subdivision.

A problem arises when placing these Steiner points near vertices of the face since the shortest possible segment becomes infinitesimal in length. A similar issue was encountered by Kenyon and Kenyon [7] and Das and Narasimhan [3] during their work on rectifiable curves on the plane and in 3-space, respectively. The problem arises since the distance between adjacent Steiner points, in the near vicinity of a vertex, would have to be infinitesimal requiring an infinite number of Steiner points. We address this problem by constructing spheres around the vertices which have a very small radius (at most ϵ times the shortest distance from the vertex to an edge that is not incident to the vertex). The graph construction procedure never adds Steiner points within these spheres centered around each vertex of the polyhedron. This allows us to put a lower bound on the length of the smallest possible edge that passes between two adjacent Steiner points and hence we are able to add a finite number of Steiner points. As a result, if the shortest path passes through one of these spheres, the approximate path may pass through the vertex of the polyhedra, corresponding to the center of the sphere.

We show that there exist paths in this graph with costs that are within $(1 + \epsilon)$ times the shortest path costs. For the purpose of simplifying the proofs, we actually show that the approximation is within the bound of $(1 + \frac{3-2\epsilon}{1-2\epsilon}\epsilon)$ times the shortest path length in the unweighted scenario and within the bound of $(1 + (2 + \frac{2W}{(1-2\epsilon)w})\epsilon)$ times the shortest cost in the weighted scenario where $0 < \epsilon < \frac{1}{2}$ and $\frac{W}{w}$ is the largest to smallest weight ratio of the faces of $\mathcal{P}$. The desired ϵ-approximation is achieved by dividing ϵ by $\frac{3-2\epsilon}{1-2\epsilon}$ or $(2 + \frac{2W}{(1-2\epsilon)w})$ for the unweighted and weighted case, respectively. We can simplify the bounds of our algorithm when $\epsilon < 1/6$. The bounds become $(1 + 4\epsilon)$ and $(1 + (2 + 3\frac{W}{w})\epsilon)$ for the unweighted and weighted case, respectively. The running time of our algorithm is the cost for computing the graph G plus that of running a shortest path algorithm in G. The graph consists of $|V| = nm$ vertices and $|E| = nm^2$ edges where $m = O(\log_\delta(|L|/r))$, $|L|$ is the length of the longest edge, r is ϵ times the minimum distance from any vertex to the boundary of the union of its incident faces (denoted as minimum height h of any face), and $\delta \geq 1 + \epsilon \sin \theta$, where θ is the minimum angle between any two adjacent edges of $\mathcal{P}$.

We also provide an algorithm to compute a subgraph G^* of G, on the same vertex set as that of G, but with only $O(nm \log nm)$ edges. G^* has the property that for any edge $\overline{uv} \in G$, there exists a path in G^* whose length is at most $(1 + \epsilon)\|\overline{uv}\|$. This results in an ϵ-approximation algorithm for the shortest path problem and it runs in $O(nm \log nm)$ time. (To study the entries in Table 1, set $h = \Omega(1/N)$ and $\sin \theta = \Omega(1/N^2)$, where the vertices are assumed to have integer coordinates bounded by N).

Our analysis reveals the exact relationship between the geometric parameters and the algorithm's running time. The dependence on geometric parameters is an interesting feature of several approximation geometric algorithms. Many researchers have advocated the use of geometric parameters in analyzing the performance of geometric algorithms, and our result indicates that if the geometric parameters are "well-behaved" then the asymptotic complexity of our algorithms is several orders of magnitude better than existing ones. One of the conclusions from our study is that while studying the performance of geometric algorithms, geometric parameters (e.g. fatness, density, aspect ratio, longest, closest) should not be ignored, and in fact it could potentially be very useful to express the performance that includes the relevant geometric parameters.

2 Preliminaries

Let s and t be two vertices of a triangulated polyhedral surface $\mathcal{P}$ with n faces. A weight $w_i > 0$ is associated with each face $f_i \in \mathcal{P}$ such that the cost of travel through f_i is the distance traveled times w_i. Define W and w to be the maximum and minimum weight of all $w_i, 1 \leq i \leq n$, respectively.

Property 1. An edge of $\mathcal{P}$ cannot have a weight greater than its adjacent faces.

Let L be the longest edge of $\mathcal{P}$. Let $\pi(s,t)$ be a shortest Euclidean length path between s and t that remains on $\mathcal{P}$ with path length $|\pi(s,t)|$. Let $s_1, s_2, ..., s_k$ be the segments of $\pi(s,t)$ passing through faces $f_1, f_2, ..., f_k$. Similarly, define $\Pi(s,t)$ to be a shortest weighted cost path between s and t that remains on $\mathcal{P}$ with weighted cost denoted as $\|\Pi(s,t)\|$. Define $G(V,E) = G_1 \cup G_2 \cup ... G_n$ to be a graph such that $G_i, 1 \leq i \leq n$ is a subgraph created on face f_i with vertices lying on edges of f_i and edges of G_i lying across face f_i. Let $E(G_i)$ represent the edges of G_i and $V(G_i)$ represent the vertices of G_i. Let $\pi'(s,t) = s'_1, s'_2, ..., s'_k$ be a path in G passing through the same faces as $\pi(s,t)$ with length $|\pi'(s,t)|$. Similarly let $\Pi'(s,t)$ be a path in G passing through the same faces as $\Pi(s,t)$ with weighted cost $\|\Pi'(s,t)\|$.

Let v be a vertex of $\mathcal{P}$. Define h_v to be the minimum distance from v to the boundary of the union of its incident faces. Define a polygonal cap C_v, called a *sphere*, around v, as follows. Let $r_v = \epsilon h_v$ for some $0 < \epsilon$. Let r be the minimum r_v over all v. Let vuw be a triangulated face incident to v. Let u' (w') be at the distance of r_v from v on vu (vw). This defines a triangular sub-face $vu'w'$ of vuw. The sphere C_v around v consists of all such sub-faces incident at v.

Property 2. The distance between any two spheres C_{v_a} and C_{v_b} is greater than $(1 - 2\epsilon)h_{v_a}$.

Define θ_v to be the minimum angle (measured in 3D) between any two edges of $\mathcal{P}$ that are incident to v. Let θ be the minimum θ_v. A weighted path may *critically use* an edge of $\mathcal{P}$ by traveling along it and then reflecting back into the face [11]. We distinguish between two types of path segments of a shortest path: 1) *face-crossing* segments which cross a face and do not critically use an edge, and 2) *edge-using* segments which lie along an edge (critically using it). In the unweighted domain, edge-using segments span the entire length of an edge in $\mathcal{P}$ and a face can only be crossed once by a shortest path [14]. However, in the weighted domain, a face may be crossed more than once and so a weighted shortest path may have $\theta(n^2)$ segments, see [11].

3 An ϵ - Approximation Scheme

This section presents the approximation scheme by first describing the computation of the graph G, which discretize the problem. The construction of G

depends on the choice of ϵ, so we assume that a positive $\epsilon < \frac{1}{2}$ has been chosen and is fixed. A shortest path in G between s and t will be the ϵ-approximation path $\pi'(s,t)$ (or $\Pi(s,t)$) that we report.

Our proof strategy is as follows. We present a construction to show that there exists a path, $\pi'(s,t)$ (respectively, $\Pi'(s,t)$), between s and t in G, with cost at most $(1 + \frac{3-2\epsilon}{1-2\epsilon}\epsilon)$ (respectively, $(1 + (2 + \frac{2W}{(1-2\epsilon)w})\epsilon))$ times the cost of $\pi(s,t)$ (respectively, $\Pi(s,t)$). Consider a shortest path $\Pi(s,t)$ in $\mathcal{P}$. It is composed of several segments which go through faces, and/or along edges, and/or through spheres around a vertex. For segments of $\Pi(s,t)$ that are not completely contained inside spheres, we show that there exists an appropriate edge in the graph. For segments that are lying inside a sphere, we use a "charging" argument. (Due to the lack of space, proofs are omitted; see [9] for details.)

3.1 An Algorithm to Compute the Graph

For each vertex v of face f_i we do the following: Let e_q and e_p be the edges of f_i incident to v. First, place Steiner points on edges e_q and e_p at distance r_v from v; call them q_1 and p_1, respectively. By definition, $|\overline{vq_1}| = |\overline{vp_1}| = r_v$. Define $\delta = (1 + \epsilon \cdot \sin \theta_v)$ if $\theta_v < \frac{\pi}{2}$, otherwise $\delta = (1 + \epsilon)$. We now add Steiner points $q_2, q_3, ..., q_{\mu_q-1}$ along e_q such that $|\overline{vq_j}| = r_v\delta^{j-1}$ where $\mu_q = \log_\delta(|e_q|/r_v)$. Similarly, add Steiner points $p_2, p_3, ..., p_{\mu_p-1}$ along e_p, where $\mu_p = \log_\delta(|e_p|/r_v)$. Define $\mathrm{dist}(a, e)$ as the minimum distance from a point a to an edge e. The segment from a to e will be a perpendicular to e. This strategy creates sets of Steiner points along edges e_q and e_p (see Figure 1a).

Claim 3.11. $|\overline{q_iq_{i+1}}| = \epsilon \cdot \mathrm{dist}(q_i, e_p)$ and $|\overline{p_jp_{j+1}}| = \epsilon \cdot \mathrm{dist}(p_j, e_q)$ where $0 < i < \mu_q$ and $0 < j < \mu_p$.

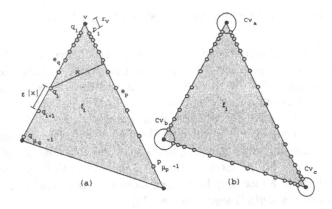

Fig. 1. a) Placement of Steiner points on the edges of f_i that are incident to vertex v. b) Results of merging Steiner points along edges.

Since we have added Steiner points based on the minimum angle θ_v about v, we obtain "concentric parallel wavefronts" centered at v consisting of Steiner point layers along the incident edges of v. Since this construction is made for each vertex of a face f_i, there will be two overlapping sets of Steiner points on each edge of f_i. To eliminate this overlap, we reduce the number of Steiner points on each edge. If two sets of Steiner points on an edge originate from the endpoints of an edge e, we determine the point on e where the interval sizes from each set are equal and eliminate all larger intervals. Intuitively, intervals are eliminated from one set if there are small intervals in the other set that overlap with it (see Figure 1b). The vertices of G_i will be Steiner points as well as the vertices of $\mathcal{P}$ defining f_i. The edges of G_i form a complete graph on its vertices. The graph G is defined to be the union $G_1 \cup G_2 \cup ...G_n$.

Claim 3.12. G is connected.

Claim 3.13. At most $m \leq 2(1 + \log_\delta(|L|/r))$ Steiner points are added to each edge of f_i, for $1 \leq i \leq n$.

Claim 3.14. G has $O(n \log_\delta(|L|/r))$ vertices and $O(n(\log_\delta(|L|/r))^2)$ edges.

Theorem 1. Let $0 < \epsilon < \frac{1}{2}$. Let $\mathcal{P}$ be a simple polyhedron with n faces and let s and t be two of its vertices. An approximation $\pi'(s,t)$ of a Euclidean shortest path $\pi(s,t)$ between s and t can be computed such that $|\pi'(s,t)| \leq (1 + \frac{3-2\epsilon}{1-2\epsilon}\epsilon)|\pi(s,t)|$. An approximation $\Pi'(s,t)$ of a weighted shortest path $\Pi(s,t)$ between s and t can be computed such that $\|\Pi'(s,t)\| \leq (1 + (2 + \frac{2W}{(1-2\epsilon)w})\epsilon)\|\Pi(s,t)\|$. The approximations can be computed in $O(mn \log mn + nm^2)$ time where $m = \log_\delta \frac{|L|}{r}$, and $\delta = (1 + \epsilon \sin \theta)$.

Proof. For both cases, we show that there exists a path in G that satisfies the claimed bounds using Lemma 2 and Lemma 4, respectively. Dijkstra's algorithm will either compute this path or a path with equal or better cost, and therefore the path computed by Dijkstra's algorithm as well satisfies the claimed approximation bounds. The running time of the algorithm follows from the size of the graph as stated in Claim 3.14. A variant of Dijkstra's algorithm using Fibonacci heaps [4] is employed to compute the path in the stated time bounds. □

3.2 Proof of Correctness

Consider a subgraph G_j, $1 \leq j \leq n$, as defined above. Let v be a vertex of a face f_j with edges e_p and e_q incident to v. We need the following technical lemma.

Lemma 1. Let s_i be the smallest segment contained within f_j such that one endpoint of s_i intersects e_q between q_i and q_{i+1} and the other endpoint intersects e_p. It holds that $|\overline{q_i q_{i+1}}| \leq \epsilon|s_i|$. Furthermore, if $\theta_v < \frac{\pi}{2}$ then s_i is a perpendicular bisector to e_p and if $\theta_v \geq \frac{\pi}{2}$ then $|s_i| \geq |\overline{vq_i}|$.

Let s_i be a segment of $\pi(s,t)$ (or $\Pi(s,t)$) crossing face f_i. Each s_i, must be of one of the following types:

    i) $s_i \cap C_v = \emptyset$, ii) $s_i \cap C_v = $ subsegment of s_i, or iii) $s_i \cap C_v = s_i$.

Let $C_{\sigma_1}, C_{\sigma_2}, ..., C_{\sigma_\kappa}$ be a sequence of spheres (listed in order from s to t) intersected by type ii) segments of $\pi(s,t)$ such that $C_{\sigma_j} \neq C_{\sigma_{j+1}}$. Now define subpaths of $\pi(s,t)$ (and $\Pi(s,t)$) as being one of two kinds:

Definition 1. <u>Between-sphere subpath:</u> *A path consisting of a type ii) segment followed by zero or more consecutive type i) segments followed by a type ii) segment. These subpaths will be denoted as* $\pi(\sigma_j, \sigma_{j+1})$ *($\Pi(\sigma_j, \sigma_{j+1})$ for weighted case) whose first and last segments intersect C_{σ_j} and $C_{\sigma_{j+1}}$, respectively. We will also consider paths that begin or/and end at a vertex to be a degenerate case of this type of path containing only type i) segments.*

Definition 2. <u>Inside-sphere subpath:</u> *A path consisting of one or more consecutive type iii) segments all lying within the same C_{σ_j}; these are denoted as $\pi(\sigma_j)$ ($\Pi(\sigma_j)$ for weighted case).*

Note that inside-sphere subpaths of $\pi(s,t)$ (and $\Pi(s,t)$) always lie between two between-sphere subpaths. That is, $\pi(\sigma_j)$ lies between $\pi(\sigma_{j-1}, \sigma_j)$ and $\pi(\sigma_j, \sigma_{j+1})$.

Claim 3.21. *Let s_i be a type i) segment with one endpoint between Steiner points q_j and q_{j+1} on edge e_q of a face f_i and the other endpoint between Steiner points p_k and p_{k+1} on edge e_p of f_i.*
Then $\max(\min(|\overline{q_j p_k}|, |\overline{q_j p_{k+1}}|), \min(|\overline{q_{j+1} p_k}|, |\overline{q_{j+1} p_{k+1}}|)) \leq (1+\epsilon)|s_i|.$

Claim 3.22. *Let s_i be a type ii) segment crossing edge e_q of f_i between Steiner points q_j and q_{j+1} and crossing e_p between v and Steiner point p_1, where $j \geq 1$ and v is the vertex common to e_q and e_p. Then $|\overline{q_1 q_j}|$ and $|\overline{q_1 q_{j+1}}|$ are less than $(1+\epsilon)|s_i|.$*

Bounding the Unweighted Approximation: We first describe the construction of an approximate path $\pi'(s,t)$ in G, given a shortest path $\pi(s,t)$ in $\mathcal{P}$. Consider a between-sphere subpath $\pi(\sigma_j, \sigma_{j+1})$, which consists of type i) and type ii) segments only. First examine a type i) segment s_i of $\pi(\sigma_j, \sigma_{j+1})$ that passes through edges e_q and e_p of face f_i. Assume s_i intersects e_q between Steiner points q_j and q_{j+1} and also intersects e_p between Steiner points p_k and p_{k+1}, where $j, k \geq 1$. The approximated path is chosen such that it enters face f_i through Steiner point q_j or q_{j+1}. W.l.o.g. assume that the approximated path enters f_i at q_j. Choose s_i' to be the shortest of $\overline{q_j p_k}$ and $\overline{q_j p_{k+1}}$. It is easily seen that $\pi'(\sigma_j, \sigma_{j+1})$ is connected since adjacent segments s_{i-1}' and s_{i+1}' share an endpoint (i.e., a Steiner point).

Now examine a type ii) segment of $\pi(\sigma_j, \sigma_{j+1})$; this can appear as the first or last segment. W.l.o.g. assume that it is the first segment. Let this segment enter f_i between Steiner points q_j and q_{j+1} and exit between vertex v_{σ_j} and Steiner point p_1 on e_p. Let $s_i' = \overline{q_1 q_j}$ (if s_i is the last segment, then we either choose s_i' to be $\overline{q_1 q_j}$ or $\overline{q_1 q_{j+1}}$ depending on at which Steiner point the approximated path up to f_i enters f_i). It is easily seen that the combination of these approximated segments forms a connected chain of edges in G which we will call $\pi'(\sigma_j, \sigma_{j+1})$. One crucial property of $\pi'(\sigma_j, \sigma_{j+1})$ is that it begins at a point where C_{σ_j} intersects an edge of $\mathcal{P}$ and ends at a point where $C_{\sigma_{j+1}}$ intersects an edge of $\mathcal{P}$.

Consider two consecutive between-sphere subpaths of $\pi(s,t)$, say $\pi'(\sigma_{j-1}, \sigma_j)$ and $\pi'(\sigma_j, \sigma_{j+1})$. They are disjoint from one another, however, the first path ends at sphere C_{σ_j} and the second path starts at C_{σ_j}. Join the end of $\pi'(\sigma_{j-1}, \sigma_j)$ and the start of $\pi'(\sigma_j, \sigma_{j+1})$ to vertex v_{σ_j} by two segments (which are edges of G). These two segments together will be denoted as $\pi'(\sigma_j)$. This step is repeated for each consecutive pair of between-sphere subpaths so that all subpaths are joined to form $\pi'(s, t)$. (The example of Figure 2 shows how between-sphere subpaths are connected to inside-sphere subpaths.) Constructing a path in this manner results in a continuous path that lies on the surface of $\mathcal{P}$.

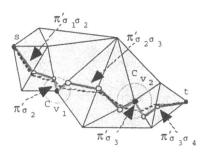

Fig. 2. An example showing the between-sphere and inside-sphere subpaths that connect to form the approximated path $\pi(s, t)$.

Claim 3.23. Let $\pi'(\sigma_{j-1}, \sigma_j)$ be a between-sphere subpath of $\pi'(s, t)$ corresponding to an approximation of $\pi(\sigma_{j-1}, \sigma_j)$ where $1 < j \leq \kappa$. Then $|\pi'(\sigma_j)| \leq \frac{2\epsilon}{1-2\epsilon}|\pi(\sigma_{j-1}, \sigma_j)|$, where $0 < \epsilon < \frac{1}{2}$.

Proof. From Property 2, the distance between $C_{\sigma_{j-1}}$ and C_{σ_j} must be at least $(1 - 2\epsilon)h_{v_{\sigma_j}}$. Since $\pi(\sigma_{j-1}, \sigma_j)$ is a between-sphere subpath, it intersects both $C_{\sigma_{j-1}}$ and C_{σ_j}. Thus $|\pi(\sigma_{j-1}, \sigma_j)| \geq (1 - 2\epsilon)h_{v_{\sigma_j}}$. By definition, $\pi'(\sigma_j)$ consists of exactly two segments which together have length satisfying $|\pi'(\sigma_j)| = 2r_{v_{\sigma_j}} = 2\epsilon h_{v_{\sigma_j}}$. Thus, $|\pi(\sigma_{j-1}, \sigma_j)| \geq \left(\frac{1-2\epsilon}{2\epsilon}\right)|\pi'(\sigma_j)|$ which can be re-written as $|\pi'(\sigma_j)| \leq \frac{2\epsilon}{1-2\epsilon}|\pi(\sigma_{j-1}, \sigma_j)|$. □

Lemma 2. If $\pi(s, t)$ is a shortest path in $\mathcal{P}$, where s and t are vertices of $\mathcal{P}$ then there exists an approximated path $\pi'(s, t) \in G$ for which $|\pi'(s, t)| \leq \left(1 + \frac{3-2\epsilon}{1-2\epsilon}\epsilon\right)|\pi(s, t)|$, where $0 < \epsilon < \frac{1}{2}$.

Proof. Using the results of Claim 3.22 and Claim 3.23, we can "charge" the cost of each inside-sphere subpath $\pi'(\sigma_j)$ to the between-sphere subpath $\pi'(\sigma_{j-1}, \sigma_j)$ as follows: $|\pi'(\sigma_{j-1}, \sigma_j)| + |\pi'(\sigma_j)| \leq (1+\epsilon)|\pi(\sigma_{j-1}, \sigma_j)| + \left(\frac{2\epsilon}{1-2\epsilon}\right)|\pi(\sigma_{j-1}, \sigma_j)| = \left(1 + \frac{3-2\epsilon}{1-2\epsilon}\epsilon\right)|\pi(\sigma_{j-1}, \sigma_j)|$. The union of all subpaths $\pi'(\sigma_{j-1}, \sigma_j)$ and $\pi'(\sigma_j)$ form $\pi'(s, t)$ where $2 \leq j \leq \kappa$. Hence, we have bounded $|\pi'(s, t)|$ w.r.t. the between-sphere subpaths of $\pi(s, t)$. Therefore
$$|\pi'(s, t)| \leq \left(1 + \frac{3-2\epsilon}{1-2\epsilon}\epsilon\right)\sum_{j=2}^{\kappa}|\pi(\sigma_{j-1}, \sigma_j)| \leq \left(1 + \frac{3-2\epsilon}{1-2\epsilon}\epsilon\right)|\pi(s, t)|.$$
□

Bounding the Weighted Approximation: Given a shortest path $\Pi(s,t)$, we construct a path $\Pi'(s,t)$ in a similar manner as in the unweighted scenario. However, we must consider the approximation of edge-using segments since they may no longer span the full length of the edge which they are using. Consider an edge-using segment s_i of $\Pi(s,t)$ on edge e_p of $\mathcal{P}$ with endpoints lying in Steiner point intervals $[p_y, p_{y+1}]$ and $[p_{u-1}, p_u]$ along e_p, where $y < u$. Let s_{i-1} and s_{i+1}, respectively, be the two crossing segments representing the predecessor and successor of s_i in the sequence of segments in $\pi(s,t)$. We will assume that two such edges exist although it is possible that s_{i-1} and s_i meet at a vertex of $\mathcal{P}$; which can easily be handled as well. We choose an approximation s'_i of s_i to be one of $\overline{p_y p_{u-1}}$ or $\overline{p_{y+1} p_{u-1}}$ depending on whether s'_{i-1} intersects e_p at p_y or p_{y+1}, respectively. Note that we make sure to choose s'_i so that it is connected to s'_{i-1}. Of course, s'_{i+1} will also be chosen to ensure connectivity with s'_i. In the degenerate case where $u = y + 1$, then there is no approximation for s_i. Instead, s'_{i-1} is connected directly to s'_{i+1}. In fact, Dijkstra's algorithm will never choose such a subpath since it does not make use of e_p. However, the path it does choose will be better than this, so our bound will hold for this better path as well.

Claim 3.24. *Let s_i be an edge-using segment of $\Pi(\sigma_j, \sigma_{j+1})$ and let s_{i-1} be the segment of $\Pi(\sigma_j, \sigma_{j+1})$ preceding s_i. There exists a segment s'_i of $\Pi'(\sigma_j, \sigma_{j+1})$ for which $\|s'_i\| \leq \|s_i\| + \epsilon\|s_{i-1}\|$.*

Lemma 3. *If $\Pi'(\sigma_{j-1}, \sigma_j)$ is a between-sphere subpath of $\Pi'(s,t)$ corresponding to an approximation of $\Pi(\sigma_{j-1}, \sigma_j)$ then $\|\Pi'(\sigma_{j-1}, \sigma_j)\| \leq (1+2\epsilon)\|\Pi(\sigma_{j-1}, \sigma_j)\|$.*

Claim 3.25. *Let $\Pi'(\sigma_{j-1}, \sigma_j)$ be a between-sphere subpath of $\Pi'(s,t)$ corresponding to an approximation of $\Pi(\sigma_{j-1}, \sigma_j)$ then $\|\Pi'(\sigma_j)\| \leq \frac{2\epsilon W}{(1-2\epsilon)w}\|\Pi(\sigma_{j-1}, \sigma_j)\|$ where $0 < \epsilon < \frac{1}{2}$.*

We have made the assumption that $\Pi'(\sigma_j)$ consists of segments passing through faces that have weight W. Although this may be true in the worst case, we could use the maximum weight of any face adjacent to v_{σ_j}, which typically would be smaller than W. In addition, we have assumed that $\Pi'(\sigma_{j-1}, \sigma_j)$ traveled through faces with minimum weight. We could determine the smallest weight of any face through which $\Pi'(\sigma_{j-1}, \sigma_j)$ passes and use that in place of w. This would lead to a better bound.

Lemma 4. *If $\Pi(s,t)$ is a shortest weighted path in $\mathcal{P}$, where s and t are vertices of $\mathcal{P}$ then there exists an approximated path $\Pi'(s,t) \in G$ such that $\|\Pi'(s,t)\| \leq (1 + (2 + \frac{2W}{(1-2\epsilon)w})\epsilon)\|\Pi(s,t)\|$ where $0 < \epsilon < \frac{1}{2}$.*

Proof. Using the results of Claim 3.25 and Lemma 3, it can be shown that
$$\|\Pi'(\sigma_{j-1}, \sigma_j)\| + \|\Pi'(\sigma_j)\| \leq (1 + (2 + \frac{2W}{(1-2\epsilon)w})\epsilon)\|\Pi(\sigma_{j-1}, \sigma_j)\|.$$
This essentially "charges" the length of an inside-sphere subpath to a between-sphere subpath. The union of all such subpaths form $\Pi'(s,t)$. This allows us to approximate $\Pi'(s,t)$ within the bound of $1 + (2 + \frac{2W}{(1-2\epsilon)w})\epsilon$ times the total cost of all the between-sphere subpaths of $\Pi(s,t)$. Since $\Pi(s,t)$ has cost at least that of its between-sphere subpaths, $\|\Pi'(s,t)\| \leq (1 + (2 + \frac{2W}{(1-2\epsilon)w})\epsilon)\|\Pi(s,t)\|$. $\qquad\square$

4 Reduced Approximation Graph

We show that some of the edges of the approximation graph G can be removed, so that for the obtained graph G^*, our results hold with a slightly worse constant. Since the reduced graph has only $O(nm \log mn)$ edges the running time of the resulting algorithm will improve substantially.

The graph $G^* = (V(G), E(G^*))$ is a subgraph of G having the same vertex set and $E(G^*) \subset E(G)$. We describe the construction of G^* by describing the choice of edges for $E(G^*)$. All edges in $E(G)$ that are subsegments of the edges of $\mathcal{P}$ remain in $E(G^*)$. The vertices of $\mathcal{P}$ in G^* are adjacent only to their neighboring Steiner points. Now we consider a fixed Steiner point p and describe the edges of G^* incident to p. Assume that p lies on an edge e of $\mathcal{P}$. By our construction of G the point p is connected to all Steiner points that lie on one of the four (or less) edges of $\mathcal{P}$ sharing a face (triangle) with e. We describe the construction on one of these edges, say e_1, and let $q_1, \ldots q_k$ be the Steiner points on e_1. Let M be the point closest to p on interval (q_1, q_k). The edges $\overline{pq_1}$, and $\overline{pq_k}$ are in $E(G^*)$. We choose the edges joining p with points in the subintervals (q_1, M) and (M, q_k) as follows: Consider the interval (M, q_k) and define a sequence of points $x_0, x_1, \ldots, x_\kappa$ in this interval, so that $|\overline{x_{i-1} x_i}| = \varepsilon |\overline{px_{i-1}}|$. Observe that there is at least one Steiner point in each of the intervals (x_{i-1}, x_i), for $i = 1, \ldots, \kappa$. Now, for $i = 1, \ldots \kappa$, we denote by q^i the Steiner point closest to x_{i-1} in the interval (x_{i-1}, x_i) and define $\overline{pq^i}$ to be edges of G^*. By the same procedure we define the subset of the Steiner points in (M, q_1) and connect p to the points in this subset. Omitting the technical proof, we claim that the out-degree of any Steiner point is $O(\log m)$; hence G^* has $O(nm \log mn)$ edges and any edge e in G can be approximated by a path e^* in G^* so that $\|e^*\|$ is an ϵ- approximation of $\|e\|$. The result is summarized in the following theorem:

Theorem 2. *An ϵ-approximate weighted shortest path between two vertices on a polytope consisting of n triangular faces can be computed in $O(nm \log mn)$ time, where m is a constant that depends upon ϵ and the geometric parameters as discussed before.*

5 Conclusions and Ongoing Work

We have presented algorithms to compute ϵ-approximate paths on weighted polyhedra. More specifically, the algorithms compute paths from the source vertex s to all vertices, Steiner points which are introduced on edges of the polyhedron. The techniques described in this paper can be used to derive algorithms for shortest path queries, as discussed in [1]. An alternative approach, which we are investigating, is to compute the relevant portion of the subgraphs G_i on the fly. It is clear that in Dijkstra's algorithm when the current vertex v (with least cost) explores the edges incident to it, we don't have to explore all of them because of the following: suppose the approximate path to v is through an edge $\overline{uv}$, then from v we need to explore those edges which obey Snell's law with respect to $\overline{uv}$. We suspect that the total number of edges that needs to be explored with this modification will be substantially lower. Moreover, we do not have to sacrifice the accuracy of the path obtained.

References

1. L. Aleksandrov, M. Lanthier, A. Maheshwari and J.-R. Sack, "An ϵ-Approximation Algorithm for Weighted Shortest Path Queries on Polyhedral Surfaces", to appear *14th European Workshop on Computational Geometry*, Barcelona, Spain, 1998.

2. J. Choi, J. Sellen and C.K. Yap, "Approximate Euclidean Shortest Path in 3-Space", *Proc. 10th Annual Symp. on Computational Geometry*, 1994, pp. 41-48.

3. G. Das and G. Narasimhan, "Short Cuts in Higher Dimensional Space", *Proceedings of the 7th Annual Canadian Conference on Computational Geometry*, Québec City, Québec, 1995, pp. 103-108.

4. M.L. Fredman and R.E. Tarjan, "Fibonacci Heaps and Their Uses in Improved Network Optimization Algorithms", *J. ACM*, **34**(3), 1987, pp.596-615.

5. J. Goodman and J. O'Rourke, Eds., *Handbook of Discrete and Computational Geometry*, CRC Press LLC, Chapter 24, 1997, pp. 445-466.

6. P. Johansson, "On a Weighted Distance Model for Injection Molding", Linköping Studies in Science and Technology, Thesis no. 604 LiU-TEK-LIC-1997:05, Division of Applied Mathematics, Linköping University, Linköping, Sweden, 1997.

7. C. Kenyon and R. Kenyon, "How To Take Short Cuts", *Discrete and Computational Geometry*, Vol. 8, No. 3, 1992, pp. 251-264.

8. M. Lanthier, A. Maheshwari and J.-R. Sack, "Approximating Weighted Shortest Paths on Polyhedral Surfaces", *Proceedings of the 13th Annual ACM Symposium on Computational Geometry*, 1997, pp. 274-283.

9. M. Lanthier, "Shortest Path Problems on Polyhedral Surfaces", *Ph.D. Thesis in progress*, School of Computer Science, Carleton University, Ottawa, Canada, 1998.

10. C. Mata and J. Mitchell, "A New Algorithm for Computing Shortest Paths in Weighted Planar Subdivisions", *Proceedings of the 13th Annual ACM Symposium on Computational Geometry*, 1997, pp. 264-273.

11. J.S.B. Mitchell and C.H. Papadimitriou, "The Weighted Region Problem: Finding Shortest Paths Through a Weighted Planar Subdivision", *Journal of the ACM*, **38**, January 1991, pp. 18-73.

12. C.H. Papadimitriou, "An Algorithm for Shortest Path Motion in Three Dimensions", *Information Processing Letters*, **20**, 1985, pp. 259-263.

13. J.-R. Sack and J. Urrutia Eds., *Handbook on Computational Geometry*, Elsevier Science B.V., to appear.

14. M. Sharir and A. Schorr, "On Shortest Paths in Polyhedral Spaces", *SIAM Journal of Computing*, **15**, 1986, pp. 193-215.

15. Paradigm Group Webpage, School of Computer Science, Carleton University, http://www.scs.carleton.ca/~gis.

Facility Location with Dynamic Distance Functions

(Extended Abstract)

Randeep Bhatia[1] [*], Sudipto Guha[2] [**], Samir Khuller[1] [***], and Yoram J. Sussmann[1] [†]

[1] Dept. of Computer Science, University of Maryland, College Park, MD 20742. Email addr: {`randeep,samir,yoram`}`@cs.umd.edu`
[2] Dept. of Computer Science, Stanford University. Email addr: `sudipto@cs.stanford.edu`

Abstract. Facility location problems have always been studied with the assumption that the edge lengths in the network are *static* and do not change over time. The underlying network could be used to model a city street network for emergency facility location/hospitals, or an electronic network for locating information centers. In any case, it is clear that due to traffic congestion the traversal time on links *changes* with time. Very often, we have estimates as to how the edge lengths change over time, and our objective is to choose a set of locations (vertices) as centers, such that at *every* time instant each vertex has a center close to it (clearly, the center close to a vertex may change over time). We also provide approximation algorithms as well as hardness results for the K-center problem under this model. This is the first comprehensive study regarding approximation algorithms for facility location for good time-invariant solutions.

1   Introduction

Previous theoretical work on facility location typically has addressed situations in which we want to locate facilities in a network and optimize an objective function. The edge lengths in the network (or distances) are typically assumed to be *static*. In practice however, edge lengths are not static. For example, in emergency facility location, the transit time may be a function of the traffic load at the current time. The same is true for locating information centers in

---

[*] Also with LCC Inc., 7925 Jones Branch Drive, McLean VA 22102.
[**] This work was done while this author was at the University of Maryland and his research was supported by NSF Research Initiation Award CCR-9307462.
[***] Research supported by NSF Research Initiation Award CCR-9307462, and NSF CAREER Award CCR-9501355.
[†] Research supported by NSF CAREER Award CCR-9501355.

networks. In this paper, we define a model of dynamic distance functions and study approximation algorithms for the K-center problem, a fundamental facility location problem, under this general model. Although in reality edge lengths may behave unpredictably, we often have estimates as to how they behave as a function of time at the macro level. In particular, the transit times in a city may oscillate periodically during the day. We will therefore assume that we have some knowledge of the behavior of edge lengths.

Our objective in this paper is to study the problem of placing facilities in such a way that *at all possible times we meet our objective criteria effectively*. We have to choose placements for the facilities, to minimize the distance d so that each node has a facility within distance d *at all possible times*.

The dynamic edge length model is a much more realistic model for studying many fundamental network problems. In this paper we initiate this study for the facility location problem and leave open a host of other problems. For example it is natural to study the problem of finding a spanning tree of the network whose maximum weight over time is minimum. Ravi and Goemans [21] studied this problem in the context of bicriterion approximation and their results imply a $1+\epsilon$ approximation for this problem when there are only two "time-slots". Similar bicriterion approximation results are known for the shortest path problem [27, 8]. Some of the other fundamental problems, such as finding a matching whose maximum weight over time is minimized, are completely open. Our experience with this general model has convinced us that even the simplest network problems are quite complex when studied under this model. For example, many of the problems which can be solved in polynomial time on a static network are NP-complete in the general model even when there are two time-slots. In addition, the techniques used for solving these problems on a static network do not easily extend to the general model.

We first discuss the basic K-center problem and then discuss our specific model and results.

K-Center Problem:

The basic K-center problem is a fundamental facility location problem [23, 7, 12, 10, 11] and is defined as follows: given an undirected edge-weighted graph $G = (V, E)$ find a subset $S \subseteq V$ of size at most K such that each vertex in V is "close" to some vertex in S. More formally, the objective function is defined as follows:

$$\min_{S \subseteq V, \, |S| \le K} \, \max_{u \in V} \, \min_{v \in S} d(u, v)$$

where d is the distance function defined by the underlying edge weights and hence satisfies the triangle inequality. For convenience we may assume that the graph is a complete graph, where each edge length is simply the distance between the corresponding vertices. For example, one may wish to install K fire stations and minimize the maximum distance (response time) from a location to its closest fire station. Some applications are mentioned in [17, 18]. The problem is known to be NP-hard [6].

1.1 Basic Notation

An approximation algorithm with a factor of ρ, for a minimization problem, is a polynomial time algorithm that guarantees a solution with cost at most ρ times the optimal solution. Approximation algorithms for the basic K-center problem [7, 10] that achieve an approximation factor of 2 are known to be optimal [7, 12, 10, 11]. Several approximation algorithms are known for interesting generalizations of the basic K-center problem as well [5, 11, 20, 1, 16, 15, 3, 14]. The generalizations include cases when each node v has an associated cost $c(v)$ for placing a center on it, and rather than limiting the number of centers, we have a limited budget K [11, 20, 26]. In this case we want to find a subset $S \subseteq V$ with the property $c(S) = \sum_{s \in S} c(s) \leq K$, with the objective function

$$\min_{S \subseteq V \ c(S) \leq K} \max_{u \in V} \min_{v \in S} d(u, v).$$

Other generalizations include cases where each node v has a weight $w(v)$ associated with it and we want to minimize the weighted distance from a node to its closest center [13, 4, 5, 20, 26]. This models the situation where nodes have a non-uniform importance level, and higher weight nodes are required to be closer to centers than lower weight nodes. More formally we want to find a subset $S \subseteq V$ of size at most K, where the objective function is

$$\min_{S \subseteq V \ |S| \leq K} \max_{u \in V} \min_{v \in S} w(u) d(u, v).$$

This does lead to an asymmetric notion of distance even though the function d is symmetric.

For the basic K-center problem, $w(v) = c(v) = 1$ for every node v. For the weighted (cost) problems $c(v) = 1$ ($w(v) = 1$) for every node v. For the most general version of the K-centers problem, we have both cost and weight functions defined on the vertices.

We address the case when the distance function *changes* with time. Assume that $d^t(u, v)$ is the distance from u to v at time t. We assume that this function is specified in some convenient manner (for example, it could be specified as a piecewise linear function, or as a step function). The function $d^t(u, v)$ could also be specified simply as a set of points $(t, d^t(u, v))$. These could be obtained for example by "sampling" transit times at different times of day. Our objective function (for the basic K-center problem) is

$$\min_{S \subseteq V \ |S| \leq K} \max_{u \in V} \max_{t} \min_{v \in S} d^t(u, v).$$

We also study other variants/generalizations of the K-center problem under this model. These results may be found in the full version of our paper [2].

A time-slot is defined to be those instants of time over which all edge lengths are invariant. We assume that time can be partitioned into T time-slots. Note that each time-slot t can be associated with a static distance function d^t, which is assumed to be distinct for each time-slot. Let β be the smallest value such that for any edge e_i, $\frac{\max_t d^t(e_i)}{\min_t d^t(e_i)} \leq \beta$. We call β the *traffic-load factor*.

A *dominating set* in a graph (digraph) is a subset S of vertices with the property that every vertex is either in S or is adjacent (has an edge) to some vertex in S. A *matching* M in a graph is a subset of edges that do not share a vertex in common. An *edge cover* S in a graph is a subset of edges such that every vertex is incident to at least one edge in S. A minimum-cost edge cover can be found in polynomial time [24, pages 580–583].

1.2 Our Results

We study the basic K-center problem as well as several variations of this problem under this model. We provide constant-factor approximation algorithms for all these problems when there are two time-slots (this models the situation of "rush hour" and "non-rush hour" travel times). For example, we could declare 7am to 9am and 4pm to 6pm as "rush hour" and all other times as "non-rush hour". (The rush hour could happen many times during the day. What is important is that there are only two distinct distance functions to consider, one for rush hour and one for non-rush hour.)

We show that even under the simple time-slot model for varying distances, if there are more than two time-slots then no constant approximation is possible. We also provide approximation algorithms for several variations of the K-center problem for arbitrarily many time-slots, including weights and costs, with factors that are close to the best possible unless $P = NP$. These results are summarized in Table 1. The only known lower bound for any of the K-center problems is 2, the lower bound for the basic K-center problem. It can be shown that the K-center problem with weights and costs is a generalization of the K-suppliers problem, for which a lower bound of 3 due to Karloff appears in [11]. For two time-slots with costs, all factors achieved in this paper match the best known factor for static distance functions. **We also study other variants/generalizations of the K-center problem under this model. These results may be found in the full version of our paper [2].**

Problem	2 time-slots		T time-slots	
	factor	lower bound	factor	lower bound
Basic + weights	3	2 [12]	$1 + \beta$	$\max(2, \beta)$
Basic + weights + costs	3	3 [26]	$1 + 2\beta$	$\max(3, \beta)$

* β is the maximum ratio of an edge's greatest length to its shortest length

Table 1. Results

We can solve all of the above problems in a unified manner using matching techniques. The algorithms for arbitrary time-slots are based on an extension of the Hochbaum-Shmoys method [10, 11].

Recently, we have learned that Hochbaum and Pathria [9] motivated by [22] have obtained a factor 3 approximation for the basic K-center problem as well (for 2 time-slots).

2 Hardness of Approximation for 3 time-slots

Lemma 1. *With three time-slots and no restriction on how the distance function can change with time, no approximation ratio for the K-center problem is possible unless $P = NP$.*

Proof. For contradiction let us assume that a polynomial time ρ-approximation algorithm exists, for some constant ρ. We show that we can use this algorithm to solve the 3-dimensional matching problem [6] in polynomial time.

Let the instance of the 3-dimensional matching problem ($3DM$) be the three sets A, B, C and a set of m ordered triples over $A \times B \times C$, where $|A| = |B| = |C| = K$. For every triple u let $u(i)$ denote its ith component. Note that $u(1) \in A, u(2) \in B$ and $u(3) \in C$. We create a graph G with m vertices, one for each ordered triple. G has an edge between every pair of vertices. We set $T = 3$. In the following we use the term vertices and triples interchangeably, as there is a one to one correspondence between them. We set:

$$d^i(u, v) = \begin{cases} 1 & \text{if } u(i) = v(i) \\ \rho + \epsilon & \text{otherwise} \end{cases}$$

First note that any solution to the $3DM$ instance corresponds to K centers which cover the vertices of G at distance at most one, in every time-slot. Hence if the ρ-approximation algorithm returns a solution of cost more than ρ then the $3DM$ instance has no solution. Let S be the solution of cost at most ρ returned by the ρ-approximation algorithm. We show that S is also a solution to the $3DM$ instance. Let $a \in A$. Let u be a triple for which $u(1) = a$ (if such a triple does not exist then we already know that the $3DM$ instance has no solution). Since u must be covered at distance at most ρ at time-slot 1, and since all the edges at any time-slot are of length 1 or $\rho + \epsilon$, there exists a triple $v \in S$ such that $v(1) = a$. Similarly for any $b \in B$ ($c \in C$) there exists a triple $u \in S$ such that $u(2) = b$ ($u(3) = c$). Also since $|S| \leq K$, S is a solution to the $3DM$ instance.

From the proof it is easy to conclude the following.

Corollary 1. *Unless $P = NP$ no polynomial time algorithm achieves a factor of $\beta - \epsilon$, for any $\epsilon > 0$, where β is the traffic-load factor.*

3 Approximation for 2 time-slots

In this section we present algorithms that find approximate solutions for several generalizations of the K-center problem with two time-slots.

We are given a graph $G = (V, E)$, and functions d^1, d^2 from the edge set E into $\mathcal{R}$, which denote the distances in the first and the second time-slots respectively.

High-Level Description:

We will use the standard approach of constructing a threshold graph as pioneered by Hochbaum and Shmoys [10,11]. We describe the algorithm for the most general case, when each node has a weight and an associated cost to construct centers.

The algorithm operates as follows. Consider the following set of weighted edge lengths:

$$\{w(u_i)d^t(u_i, u_j), w(u_j)d^t(u_j, u_i) \mid 1 \leq t \leq 2, 1 \leq i < j \leq |V|\}$$

Let $\ell_1, \ell_2, \ldots, \ell_p$ be the sorted list of these weighted edge lengths in increasing order.

The algorithm fixes a distance threshold $\delta = \ell_i$ for increasing values of i, and then considers the directed graphs G_δ^t one for each time slot $t = 1, 2$. The threshold graph G_δ^t includes only those edges whose weighted length in time slot t is at most the threshold δ. More formally $G_\delta^t = (V, E_\delta^t)$, where E_δ^t is the set of edges (u, v) with the property that $w(u)d^t(u, v) \leq \delta$. Note that if there is a solution to the original problem with distance δ, then there is a subset of vertices S, with cost at most K, such that S is a solution for each graph G_δ^t under the distance function d^t. The algorithm independently finds two solutions one for each threshold graph. Note that since the edge weights of the threshold graphs do not change with time, the algorithm can invoke existing approximation algorithms for solving the static version of the problem, for each threshold graph. The algorithm now uses the two solutions found above to reduce the problem to a minimum cost edge cover problem in an auxiliary graph to find a solution with cost K. The last step increases the approximation factor.

Note that the algorithm will search for the smallest threshold value for which it finds a solution with cost at most K. This can be done by linear or binary search on the list of edge lengths.

We illustrate this general technique in the following sections.

3.1 Basic K-centers

We find potential center locations in each graph G_δ^t for $t = 1, 2$, using the Hochbaum-Shmoys method[1] as follows. Assume that all vertices in G_δ^t are un-

---

[1] Strictly speaking, their method does not address the case when nodes have weights, but can be easily extended to handle node weights [20].

marked initially and $M_t = \emptyset$. (M_t is a set of vertices that are marked as a potential set of locations for placing centers for time slot t.) We repeatedly pick an unmarked vertex v with maximum weight and place a center on it, and also add it to M_t. We mark all vertices within weighted distance 2δ from v. Specifically, we mark all vertices x such that $w(x)d^t(x, v) \leq 2\delta$.

We now find final center locations using the potential centers in M_1 and M_2. We construct a new auxiliary graph G' on the vertex sets M_1 and M_2. A vertex in each set contributes to a distinct node in G'.

We define *neighborhood* as follows: Vertex z is in the neighborhood of $x \in M_t$ if $w(x)d^t(x, z) \leq \delta$. (Note that by this definition, a vertex is in its own neighborhood.)

Note that by construction a vertex can be in the neighborhood of only one vertex in M_t. If it was in the neighborhood of two nodes in M_t, say of x and y, then the first node to be added to M_t would mark the other node. (If $w(x) \geq w(y)$ then x is chosen before y and since $w(y)d^t(x, y) \leq w(y)d^t(y, z) + w(x)d^t(x, z) \leq 2\delta$, y will get marked when x is chosen.)

If a vertex z belongs to the neighborhood of $x \in M_1$ and $y \in M_2$, create an edge $e_z = (x, y)$ in G'. If a vertex z belongs to the neighborhood of $x \in M_1$ (M_2) but of no vertex in M_2 (M_1), create a self loop $e_z = (x, x)$ in G'. The cost of edge e_z is $c(z)$, and its label is z.

Find a minimum cost edge cover C of this graph G', and let V_C be the vertices corresponding to the labels on the edges in C. If $c(V_C) \leq K$, return V_C as the solution. Otherwise repeat the above procedure for a higher threshold value.

Theorem 1. *The above algorithm yields a solution of distance at most 3 times the distance for the optimal placement of centers.*

Proof. Assume there is a placement of centers with total cost at most K, such that every vertex has weighted distance at most δ from its nearest center in either time-slot. Let the set of centers be OPT. The placement of the centers has to be such that each center in OPT appears in the neighborhood in G_δ^1 (G_δ^2) of only one node in M_1 (M_2).

For each center $z \in OPT$, choose the edge e_z in G'. This set of edges covers all the nodes in G', guaranteeing the existence of an edge cover of cost at most K.

The distance bound follows from the fact that in each time-slot t and for every vertex $v \in V$, there is a vertex u in M_t within weighted distance at most 2δ from v (by the maximality of M_t). Since u is covered by some edge e_z in C, it is at most weighted distance δ from the node z in V_C that corresponds to the edge e_z. Thus $w(v)d^i(v, z) \leq w(v)d^i(v, u) + w(u)d^i(u, z) \leq 2\delta + \delta = 3\delta$. (Here we use the fact that the algorithm chooses an unmarked vertex of maximum weight.) Hence all the vertices in G are at most distance 3δ from some node in V_C in each time-slot.

Note: we have constructed an example (omitted here due to space constraints) that shows that the performance bound of 3 is tight for our algorithm.

Remark. We showed above how to obtain a solution to the K-center problem with two time-slots whose distance is at most 3 times the optimal. However, it is possible that one time-slot will have many short edges, in which case the value of δ that we use could be much larger than necessary *for that time-slot.* For example, if one time-slot represents rush hour and the other represents non-rush hour, then distances will be much shorter in the second time-slot. We cannot avoid long distances during rush hour, but we would like everyone to have a center nearby (in terms of travel time) during non-rush hour, since there is no reason not to.

We can guarantee this by using two different values δ_1 and δ_2, one for each time-slot, and solving each time-slot separately as before. To combine the solutions, we make the definition of *neighborhood* depend on the time-slot. That is, we say that vertex z is in the neighborhood of $x \in M_t$ if $w(x)d^t(x, z) \leq \delta_t$. Thus in each time slot t the weighted distance is at most $3\delta_t$.

4 Approximation for bounded variance on edge-lengths

In this section we consider the case when there are arbitrarily many time-slots. We restrict our attention to instances of the problem in which the ratio of an edge's maximum length to its minimum length is bounded by some constant β. This section considers the case of distinct time-slots with edge lengths given by a step function.

4.1 Basic K-centers

For each edge e_i we assume that

$$\max_t d^t(e_i) \leq \beta \cdot \min_t d^t(e_i).$$

We first give an algorithm for the weighted version (weights on the nodes), and then show how to modify it when there are costs associated with building centers as well. We note that getting an approximation factor of 2β is trivial by a direct extension of the algorithm for the K-center problem with static distance functions. We present a better algorithm with an approximation ratio of $1 + \beta$.

List all the weighted edge lengths

$$\{w(u_i)d^t(u_i, u_j), w(u_j)d^t(u_j, u_i) \mid 1 \leq t \leq T, 1 \leq i < j \leq |V|\}$$

Let $\ell_1, \ell_2, \ldots, \ell_p$ be the sorted list of these weighted edge lengths in increasing order. We will use the standard approach of constructing a threshold graph as pioneered by Hochbaum and Shmoys [10, 11].

The main idea behind the algorithm is to fix a distance threshold $\delta = \ell_i$ for increasing values of i, and to then consider the directed graph G_δ^t. For each time-slot t, let $G_\delta^t = (V, E_\delta^t)$, where E_δ^t is the set of edges (v, u) with the property that $w(v)d^t(v, u) \leq \delta$. (Note that by this definition, the self-loop (v, v) is also included.) If there is an optimal solution with optimal weighted distance δ, then there is an optimal subset S^* of K vertices that forms a dominating set for each graph G_δ^t. (In other words, at any point of time, each vertex is either in the dominating set or has an edge to some vertex in the dominating set.)

The algorithm works as follows: consider a vertex v. Assume that v has an edge to $u \in S^*$ at time-slot t. If we place a center at v for time-slot t, we would like to "cover" all the vertices that are covered by u during any time-slot. If we pick the heaviest unmarked vertex (at some time-slot t) at each step, we can cover all unmarked vertices that can cover it. By our choice of v, all the vertices covered by u can reach v by using at most two edges: one edge from E_δ^t and one edge from $E_\delta^{t'}$ for any time-slot t'. So we "mark" v together with all vertices w such that w can reach v by using at most two such edges. We mark the nodes for the time-slots during which they are covered: the goal is for every node to be marked for every time-slot. This guarantees that we cover each vertex within a distance of $(1 + \beta)\delta$. The algorithm is shown in Fig. 1.

Theorem 2. *The above algorithm returns a solution to the weighted time-varying facility location problem with an approximation ratio of $1 + \beta$, where $\beta = \max_e \times \frac{\max_t d^t(e)}{\min_t d^t(e)}$.*

Proof. Consider a vertex $x \notin S$. Let $1 \leq t \leq T$ and $marked^t(x)$ be set TRUE for the first time in the WHILE loop when a center is placed at $v \in S$ by the algorithm. Note that $w(x) \leq w(v)$. We have to argue that for each vertex that is marked for time-slot t, we can guarantee that x is within a weighted distance $(1 + \beta)\delta$ of $v \in S$ in time slot t. There are a few cases, and the proof for each case is shown as part of the pseudo-code in comments.

Since when the algorithm terminates all nodes not in S are marked for every time-slot, S dominates all vertices, in every time-slot, within a weighted distance of $(1 + \beta)\delta$.

It remains to show that if $|S| > K$, then the optimal solution using K centers has weighted distance strictly greater than δ.

Consider a vertex $v \in S$ and a vertex u in the optimal solution that covers v at some time t (u could be v). We will show that the selection of vertex v as a center, along with the choice t for time-slot, causes any node w covered by u at *any* time t', to be marked for t'. We have edge $(v, u) \in E_\delta^t$ and edge $(w, u) \in E_\delta^{t'}$.

Time-Invariant Bounded K-centers(G, δ).
1   $S = \emptyset$.
2   for $1 \leq t \leq T$
3   for all v
4           $marked^t(v) = $ FALSE.
5   while $\exists v \notin S, t \in [1, T]$ with $marked^t(v) = $ FALSE do
6   let v, t be such a pair for which $w(v)$ is maximized.
7   create center at v and set $S = S \cup v$.
8   for $(v, u) \in E_\delta^t$
9   for $1 \leq t' \leq T$
10   set $marked^{t'}(u) = $ TRUE.
                (* $w(u)d^{t'}(u, v) \leq w(v)\beta d^t(u, v) \leq \beta\delta$. *)
11   for $(w, u) \in E_\delta^{t'}$
12   set $marked^t(w) = $ TRUE.
                    (* $w(w)d^t(w, v) \leq w(w)d^t(w, u) + w(w)d^t(u, v) \leq \beta\delta + \delta$. *)
13   set $marked^{t'}(w) = $ TRUE.
                    (* $w(w)d^{t'}(w, v) \leq w(w)d^{t'}(w, u) + w(w)d^{t'}(u, v) \leq \delta + \beta\delta$. *)
14   for $1 \leq t' \leq T$
15   for $(v, u) \in E_\delta^{t'}$
16   set $marked^t(u) = $ TRUE.
                (* $w(u)d^t(u, v) \leq w(v)\beta d^{t'}(u, v) \leq \beta\delta$. *)
17   set $marked^{t'}(u) = $ TRUE.
                (* $w(u)d^{t'}(u, v) \leq w(v)d^{t'}(u, v) \leq \delta$. *)
18   for $(w, u) \in E_\delta^t$
19   set $marked^t(w) = $ TRUE.
                    (* $w(w)d^t(w, v) \leq w(w)d^t(w, u) + w(w)d^t(u, v) \leq \delta + \beta\delta$. *)
20   set $marked^{t'}(w) = $ TRUE.
                    (* $w(w)d^{t'}(w, v) \leq w(w)d^{t'}(w, u) + w(w)d^{t'}(u, v) \leq \beta\delta + \delta$. *)

Fig. 1. Algorithm for K-centers with bounded variance on edge lengths

Therefore vertex w is marked for time-slot t'. Since the selection of vertex v as a center, along with the choice t for time-slot, causes all vertices that are covered by u in any time-slot t' to be marked for time-slot t', the size of the optimal solution is at least $|S|$.

The cost case. Here we assume that vertices have cost and we have a limited budget to spend on the centers. Our algorithm for the cost case first runs the algorithm for the weighted case to get the set S of centers, and then just shifts each center in S to a node with the least cost, among all nodes that it has edges to (including itself). Let S' be the resulting set of centers.

Theorem 3. *The above algorithm returns a solution to the time-varying facility location problem with costs, with an approximation ratio of $1 + 2\beta$, where $\beta = \max_e \frac{\max_t d^t(e)}{\min_t d^t(e)}$.*

Proof. Note that by the proof of Thm. 2, in every time-slot each node is within a distance $(1 + \beta)\delta + \beta\delta$ of some node in S'. Also by the proof of Theorem 2, for each $v \in S$ there is a distinct optimal center x_v, which covers v in some time-slot. Hence if v is shifted to v' by the algorithm then $c(v') \leq c(x_v)$. Summing over all the nodes in S we get that the cost of the nodes in S' is a lower bound on the optimal cost.

Acknowledgments: We thank the anonymous referees for useful comments.

References

1. J. Bar-Ilan, G. Kortsarz and D. Peleg, "How to allocate network centers", *J. Algorithms*, 15:385–415, (1993).
2. R. Bhatia, S. Guha, S. Khuller and Y. J. Sussmann, "Facility location with dynamic distance functions", Technical Report CS-TR-3834, UMIACS-TR-97-70, Univ of Maryland (1997).
3. S. Chaudhuri, N. Garg, and R. Ravi, "Best possible approximation algorithms for generalized k-Center problems", Technical Report MPI-I-96-1-021, Max-Planck-Institut für Informatik, Im Stadtwald, 66123 Saarbrücken, Germany, (1996).
4. Z. Drezner, "The p-centre problem—heuristic and optimal algorithms", *J. Opl. Res. Soc.* Vol 35:741–748, (1984).
5. M. Dyer and A. M. Frieze, "A simple heuristic for the p-center problem", *Operations Research Letters*, Vol 3:285–288, (1985).
6. M. R. Garey and D. S. Johnson, *Computers and Intractability: A guide to the theory of NP-completeness*, Freeman, San Francisco, 1978.
7. T. Gonzalez, "Clustering to minimize the maximum inter-cluster distance", *Theoretical Computer Science*, Vol 38:293–306, (1985).
8. R. Hassin, "Approximation schemes for the restricted shortest path problems", *Mathematics of Operations Research*, Vol 17:36-42, No 1. Feb. 1992.
9. D. Hochbaum, personal communication, Oct (1996).
10. D. Hochbaum and D. B. Shmoys, "A best possible heuristic for the k-center problem", *Mathematics of Operations Research*, Vol 10:180–184, (1985).
11. D. Hochbaum and D. B. Shmoys, "A unified approach to approximation algorithms for bottleneck problems", *Journal of the ACM*, Vol 33(3):533–550, (1986).
12. W. L. Hsu and G. L. Nemhauser, "Easy and hard bottleneck location problems", *Discrete Applied Mathematics*, Vol 1:209–216, (1979).
13. O. Kariv and S. L. Hakimi, "An algorithmic approach to network location problems. I: The p-centers", *SIAM J. Appl. Math*, Vol 37:513–538, (1979).
14. S. Khuller, R. Pless, and Y. J. Sussmann, "Fault tolerant K-Center problems", *Proc. of the 3rd Italian Conference on Algorithms and Complexity*, LNCS 1203, pages 37–48, (1997).

15. S. Khuller and Y. J. Sussmann, "The capacitated K-Center problem", *Proc. of the 4th Annual European Symposium on Algorithms*, LNCS 1136, pages 152–166, (1996).

16. S. O. Krumke, "On a generalization of the p-center problem", *Information Processing Letters*, Vol 56:67–71, (1995).

17. H. L. Morgan and K. D. Levin, "Optimal program and data locations in computer networks", *Communications of the ACM*, Vol 20:315–322, (1977).

18. K. Murthy and J. Kam, "An approximation algorithm to the file allocation problem in computer networks", *Proc. of the 2nd ACM Symposium on Principles of Database Systems*, pages 258–266, (1983).

19. R. Panigrahy, "An $O(\log n)$ approximation algorithm for the asymmetric p-center problem", *manuscript*, 1995.

20. J. Plesnik, "A heuristic for the p-center problem in graphs", *Discrete Applied Mathematics*, Vol 17:263–268, (1987)

21. R. Ravi, M. X. Goemans, "The constrained minimum spanning tree problem", *SWAT 1996*, 66-75.

22. D. Serra and V. Marianov, "The P-median problem in a changing network: The case of Barcelona", paper presented at the *International Symposium in Locational Decisions VII, (ISOLDE)*, Edmonton, Canada, (1996).

23. C. Toregas, R. Swain, C. Revelle and L. Bergman, "The location of emergency service facilities", *Operations Research*, Vol 19:1363–1373, (1971).

24. J. Van Leeuwen, Ed., *Handbook of Theoretical Computer Science, Vol. A: Algorithms and Complexity*, The MIT Press/Elsevier, 1990.

25. S. Vishwanathan, "An $O(\log^* n)$ approximation algorithm for the asymmetric p-Center problem", *Proc. of the 7th Annual ACM-SIAM Symposium on Discrete Algorithms*, pages 1–5, (1996).

26. Q. Wang and K. H. Cheng, "A heuristic algorithm for the k-center problem with cost and usage weights", *Proc. of the Twenty-Eighth Annual Allerton Conference on Communication, Control, and Computing*, pages 324–333, (1990).

27. A. Warburton, "Approximation of pareto optima in multiple-objective, shortest path problems", *Operations Research* Vol 35:70–79, (1987).

An Approximation Scheme for Bin Packing with Conflicts [*]

Klaus Jansen[1]

IDSIA, Corso Elvezia 36, 6900 Lugano, Switzerland,
klaus@idsia.ch

Abstract. In this paper we consider the following bin packing problem with conflicts. Given a set of items $V = \{1, \ldots, n\}$ with sizes $s_1, \ldots, s_n \in (0, 1]$ and a conflict graph $G = (V, E)$, we consider the problem to find a packing for the items into bins of size one such that adjacent items $(j, j') \in E$ are assigned to different bins. The goal is to find an assignment with a minimum number of bins.

This problem is a natural generalization of the classical bin packing problem. We propose an asymptotic approximation scheme for the bin packing problem with conflicts restricted to d-inductive graphs with constant d. This graph class contains trees, grid graphs, planar graphs and graphs with constant treewidth. The algorithm finds an assignment for the items such that the generated number of bins is within a factor of $(1 + \epsilon)$ of optimal, and has a running time polynomial both in n and $\frac{1}{\epsilon}$.

1 Introduction

1.1 Problem Definition

In this paper we consider the following bin packing problem with conflicts. The input I of the problem consists of an undirected graph $G = (V, E)$ with a set of items $V = \{1, \ldots, n\}$ and sizes $s_1, \ldots, s_n$. We assume that each item size is a rational number in the interval $(0, 1]$. The problem is to partition the set V of items into independent sets or bins $U_1, \ldots, U_m$ such that $\sum_{i \in U_j} s_i \leq 1$ for each $1 \leq j \leq m$. The goal is to find a conflict-free packing with a minimum number m of bins. For any instance $I = (G = (V, E), (s_1, \ldots, s_n))$, let $SIZE(I) = \sum_{i=1}^{n} s_i$ denote the total size of the n items, and let $OPT(I)$ denote the minimum number of unit size bins needed to pack all items without conflicts. For graph classes not defined in this paper we refer to [8].

One application of the problem is the assignment of processes to processors. In this case, we have a set of processes (e.g. multi media streams) where some of the processes are not allowed to execute on the same processor. This can be for

---

[*] This research was done while the author was associated with the MPI Saarbrücken and was supported partially by the EU ESPRIT LTR Project No. 20244 (ALCOM-IT) and by the Swiss Office Fédéral de l'éducation et de la Science project n 97.0315 titled "Platform".

reason of fault tolerance (not to schedule two replicas of the same process on the same cabinet) or for efficiency purposes (better put two cpu intensive processes on different processors). The problem is how to assign a minimum number of processors for this set of processes. A second application is given by storing versions of the same file or a database. Again, for reason of fault tolerance we would like to keep two replicas / versions of the same file on different file server.

Another problem arises in load balancing the parallel solution of partial differential equations (pde's) by two-dimensional domain decomposition [2, 1]. The domain for the pde's is decomposed into regions where each region corresponds to a subcomputation. The subcomputations are scheduled on processors so that subcomputations corresponding to regions that touch at even one point are not performed simultaneously. Each subcomputation j requires one unit of running time and s_j gives the amount of a given resource (e.g. number of used processors or the used storage). The goal of the problem is to find a schedule with minimum total completion time. In general, the created conflict graphs are nonplanar. But if the maximum number of regions touching at a single point is constant, then the mutual exclusion graph is d-inductive with constant d. Other applications are in constructing school course time tables [17] and scheduling in communication systems [10].

1.2 Results

If E is an empty set, we obtain the classical bin-packing problem. Furthermore, if $\sum_{j \in V} s_j \leq 1$ then we obtain the problem to compute the chromatic number $\chi(G)$ of the conflict graph G. This means that the bin packing problem with conflicts is NP-complete even if $E = \emptyset$ or if $\sum_{j \in V} s_j \leq 1$. We notice that no polynomial time algorithm has an absolute worst case ratio smaller than 1.5 for the bin packing problem, unless $P = NP$. This is obvious since such an algorithm could be used to solve the partition problem [7] in polynomial time. For a survey about the bin packing problem we refer to [4]. The packing problem for an arbitrary undirected graph is harder to approximate, because Feige and Kilian [5] proved that it is hard to approximate the chromatic number to within $\Omega(|V|^{1-\epsilon})$ for any $\epsilon > 0$, unless $NP \subset ZPP$.

In [1,3] the bin packing problem with conflicts and with unit-sizes ($s_j = \frac{1}{\ell}$ for each item $j \in V$) was studied. Baker and Coffman called this packing problem (with unit sizes) Mutual Exclusion Scheduling (short: MES). In [3] the computational complexity of MES was studied for different graph classes like bipartite graphs, interval graphs and cographs, arbitrary and constant numbers m of bins and constant ℓ. Lonc [16] showed that MES for split graphs can be solved in polynomial time. Baker and Coffman [1] have proved e.g. that forest can be scheduled optimally in polynomial time and have investigated scheduling of planar graphs resulting from a two-dimensional domain decomposition problem. A linear time algorithm was proposed in [14] for MES restricted to graphs with constant treewidth and fixed m. Furthermore, Irani and Leung [10] have studied on-line algorithms for interval and bipartite graphs.

In [12], we have proposed several approximation algorithms A with constant absolute worst case bound $A(I) \leq \rho \cdot OPT(I)$ for the bin packing problem with conflicts for graphs that can be colored with a minimum number of colors in polynomial time. Using a composed algorithm (an optimum coloring algorithm and a bin packing heuristic for each color set), we have obtained an approximation algorithm with worst case bound ρ between 2.691 and 2.7. Furthermore, using a precoloring method that works for e.g. interval graphs, split graphs and cographs we have an algorithm with bound 2.5. Based on a separation method we have got an algorithm with worst case ratio $2 + \epsilon$ for cographs and graphs with constant treewidth.

A d-inductive graph introduced in [9] has the property that the vertices can be assigned distinct numbers $1, \ldots, n$ in such a way that each vertex is adjacent to at most d lower numbered vertices. We assume that an order $v_1, \ldots, v_n$ is given such that $|\{v_j | j < i, \{v_j, v_i\} \in E\}| \leq d$ for each $1 \leq i \leq n$. We notice that such an order (if one exists) can be obtained in polynomial time. In other words, the problem to decide whether a graph is d - inductive can be solved in polynomial time [11]. If G is d-inductive then there must be a vertex v in G of degree at most d. We can take $v_n = v$ and can compute recursively an order $v_1, \ldots, v_{n-1}$ for the remaining induced subgraph $G[V \setminus \{v_n\}]$. Since the property d-inductive is hereditary for induced subgraphs of G, we cannot reach a subgraph with all vertices of degree larger than d.

It is clear that $d + 1$ is an upper bound on the chromatic number of any d-inductive graph and that a $(d + 1)$-coloring can be computed in polynomial time using the ordering. As examples, planar graphs are 5-inductive and graphs with constant treewidth k are k-inductive. Moreover, trees are 1-inductive and grid graphs are 2-inductive.

The goal is to find an algorithm A with a good asymptotic worst case bound, that means that

$$limsup_{OPT(L) \to \infty} \frac{A(I)}{OPT(I)}$$

is small. In this paper, we give an asymptotic approximation scheme for the bin packing problem with conflicts restricted to d-inductive graphs. Our main result is the following:

Theorem 1. *There is an algorithm A which, given a set of items $V = \{1, \ldots, n\}$ with sizes $s_1, \ldots, s_n \in (0, 1]$, a d-inductive graph $G = (V, E)$ with constant d and a positive number ϵ, produces a packing of the items without conflicts into at most*

$$A_\epsilon(I) \leq (1 + \epsilon)OPT(I) + O(\frac{1}{\epsilon^2})$$

bins. The time complexity of A is polynomial in n and $\frac{1}{\epsilon}$.

2 The Algorithm

In the first step of our algorithm, we remove all items with sizes smaller than $\delta = \frac{\epsilon}{2}$ and consider a restricted bin packing problem as proposed by Fernandez de la Vega and Lueker [6].

2.1 Restricted Bin Packing

For all $0 < \delta < 1$ and positive integers m, the *restricted bin packing* problem $RBP[\delta, m]$ is defined as the bin packing problem (without considering conflicts) restricted to instances where the item sizes take on at most m distinct values and each item size is at least as large as δ. An input instance for $RBP[\delta, m]$ can be represented by a multiset $M = \{n_1 : s_1, n_2 : s_2, \ldots, n_m : s_m\}$ such that $1 \geq s_1 > s_2 > \ldots > s_m \geq \delta$. Furthermore, a packing of a subset of items in a unit size bin B is given also by a multiset $B = \{b_1 : s_1, b_2 : s_2, \ldots, b_m : s_m\}$ such that b_i is the number of items of size s_i that are packed into B. For fixed M, a feasible packing B^t is denoted by a m-vector (called *bin type*) $(b_1^t, \ldots, b_m^t)$ of non-negative integers such that $\sum_{i=1}^{m} b_i^t s_i \leq 1$. Two bins packed with items from M have the same *bin type* if the corresponding packing vectors are identical. Using the parameter $\delta = \frac{\epsilon}{2}$, the number of items in a bin is bounded by a constant $\sum_{i=1}^{m} b_i^t \leq \lfloor \frac{1}{\delta} \rfloor = \lfloor \frac{2}{\epsilon} \rfloor$.

Given a fixed set $S = \{s_1, \ldots, s_m\}$, the collection of possible bin types is fully determined and finite. The number q of bin types with respect to S can be bounded by $\binom{m+\ell}{\ell}$ where $\ell = \lfloor \frac{1}{\delta} \rfloor$ [6]. Fernandez de la Vega and Lueker [6] used a linear grouping method to obtain a restricted bin packing instance $RBP[\delta, m]$ with fixed constant $m = \lfloor \frac{n'}{k} \rfloor$ where n' is the number of large items (greater than or equal to δ) and where $k = \lceil \frac{\epsilon^2 n'}{2} \rceil$. Since the number q of bin types is also a fixed constant, these bin packing instances can be solved in polynomial time using a integer linear program [6].

A solution x to an instance M of $RBP[\delta, m]$ is a q-vector of non negative integers $(x_1, \ldots, x_q)$ where x_t denotes the number of bins of type B^t used in x. A q-vector is *feasible*, if and only if

$$\sum_{t=1}^{q} x_t b_i^t = n_i \; \forall 1 \leq i \leq m$$
$$x_t \in \mathbb{N}_0 \qquad \forall 1 \leq t \leq q$$

where n_i is the number of items of size s_i. We get the integer linear program using the (in-)equalities above and the objective function $\sum_{t=1}^{q} x_t$ (the number of bins in the solution). Let $LIN(I)$ denote the value of the optimum solution for the corresponding linear program. Karmarkar and Karp [15] gave a polynomial time algorithm A for the restricted bin packing problem such that

$$A(I) \leq LIN(I) + 1 + \frac{m+1}{2}.$$

Their algorithm runs in time polynomial in n and $\frac{1}{\epsilon}$ and produces an integral solution for the large items with at most m non-zero components (or bin types)

x_t. Considering the linear grouping method in [6], we get k additional bins with one element. In total, the number of bins generated for the instance J with the large items is at most

$$OPT(J) + 1 + \frac{m+1}{2} + k$$

where k is at most $\epsilon \cdot OPT(J) + 1$.

2.2 Generation a Solution without Conflicts

The algorithm of Karmarkar and Karp generates a packing of the large items into bins, but with some possible conflicts between the items in the bins. In this subsection, we show how we can modify the solution to get a conflict - free packing for the large items. In the following, we consider a non-zero component x_t in the solution of the algorithm A of Karmarkar and Karp. The idea of our algorithm is to compute for a packing with x_t bins of type B^t a conflict free packing that uses at most a constant number of additional bins.

Let $C_1, \ldots, C_{x_t} \subset V$ be a packing into x_t bins of type t. Each set of items C_i can be packed into a bin of type $B^t = (b_1^t, \ldots, b_m^t)$. That means that

$$|\{v \in C_i | s_v = s_j\}| \leq b_j^t$$

for each $1 \leq j \leq m$ and each $1 \leq i \leq x_t$. We may assume that the cardinalities $|\{v \in C_i | s_v = s_j\}|$ are equal to b_j^t for each set C_i; otherwise we insert some dummy items. We define with $M = \sum_{j=1}^m b_j^t$ the number of items in each bin of type B^t. Notice that the number M of items in a bin is bounded by the constant $\lfloor \frac{2}{\epsilon} \rfloor$.

Next, we consider the subgraph of G induced by the vertex set $C_1 \cup \ldots \cup C_{x_t}$ and label the vertices in this subgraph with $1, \ldots, M$ as follows. We sort the items in each set C_i in non-increasing order of their sizes and label the corresponding vertices (in this order) by $1, \ldots, M$. Two items x, y with the same label (possibly in two different sets C_x and C_y) have the same size, and one largest item in each set C_i gets label 1. Let $\ell(v)$ be the the label of v. The key idea is to compute independent sets U by a greedy algorithm with the property that

$$\{\ell(v) | v \in U\} = \{1, \ldots, M\}.$$

Each independent set U with this property can be packed into one bin. Moreover, the bin type of a packing for such a set U is again B^t. In general, the problem to find an independent set U with different labels $1, \ldots, M$ is NP-complete even in a forest.

Theorem 2. *The problem to find an independent set U with labels $1, \ldots, M$ in a labelled forest $W = (V, E)$ with $\ell : V \to \{1, \ldots, M\}$ is NP-complete.*

Proof. By a reduction from a satisfiability problem. Let $\alpha = c_1 \wedge \ldots \wedge c_m$ be a formula in conjunctive normal form, with two or three literals for each clause $c_i = (y_{i1} \vee y_{i2} \vee y_{i3})$ or $c_i = (y_{i1} \vee y_{i2})$ and with $y_{ij} \in \{x_1, \bar{x}_1, \ldots, x_n, \bar{x}_n\}$. We may assume (see [13]) that each variable x_k occurs either

(a) once unnegated and twice negated or
(b) once negated and twice unnegated.

We build a forest W with vertex set $V = \{a_{ij}|y_{ij}$ is in $c_i, 1 \le i \le m\}$ and labelling $\ell(a_{ij}) = i$ for $1 \le i \le m$. The edge set E is given as follows: for each variable x_k connect the vertices a_{ij} and $a_{i'j'}$, if and only if $y_{ij} = x_k$, $y_{i'j'} = \bar{x}_k$ and $(i, j) \ne (i', j')$. Using the property of the variables above, W forms a forest. Then, we can prove that α is satisfiable, if and only if there is an independent set of size m with labels $1, \ldots, m$ $\qquad\square$

Our method is based on the following idea: if we have enough vertices in a d-inductive graph, then we can find an independent set with labels $1, \ldots, M$ in an efficient way.

Lemma 1. *Let $G = (V, E)$ be a d-inductive graph with constant d and $|V| = M \cdot L$ vertices, and a labelling $\ell : V \to \{1, \ldots, M\}$ such that each label occurs exacly L times. If $L \ge d(M - 1) + 1$ then there exists an independent set U in G with labels $1, \ldots, M$.*

Proof. If $M = 1$ then we have at least one vertex with label 1. For $M \ge 2$ we choose a vertex $v \in V$ with degree $\le d$. We may assume that v is labelled with label $\ell(v) = 1$. Next, we delete all vertices in G with label 1.

Case 1: There is a vertex $v' \in \Gamma(v)$ with label $\ell(v') \ne 1$. We may assume that $\ell(v') = 2$ and that

$$|\{w \in \Gamma(v)|\ell(w) = 2\}| \ge |\{w \in \Gamma(v)|\ell(w) = i\}| \; \forall i \in \{3, \ldots, M\}.$$

Then, we delete exactly $a_2 = |\{w \in \Gamma(v)|\ell(w) = 2\}|$ vertices in G with labels $2, \ldots, M$ where we prefer the vertices in the neighbourhood of v. All vertices $w \in \Gamma(v)$ are removed after this step. In this case, we obtain a d-inductive graph $G' = (V', E')$ with $(L - a_2) \cdot (M - 1)$ vertices and labelling $\ell : V' \to \{2, \ldots, M\}$ such that each label $2, \ldots, M$ occurs exactly $L - a_2 \ge L - d$ times. Since $L' = L - a_2 \ge L - d \ge d(M - 2) + 1$, we find per induction an independent set U' in G' with labels $2, \ldots, M$. Since v is not adjacent to the vertices in U', the set $U = U' \cup \{v\}$ has the desired properties.

Case 2: All vertices $v' \in \Gamma(v)$ have label $\ell(v') = 1$, or v has no neighbour. In this case, we have directly a graph $G' = (V', E')$ with $(M - 1) \cdot L$ vertices and a labelling $\ell : V' \to \{2, \ldots, M\}$ where each label occurs exactly L times. Again, we find (by induction) an independent set U' with labels $2, \ldots, M$ that can be extended to the independent set $U = U' \cup \{v\}$ with labels $1, \ldots, M$. $\qquad\square$

The first idea is to compute for each bin type B^t and the d-inductive graph $G_t = G[C_1 \cup \cdots \cup C_{x_t}]$ with labels $1, \ldots, M_t$ (the number of labels depends on the bin type) a partition into conflict-free independent sets as follows:

(a) if G_t contains more than $M_t(d(M_t - 1) + 1)$ vertices, then we find an independent set U with labels $1, \ldots, M_t$ using the algorithm in Lemma 1,
(b) otherwise we take for each vertex v in G_t a separate bin.

Since the numbers M_t and d are fixed constants, we obtain using this idea only a constant number of additional bins. Another and better idea is to analyse the coloring problem for a labelled d-inductive graph. The first result is negative.

Theorem 3. *The problem to decide whether a forest $W = (V_W, E_W)$ with labelling $\ell : V_W \to \{1, \ldots, M\}$ can be partitioned into three independent sets U_1, U_2, U_3 where each independent set U_i contains only vertices with different labels is NP-complete.*

Proof. We use a reduction from the 3-coloring problem with no vertex degree exceeding 4, which is NP-complete, to the coloring problem for a labelled forest. Let $G = (V, E)$ be a graph with $V = \{1, \ldots, n\}$ and maximum degree 4. We substitute for each node $1, \ldots, n$ a graph H_v and construct a forest W with labelling ℓ such that G is 3-colorable if and only if W can be partitioned into three independent sets, each with different labels.

For the node substitution, we use a graph H_v with vertex set $\{v(j, k) | 1 \le j, \le 3, 1 \le k \le 4\}$ and edge set

$$\{(v(j, k), v(j', k)) | 1 \le j \ne j' \le 3, 1 \le k \le 4\}$$
$$\cup \{(v(1, k), v(2, k - 1)), (v(1, k), v(3, k - 1)) | 2 \le k \le 4\}.$$

The graph H_v has 4 outlets, labelled by $v(1, k)$. The graph H_v has the following properties:

(a) H_v is 3-colorable, but not 2-colorable,
(b) for each 3-coloring f of H_v we have $f(v(1, 1)) = f((v(1, 2)) = f(v(1, 3)) = f(v(1, 4))$,
(c) The graph K_v which arises from H_v by deleting the edges

$$\{(v(1, k), v(2, k - 1)), (v(1, k), v(3, k - 1)) | 2 \le k \le 4\}$$
$$\cup \{(v(2, k), v(3, k)) | 1 \le k \le 4\}$$

is a forest.

For each edge $e = (v, v') \in E$ of the original graph G we choose a pair of vertices $v(1, k_{e,v})$ and $v'(1, k_{e,v'})$ in the graphs K_v and $K_{v'}$ and connect these vertices. Clearly, we can choose different vertices $v(1, k_{e,v})$ and $v(1, k_{e',v})$ in K_v for different edges e, e' incident to the same vertex v. If we insert these connecting edges in the graph $\bigcup_{v \in V} K_v$, we obtain our forest W for the reduction.

Moreover, we choose a labelling $\ell : V_W \to \{1, \ldots, 5n\}$ as follows. For $v = 1, \ldots, n$ we define

$$\ell(v(1, 2)) = \ell(v(2, 1)) = \ell(v(3, 1)) = 4(v - 1) + 1,$$
$$\ell(v(1, 3)) = \ell(v(2, 2)) = \ell(v(3, 2)) = 4(v - 1) + 2,$$
$$\ell(v(1, 4)) = \ell(v(2, 3)) = \ell(v(3, 3)) = 4(v - 1) + 3,$$
$$\ell(v(2, 4)) = \ell(v(3, 4)) = 4(v).$$

The remaining vertices $v(1, 1)$ get different labels between $4n + 1$ and $5n$. Using this construction, we can prove that G is 3-colorable if and only if the union of the forest W and the disjoint union of the $5n$ complete graphs (one complete graph for each label) is 3-colorable. This proves the theorem. □

In the following, we show that the coloring problem can be approximated for labelled d-inductive graphs.

Lemma 2. *Let $G = (V, E)$ be a d-inductive graph and let $\ell : V \to \{1, \ldots, M\}$ be a labelling where each label occurs at most L times. Then, it is possible to compute a partition of G into at most $L + d$ independent sets $U_1, \ldots, U_{L+d}$ such that $|\{u \in U_j | \ell(u) = i\}| \leq 1$ for each label $i \in \{1, \ldots, M\}$ and each $1 \leq j \leq L+d$.*

Proof. Let v be a vertex with degree at most d. Per induction on $|V|$, we have a partition for $V' = V \setminus \{v\}$ into at most $L + d$ independent sets $U_1, \ldots, U_{L+d}$ with the property above. Since v has at most d neighbours and there are at most $L - 1$ other vertices with label $\ell(v)$, there is at least one independent set U_i $(1 \leq i \leq L+d)$ that does not contain a neighbour of v or a vertex with label $\ell(v)$. This implies that the partition $U_1, \ldots, U_{i-1}, U_i \cup \{v\}, U_{i+1}, \ldots, U_{L+d}$ has the desired property. □

This Lemma gives us an approximation algorithm for the coloring problem of labelled d - inductive graphs with additive factor d (since we need at least L colors). We use this approximation algorithm for each bin type B^t with non - zero component x_t. Since we have only m non - zero components x_t, the total number of bins with only conflict free items in each bin can be bounded by

$$(1 + \epsilon)OPT(J) + \frac{m+1}{2} + 2 + m \cdot d$$

where $m \leq \frac{2}{\epsilon^2}$.

2.3 Insertion of Small Items

In the following, we show how we can insert the small items into a sequence of bins with only large items. The proof contains also an algorithm to do this in an efficient way.

Lemma 3. *Let $\delta \in (0, 1/2]$ be a fixed constant. Let I be an instance of the bin packing problem with conflicts restricted to d-inductive graphs and suppose that all items of size $\geq \delta$ have been packed into β bins. Then it is possible to find (in polynomial time) a packing for I which uses at most*

$$max[\beta, (1 + 2\delta)OPT(I)] + (3d + 1)$$

bins.

Proof. The idea is to start with a packing of the large items into β bins and to use a greedy algorithm to pack the small items. Let $B_1, \ldots, B_{\beta'}$ (with $\beta' \leq \beta$) be the bins in the packing of the large items with sizes at most $1 - \delta$ and let I_δ be the set of items in I with sizes $< \delta$. First, we order all items with respect to the d-inductive graph. We obtain an order $v_1, \ldots, v_n$ of the vertices such that vertex v_i has at most d neighbours with lower index for $1 \leq i \leq n$. In this order (restricted to the small items), we try to pack the small items into the first β' bins with sizes at most $1 - \delta$. If a bin $B_i \cup \{v_j\}$ forms an independent set, then we can place v_j into the corresponding bin since the size $SIZE(B_i \cup \{v_j\}) \leq 1$. If the size of such an enlarged bin is now larger than $1 - \delta$, then we remove B_i from the list of the small bins and set $\beta' = \beta' - 1$. After this step, the remaining bins have sizes at most $1 - \delta$. It is possible that some small items v_j can not be packed into the first bins.

At the end of this algorithm, we have a list of bins $B_1, \ldots, B_{\beta''}$ with $0 \leq \beta'' \leq \beta'$ with sizes at most $1 - \delta$ and a set $I'_\delta \subset I_\delta$ of items with sizes $< \delta$ such that

(*) for each item $v_j \in I'_\delta$ and each index $1 \leq i \leq \beta''$ there exists an item $x \in B_i$ such that v_j is in conflict with x.

The other jobs in $I_\delta \setminus I'_\delta$ are placed either in one of the bins $B_1, \ldots, B_{\beta''}$ or in a bin of size now larger than $1 - \delta$. We notice that $\beta - \beta''$ bins of sizes larger than $1 - \delta$ are generated.

The key idea is to give lower and upper bounds for the number $LARGE(I, \beta'')$ of large jobs in the first β'' bins of sizes at most $1 - \delta$. Notice that we count only the large items in these bins.

$$(a) \quad LARGE(I, \beta'') \leq \beta'' \cdot \lfloor \tfrac{1}{\delta} \rfloor$$

Each large job j has size $\geq \delta$. A packing vector $B = (b_1, \ldots, b_m)$ corresponding to the bin with large numbers has the property that $\sum_{i=1}^m b_i s_i \leq 1$. This implies that $\delta \cdot \sum_{i=1}^m b_i \leq 1$ or equivalent $\sum_{i=1}^m b_i \leq \lfloor \tfrac{1}{\delta} \rfloor$ (since the numbers b_i are non-negative integers). Since we have β'' bins with at most $\lfloor \tfrac{1}{\delta} \rfloor$ large items, we get the inequality (a). This inequality holds for each undirected graph.

$$(b) \quad LARGE(I, \beta'') \geq |I'_\delta| \cdot \tfrac{\beta'' - d}{d}$$

First, we show that each small item $v_i \in I'_\delta$ can not be packed with at least $\beta'' - d$ large items $v_{j_1}, \ldots, v_{j_{\beta''-d}}$ that are in conflict with v_i and that have a larger index than i. It holds $i < j_1 < \ldots < j_{\beta''-d}$. To prove this consider the time step at which v_i can not be packed into one of the first bins with size at most $1 - \delta$. At this time step, we have at least β'' items that are in conflict with v_i (one item for each of these bins). Since v_i has at most d adjacent vertices with smaller index, there are at least $\beta'' - d$ conflict jobs with larger index. These $\beta'' - d$ items have sizes $\geq \delta$, since the other small items with larger index have not bben considered before.

In total, we get $|I'_\delta|(\beta'' - d)$ large conflict jobs in the first β'' bins. Since each large item v_j can be reached only by at most d small items with smaller index, each of these large items is counted at most d times. Therefore, we have at least $|I'_\delta| \frac{(\beta''-d)}{d}$ large items in the first β'' bins. This shows the inequality (b).

Combining the lower and upper bound, we get

$$|I_\delta'|\frac{(\beta''-d)}{d}\leq\beta''\lfloor\frac{1}{\delta}\rfloor.$$

This implies that

$$|I_\delta'|\leq\frac{d}{(\beta''-d)}\cdot\beta''\lfloor\frac{1}{\delta}\rfloor.$$

for $\beta'' > d$. If $\beta'' \geq 2d$, then we obtain the upper bound $|I_\delta'| \leq 2d\lfloor\frac{1}{\delta}\rfloor$. In this case, we compute a $(d+1)$ coloring for the items in I_δ'. Since all items $v_j \in I_\delta'$ have sizes at most δ, we can place at least $\lfloor\frac{1}{\delta}\rfloor$ of these items into one bin. Using a next fit heuristic for each color set, we obtain at most $3d+1$ bins for the items in I_δ'. In total, we obtain in this case at most $\beta + 3d + 1$ bins.

If $\beta'' \leq 2d$, we have at least $\beta - 2d$ bins with sizes larger than $1 - \delta$. The remaining jobs in I_δ' can be packed (using a $(d+1)$ - coloring and NF for each color set) such that at most $d+1$ further bins with sizes $\leq 1 - \delta$ are generated. Let $\bar{\beta}$ be the total number of generated bins with sizes $> 1 - \delta$. Then, we have the inequality

$$SIZE(I) > (1-\delta)\bar{\beta}.$$

Since $SIZE(I) \leq OPT(I)$, we have

$$\bar{\beta} < \frac{1}{1-\delta}OPT(I).$$

Since $\delta \leq \frac{1}{2}$, we get $\frac{1}{1-\delta} \leq 1 + 2\delta$. This implies that the total number of bins in this case is bounded by

$$\bar{\beta} + \beta'' + (d+1) \leq (1+2\delta)OPT(I) + (3d+1).$$

$\square$

2.4 The overall algorithm

Algorithm A_ϵ:
Input: Instance I consisting of a d-inductive graph $G = (V, E)$ with $V = \{1, \ldots, n\}$ and sizes $s_1, \ldots, s_n \in (0, 1]$.
Output: A packing of V into unit size bins without conflicts.
(1) set $\delta = \frac{\epsilon}{2}$,
(2) remove all items of size smaller than δ obtaining an instance J of the $RBP[\delta, n']$ with n' vertices,
(3) apply the algorithm of Karmarkar and Karp [15] and obtain an approximative solution for the bin packing problem without considering the conflicts and only for the large items,
(4) using the algorithm from section 2.2, modify the solution for each bin type such that each bin contains now an independent set of large items,
(5) using the algorithm from section 2.3, pack all small items (removed in step 2), and use new bins only if necessary.

3 Analysis

The total number of bins generated for the set J of large items (with item size $\geq \delta$) by the algorithm Karmarkar and Karp is bounded by

$$(1 + \epsilon)OPT(J) + \frac{1}{\epsilon^2} + 3.$$

Step (4) of our algorithm produces at most $m \cdot d$ additional bins where $m \leq \frac{2}{\epsilon^2}$. The total number of bins after step (5) is

$$max[\beta, (1 + 2\delta)OPT(I)] + (3d + 1).$$

Since $\beta \leq (1 + \epsilon)OPT(J) + \frac{2d+1}{\epsilon^2} + 3$ and $\delta = \frac{\epsilon}{2}$, we have at most

$$(1 + \epsilon)OPT(I) + \frac{2d + 1}{\epsilon^2} + 3d + 4$$

bins. Since d is a constant, this gives an approximation scheme with bound

$$A_\epsilon(I) \leq (1 + \epsilon)OPT(I) + O(\frac{1}{\epsilon^2}).$$

4 Conclusions

In this paper, we have given an asymptotic approximation scheme for the bin packing problem with conflicts restricted to d-inductive graphs with constant d. This implies an asymptotic approximation scheme for trees, grid graphs, planar graphs and graphs with constant treewidth. Our algorithm is a generalization of the algorithm by Karmarkar and Karp [15] for the classical bin packing problem.

It would be interesting to find other graph classes where this method also works. Furthermore, we would like to know whether there is an asymptotic approximation scheme or whether the bin packing problem with conflicts is MAXSNP-hard for bipartite or interval graphs.

References

1. B.S. Baker and E.G. Coffman, Mutual exclusion scheduling, *Theoretical Computer Science*, 162 (1996) 225 – 243.
2. P. Bjorstad, W.M. Coughran and E. Grosse: Parallel domain decomposition applied to coupled transport equations, in: *Domain Decomposition Methods in Scientific and Engineering Computing* (eds. D.E. Keys, J. Xu), AMS, Providence, 1995, 369 – 380.
3. H.L. Bodlaender and K. Jansen, On the complexity of scheduling incompatible jobs with unit-times, *Mathematical Foundations of Computer Science*, MFCS 93, LNCS 711, 291 – 300.
4. E.G. Coffman, Jr., M.R. Garey and D.S. Johnson, Approximation algorithms for bin-packing - a survey, in: *Approximation algorithms for NP-hard problems* (ed. D.S. Hochbaum), PWS Publishing, 1995, 49 – 106.

5. U. Feige and J. Kilian, Zero Knowledge and the chromatic number, *Conference on Computational Complexity*, CCC 96, 278 – 287.

6. W. Fernandez de la Vega and G.S. Lueker, Bin packing can be solved within $1 + \epsilon$ in linear time, *Combinatorica*, 1 (1981) 349 – 355.

7. M.R. Garey and D.S. Johnson, Computers and Intractability: A Guide to the Theory of NP-completeness, Freeman, San Francisco, 1979.

8. M.C. Golumbic, Algorithmic Graph Theory and Perfect Graphs, Academic Press, London, 1980.

9. S. Irani, Coloring inductive graphs on-line, *Algorithmica*, 11 (1994) 53 – 72.

10. S. Irani and V. Leung, Scheduling with conflicts, and applications to traffic signal control, *Symposium on Discrete Algorithms*, SODA 96, 85 – 94.

11. S. Irani, private communication.

12. K. Jansen and S. Öhring, Approximation algorithms for time constrained scheduling, *Information and Computation*, 132 (1997) 85 – 108.

13. K. Jansen, The mutual exclusion scheduling problem for permutation and comparability graphs, *Symposium on Theoretical Aspects of Computer Science*, STACS 98, LNCS 1373, 1998, 287 – 297.

14. D. Kaller, A. Gupta and T. Shermer: The χ_t - coloring problem, *Symposium on Theoretical Aspects of Computer Science*, STACS 95, LNCS 900, 409 – 420.

15. N. Karmarkar and R.M. Karp, An efficient approximation scheme for the one-dimensional bin packing problem, *Symposium on the Foundations of Computer Science*, FOCS 82, 312 – 320.

16. Z. Lonc: On complexity of some chain and antichain partition problem, *Graph Theoretical Concepts in Computer Science*, WG 91, LNCS 570, 97 – 104.

17. D. de Werra, An introduction to timetabling, *European Journal of Operations Research*, 19 (1985) 151 – 162.

Approximations for the General Block Distribution of a Matrix

Bengt Aspvall[1], Magnús M. Halldórsson[2,1], and Fredrik Manne[1]

[1] Department of Informatics, University of Bergen,
N-5020 Bergen, Norway.
{Bengt.Aspvall, Fredrik.Manne}@ii.uib.no
http://www.ii.uib.no/
[2] Science Institute, University of Iceland,
Taeknigardur, IS-107 Reykjavik, Iceland.
mmh@hi.is

Abstract. The general block distribution of a matrix is a rectilinear partition of the matrix into orthogonal blocks such that the maximum sum of the elements within a single block is minimized. This corresponds to partitioning the matrix onto parallel processors so as to minimize processor load while maintaining regular communication patterns. Applications of the problem include various parallel sparse matrix computations, compilers for high-performance languages, particle in cell computations, video and image compression, and simulations associated with a communication network.

We analyze the performance guarantee of a natural and practical heuristic based on iterative refinement, which has previously been shown to give good empirical results. When p^2 is the number of blocks, we show that the tight performance ratio is $\theta(\sqrt{p})$. When the matrix has rows of large cost, the details of the objective function of the algorithm are shown to be important, since a naive implementation can lead to a $\Omega(p)$ performance ratio. Extensions to more general cost functions, higher-dimensional arrays, and randomized initial configurations are also considered.

In comparison, Khanna et al. have shown the problem to be approximable within $O(1)$ factor [6], while Grigni and Manne have shown it to be NP-hard to approximate within any constant factor less than two [4].

1 Introduction

A fundamental task in parallel computing is the partitioning and subsequent distribution of data among processors. The problem one faces in this operation is how to balance two often contradictory aims: finding an equal distribution of the computational work and at the same time minimizing the imposed communication. In a data parallel computing environment, the running time is dominated by the processor with the maximal load, thus one seeks a distribution where the maximum load is minimized. On the other hand, blindly optimizing this factor, may lead to worse results if communication patterns are ignored.

We assume we are given data in the form of a matrix, with communication only involving adjacent items. This is typical for a large class of scientific computational problems. The partition that minimizes communication load is the *uniform partition*, or *simple block distribution*, where the n by n matrix is tiled by n/p by n/p squares. For instance, this improves on the *one-dimensional* partition, where n/p^2 columns are grouped together. However, workload, which we typically measure as the number of non-zero entries in a block, may be arbitrarily unbalanced, as non-zero entries may be highly clustered.

A partition that yields greatly improved workload is the *general block distribution*, where the blocks are arranged in an orthogonal, but unevenly spaced, grid. It can be viewed as an ordinary block partitioning of an array where one allows the dividers for one column (or row) block to be moved simultaneously. The advantage of this distribution is that it preserves both the locality of the matrix and the array-structured communication of the block distribution while at the same time allowing for different sized blocks.

If the underlying problem has a structure such that communication is local, using a rectilinear partitioning gives a simple and well structured communication pattern that fits especially well on grid connected computers. The simplicity of the general block distribution also makes it possible for compilers to schedule the communication efficiently. It has therefore been included as an approved extension for data mapping in High Performance Fortran HPF2 [5].

Applications of the general block distribution include various parallel sparse matrix computations, compilers for high-performance languages, particle in cell computations, video and image compression, and simulations associated with a communication network [2, 6, 7, 5, 10]. See [8] for a discussion of other rectilinear partitioning schemes.

Computing the optimal general block distribution was shown to be NP-hard by Grigni and Manne [4]. In fact, their proof shows that the problem is NP-hard to approximate within any factor less than 2. Khanna et al. [6] have shown the problem to be constant-factor approximable. They did not give a bound on the value of the constant attained by their algorithm, but an examination of their analysis appears to give a bound of 127. They also did not try to analyze the complexity of the algorithm, but it is specified in terms of a collection of submatrices that can be asymptotically $O(n^4)$ or square in the size of the input. They additionally indicated a simple $O(\log^2 n)$-approximate algorithm, also defined on a quadratic size collection.

The subject of the current paper is a performance analysis of a heuristic that has been considered repeatedly in the applied literature. This *iterative refinement* algorithm was given by Nicol [10], and independently by Mingozzi et. al. [9] and Manne and Sørevik [8]. It is based on iteratively improving a given solution by alternating between moving the horizontal and vertical dividers until a stationary solution is obtained. The algorithm and the general block distribution are given in Section 2.

We analyze the basic iterative refinement algorithm in Section 3.2 and find that it yields a performance ratio of $\theta(\sqrt{p})$ when the cost of each row is not a significant fraction of the whole instance. On the other hand, the performance deteriorates in instances with very heavy rows, and becomes as poor as $\theta(p)$.

In order to combat this weakness, we modify or constrain the operation of the algorithm in Section 3.3. We give two ways of modifying the objective functions of the one-dimensional subproblems that both lead to a $\theta(\sqrt{p})$ performance ratio on all instances. While complexity analysis has been omitted in this extended abstract, these versions can be implemented [1] in time $O(n^2 + p^3 \log^2 n)$ $(O(n^2 + n(p \log n)^2))$, which is linear in input size when $p \leq (n/\log n)^{2/3}$ $(p \leq \sqrt{n}/\log n)$, respectively.

We next consider the effect that starting configurations, or initial partitions, can have. In particular, a promising idea, suggested by Nicol [10], is to use random initial partitions, and possibly making multiple trials. We show in Section 3.4 that this is not beneficial, as the resulting performance ratio is $\Omega(p/\log p)$. Extensions of the results to more general cost functions and to matrices of higher dimensions are given in Section 3.5.

Our analysis here indicates that the iterative refinement algorithm has a considerably weaker worst-case behavior than is possible by polynomial-time algorithms. Nevertheless, it may be valuable especially for small to moderate values of p, which is the case in our motivating application: load balancing on parallel computers. It is also quite efficient, being sub-linear except for a simple linear-time precomputation step [1]. Finally, it is conceptually simple, natural enough to be discovered independently by at least three groups of researchers, easy to implement, and has been shown to give good results on various practical instances and test cases [10, 8].

2 The General Block Distribution

For integers a and b, let $[a, b]$ denote the interval $\{a, a+1, \ldots, b\}$.

For integers n and p, $1 \leq p \leq n$, a non-decreasing sequence $(1 = r_0, r_1, \ldots, r_p = n+1)$ of integers defines a *partition* of $[1, n]$ into the p intervals $[r_i, r_{i+1} - 1]$, for $0 \leq i < p$. For completeness, we allow empty intervals.

Definition 1 (General Block Distribution). *Given an n by n matrix A and integer p with $1 \leq p \leq n$, a general block distribution consists of a pair of partitions of $[1, n]$ into p intervals. It naturally partitions A into the p^2 contiguous blocks, or submatrices.*

A *block* is a submatrix outlined by pairs of adjacent horizontal and vertical dividers. A *column block* (*row block*) is a set of column between adjacent vertical (horizontal) dividers, respectively. A *row segment* is an intersection of a row and a block.

In a parallel environment the time spent on a computation is determined by the processor taking the longest time. The natural optimization problem is then to find a general block distribution that minimizes the maximum cost over all blocks. The cost is here taken to be the sum of of the elements in a block under the assumption that the entries of A are non-negative. The corresponding decision problem was shown in [4] to be NP-complete.

The *iterative refinement* algorithm consists of performing the following improvement step until none exists that further reduces the cost:

> With the vertical delimiters fixed, find an optimal distribution of the horizontal delimiters. Then, with the new horizontal delimiters fixed, do the same for the vertical delimiters.

Thus, the algorithm alternately performs vertical and horizontal *sweeps* until converging to a locally optimal solution. Each sweep can be viewed as a one-dimensional subproblem, for which efficient algorithms are known [3, 10, 11].

In the first vertical partition, A is partitioned optimally into p vertical intervals without the use of the horizontal delimiters. The number of iterations needed to obtain a converged solution varied between 2 and 13 in tests presented in [8].

For the remainder we may sometimes assume for convenience that we have p dividers (instead of $p-1$). Clearly this does not affect the asymptotic behavior. Note that the outlines of the matrix form additional dividers.

3   Performance Analysis

In this section, we analyze the performance guarantees of the iterative refinement algorithm and simple modifications thereof. We begin in Section 3.1 with intermediate results on the 1-D subproblem. We analyze in Section 3.2 the performance of the pure iterative refinement algorithm, which is dependent on the cost of the heaviest row of the input matrix. We then give in Section 3.3 simple modifications to the algorithm that yield better performance ratios when the input contains heavy rows. In Section 3.4 we consider strategies for the initial placement of vertical dividers, including uniform and random placement. We finally consider extensions of the problem in Section 3.5 to other cost functions and higher-dimensional arrays.

3.1   1-D Partitioning

As a tool for our studies of the general block distribution we need the following intermediate result on the one-dimensional case, i.e. how well a sequence of n non-negative numbers can be partitioned into p intervals. Let W be the sum of all the elements.

Lemma 1. *Given a positive integer p and a sequence of n non-negative numbers, a greedy algorithm yields a partition of the sequence into p intervals such that: (i) the cost of any interval excluding its last element is at most W/p, and (ii) any interval with cost more than $2W/p$ consists of a single element.*

Proof. Start the algorithm at the left end, and greedily add elements to form an interval until its cost exceeds $\frac{W}{p}$. If the cost is at most $2\frac{W}{p}$, make this the first interval, and inductively form $p-1$ intervals on the remaining array of total cost at most $W\frac{p-1}{p}$. If the cost exceeds $2\frac{W}{p}$, place a divider on both sides of the last element, forming two intervals. Then, inductively form $p-2$ intervals on the remaining array of cost at most $W\frac{p-2}{p}$. The only intervals that can have cost exceeding $2\frac{W}{p}$ are those formed by the last element added to a group, as in the second case. ∎

3.2 Pure iterative refinement

The performance ratio attained by the iterative refinement algorithm turns out to be highly dependent on the maximum cost of a row. If this cost is small, the performance is good, while it reduces to the trivial performance bound attained by the final 1-D sweep alone when the row cost is high.

Let us first notice that the cost of the optimal solution, OPT, is at least W/p^2, where W is the total cost of the matrix, since the number of blocks is p^2.

Theorem 1. *Let R denote the cost of the heaviest row, $R = \max_i \sum_j A[i,j]$. Then, the performance ratio of the pure iterative refinement algorithm equals $\theta(p)$, when $R = \theta(W/p)$, but only $\theta(\sqrt{p})$ when $R = O(W/p^{1.5})$.*

The theorem is established by the following three lemmas, along with the trivial $O(p)$ ratio obtained by a single 1-D sweep.

Lemma 2. *The performance ratio of pure iterative refinement is $\Omega(p)$ as a function of p alone.*

Proof. Consider the p^2 by p^2 matrix $A = (a_{i,j})$, where

$$a_{i,j} = \begin{cases} p^2, & \text{if } i = 1 \text{ and } j = (p-1)p+3,\ldots,p^2, \text{ or} \\ & \quad j = 1 \text{ and } i = (p-1)p+3,\ldots,p^2, \\ 1, & \text{if } i,j = p+1,\ldots,p(p-1), \text{ and} \\ 0, & \text{otherwise.} \end{cases}$$

Observe that the cost of the first column and the cost of the first row are $(p-2)p^2 = W/p$. The iterative refinement algorithm will first assign the vertical dividers to be the multiples of p. One choice compatible with the definition of the algorithm for the assignment of the horizontal dividers assigns them also the multiples of p. The cost of this partition is the cost of the heavy row, or $W/p = (p-2)p^2$, and no improvements are possible that involve either horizontal or vertical dividers but not both.

The optimal partition consists of the vertical and horizontal dividers $p(p-1)+4, p(p-1)+6,\ldots,p^2-2$, followed by $3p, 5p,\ldots,(p-1)p$. The cost of this partition is $4p^2$, for an approximation ratio of $\frac{p-2}{4}$. For the upper bound, recall that a vertical (or horizontal) partition alone achieves a performance ratio $p+1$. ∎

When the cost of each row is bounded, the algorithm performs considerably better.

Lemma 3. *Following the first (vertical) sweep of the algorithm, there exists a placement of horizontal dividers such that the cost of blocks excluding the heaviest row segments within them is at most $2(\sqrt{p}+1)OPT$.*

Proof. We show the existence of a set of horizontal dividers that achieves the bound. The algorithm, which performs optimally under the given situation, will then perform no worse.

Let O denote the set of dividers, horizontal and vertical, in some optimal 2-D solution. We say that a column block is *thick*, if at least $\sqrt{p}$ vertical dividers from O go through it. Otherwise, a column block is *thin*. The solution we construct uses the even-numbered horizontal dividers from O, as well as $\sqrt{p}/2$ dividers for each of the thick column block to minimize the cost of its blocks.

Each block from a thin column block has at most one optimal horizontal divider and $\sqrt{p}-1$ vertical dividers from O crossing the block. Hence, the cost of the block is at most $2\sqrt{p}OPT$, or within the desired bound.

Each thick column block is of cost at most W/p plus a single column. The cost of each column segment is bounded by $2OPT$, given the even-numbered horizontal dividers from O. The cost of the rest of the block, excluding the cost of the heaviest row segment, is at most W/p divided by $\sqrt{p}/2$, or $2W/p^{1.5}$. Since $OPT \geq W/p^2$, this is at most $2\sqrt{p}OPT$. Thus, blocks in thick column blocks, excluding the heaviest row segment, are of cost at most $(2\sqrt{p}+2)OPT$. ∎

The lemma holds in particular for the iterative algorithm, thus we get good bounds when row cost is small.

Corollary 1. *When each row is of cost at most $O(W/p^{1.5})$ the iterative refinement algorithm achieves a performance ratio of $O(\sqrt{p})$ in two sweeps.*

For the case of small row cost, we get lower bounds that match within a constant factor.

Lemma 4. *The performance ratio of the iterative refinement algorithm is at least $\frac{\sqrt{p}}{4}$, even when the cost of each row is less than W/p^2.*

Proof. Consider the matrix A in Figure 1. We assume that p is a square number, and let $\alpha = \sqrt{p}$. The matrix A consists of a large block of dimension $\alpha(\alpha^2 - \alpha + 1) \times \alpha(\alpha^2 - \alpha + 1)$ in the upper left corner. The value of each element in this block is $1/(\alpha^2 - \alpha + 1)$. On the diagonal below the large block there are $\alpha - 1$ blocks each of size $\alpha \times \alpha$. The value of each element in these blocks is 1.

The total cost W of A is $p^2 = \alpha^4$. The cost of each row is at most one, or W/p^2. That can be arbitrarily reduced further by making repeated copies of each column and row.

With these values the columns can be divided into $p = \alpha^2$ column blocks each consisting of α columns and of cost α^2. This is indicated by the dotted lines in Figure 1. Since each column interval has the same weight this is the initial partition that will be returned by the iterative refinement algorithm. When performing the horizontal partitioning the large block will now be regarded as having cost α^2. Thus from the horizontal point of view there are α blocks each

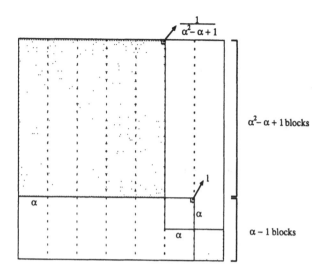

Fig. 1. The array used for showing the lower bound.

of cost α^2. Dividing each small diagonal block into α intervals will give a cost of α for each block. Similarly using α intervals on the large block divides this into blocks of cost α. Note that it is possible to achieve this bound exactly since the number of rows in the large block is $\alpha(\alpha^2\alpha + 1)$. In this way we have used α^2 row blocks and achieved a solution where each block costs $\alpha = W/p^{1.5}$ giving a perfect load balance. Thus, this is the partition the algorithm will return after the first two sweeps. Returning to the vertical delimiters we cannot improve the solution further since each column block contains a block of cost α and the algorithm terminates.

Now, consider a solution where the large block is partitioned into blocks of size at most $2\alpha \times 2\alpha$. Then the cost of each block is at most $4\frac{\alpha^2-\alpha+1/4}{\alpha^2-\alpha+1} < 4$. Using $\frac{\alpha^2}{2}$ column and row blocks one is able to cover $\alpha^3 - \frac{\alpha^2}{2}$ rows/columns, which is less than the dimension of the large block. We now have at least $\frac{\alpha^2}{2}$ row and column blocks left to use on the $\alpha - 1$ small diagonal blocks. By using $\frac{\alpha}{2}$ horizontal and vertical delimiters on each of these we get 4×4 blocks of cost 4. Thus we see that there exists a solution of overall cost at most $4 = 4W/p^2$. ∎

This bound holds even when p is as large as $n/2$. We conjecture that $\sqrt{p}/4$ is indeed the tight performance ratio of the algorithm in general, up to a lower order term.

3.3 Modified iterative refinement algorithms

The lesson learned from Lemma 2 is that one may not blindly focus only on the heaviest column/row segment in each sweep; it is essential to balance also those segments that aren't immediately causing problems. In particular, although single heavy elements (or columns/rows) can cause the maximum block cost to be large, this should not be a *carte blanche* for the remaining partition to be arbitrarily out of balance.

We present two approaches for modifying the pure iterative refinement method, which both achieve a bound of $O(\sqrt{p})$. One makes a simple modification to the objective function, and yields the desired guarantee in three sweeps. The other requires only two sweeps to obtain an $O(\sqrt{p})$-approximate solution, but diverges slightly more from the original script.

A three sweep version We use a three sweep variant of the algorithm, where the first and the third sweep are as before, but the second sweep uses the following slightly modified objective function:

> The cost of a block is the sum of all the elements in the block, excluding the heaviest row segment.

Lemma 5. *The above modified algorithm attains a performance ratio of $4\sqrt{p}+4$.*

Proof. By Lemma 3, the cost of any block after the second sweep is at most $(2\sqrt{p}+1)OPT$ plus the cost of a single row segment. We then only need to ensure that we reduce the cost of unusually heavy row segments in the third sweep, without affecting much the cost of the main parts of the blocks.

An assignment that contains every other of our previous vertical dividers, and every other of the vertical dividers from some optimal 2-D solution, ensures both: the cost of each block excluding the heaviest row segment at most doubles, while the cost of a row segment will be bounded by $2OPT$. Hence, the total cost of a block is at most $(2(2\sqrt{p}+1)+2)OPT \le (4\sqrt{p}+4)OPT$. Since such an assignment exists for the third sweep, the optimal 1-D subroutine will find a solution whose cost is no worse. ∎

A two-sweep version We now consider an algorithm that works in two sweeps, as follows:

Step 1: Find the following two sets of vertical dividers independently:
    (a) The $p/2$ dividers that minimize the maximum cost of any row segment.
    (b) Use $p/2$ dividers that minimize the maximum cost of a column block.
Step 2: Find an optimal set of horizontal dividers.

We extend the analysis of the algorithm to its *performance function*. While the performance ratio of an algorithm is only a single value, describing the worst case ratio between the heuristic and the optimal values, the performance function $\rho(OPT)$ indicates the cost of the worst solution obtained by the algorithm for each possible optimal solution cost. In many cases, this yields a more informative analysis.

First, recall that $OPT \ge W/p^2$, and thus $\rho(OPT)$ is defined only for those values of OPT. Second, consider the case when $OPT \ge 2W/p$. There is an assignment of vertical dividers so that any column block of cost more than

$2W/p$ will consists of a single column. A second sweep of horizontal dividers will then slice these separated columns optimally. Hence, $\rho(OPT) = 1$ when $OPT \geq 2W/p$.

We can generalize our analysis to show that $\rho(OPT) = O(\sqrt{\frac{W/p}{OPT}})$ for OPT in the range $[4W/p^2, 2W/p]$, providing a smoothly improving approximation bound.

Theorem 2. *The two-sweep algorithm has a performance function,*

$$\rho(OPT) = \max(O(\sqrt{\frac{W/p}{OPT}}), 1).$$

for each value of $OPT \geq 4W/p^2$.

Proof. As before, we present a particular set of horizontal dividers that achieve the bound, and thus claim that the algorithm performs no worse.

Part (a) of step one ensures that each row segment is of cost at most $2OPT$. Part (b) ensures that each column block is of cost at most $2W/p$ plus the cost of a single column (by Lemma 1).

Let $t = \sqrt{(W/p)/OPT}$. We now say that a column block is *thick*, if at least t of the optimal vertical dividers go through it, and otherwise *thin*. Observe that t is at most $\sqrt{p}$ since $OPT \geq W/p^2$.

We analyze the following set of horizontal dividers: Every other optimal horizontal divider, plus $\sqrt{p}/2$ dividers to minimize each of the at most $\sqrt{p}$ thick column blocks.

Using every other optimal horizontal dividers ensures that the cost of each column segment is at most $2OPT$, and that the cost of each thin block is at most $2tOPT$. Using t dividers to minimize the cost of blocks within each of the at most p/t thick column blocks ensures that those blocks are of cost at most $2/t$ times the cost of a column block, plus the cost of a column segment and the cost of a row segment. This is at most

$$\frac{2}{t} \cdot \frac{2W}{p} + 2OPT + 2OPT = (4t + 4)OPT.$$

In particular, this is at most $(4\sqrt{p} + 4)OPT$. ∎

This bound on the performance function can also be shown to be asymptotically tight.

3.4 Initial placement strategies

The iterative improvement method leaves open the possibility of using additional strategies for the initial placement of the vertical dividers. One approach would be to start with a *uniform placement*, with dividers at $n/p, 2n/p, \ldots, (p-1)n/p$. Nicol [10] suggests using *random placement*, where each divider is assigned a uniformly random value from 1 to n. He found this to give empirically good results. Random assignment also leaves open the possibility of repeating the whole improvement procedure, retaining the best of the resulting solutions.

Unfortunately, this approach does not improve the performance guarantee of the improvement method. In fact, with high probability, the performance ratio is decidedly worse, or $\Omega(p/\log p)$, which holds even if the procedure is repeated often. Basically, it suggests that any division strategy that is primarily based on the number of columns in each block is bound to fail. The strategy must rely on the weight of the columns. On the whole, however, we are led to the conclusion that partitioning methods that compute the horizontal and vertical dividers independently, cannot yield close to optimal approximations.

Theorem 3. *Random initial placement followed by iterative improvement has performance ratio $\Omega(p/\log p)$, expected and with high probability.*

Uniform initial placement followed by iterative improvement has performance ratio $\Theta(p)$.

The success of the algorithm on the example we shall construct depends on the size of the largest horizontal block in the initial partition. The following lemma, whose proof can be found in [1], bounds this value. Let ln denote the natural logarithm.

Lemma 6. *For a random partition of the sequence $1, 2, \ldots, n$ into p intervals, the probability that the largest interval contains at least $\alpha(n-1)\ln p/(p-1)+1$ numbers is at most $p^{-(\alpha-1)}$.*

We now prove the theorem.

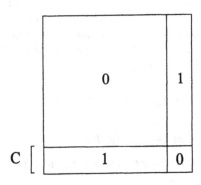

Fig. 2. An example for which a uniform or random initial assignment leads to poor performance.

Proof. We assume that $p = o(\sqrt{n})$. Let $C = \sqrt{n}$.

Consider the $n \times n$ 0/1 matrix in Figure 2. Let us refer to the rightmost C columns as the *thick vertical block*, and the lowest C rows as the *thick horizontal block*. Only the elements in the symmetric difference between the two thick blocks have a cost one; the rest are zero elements. The cost of either block, denoted by Z, is thus $C \cdot (n - C) = n^{3/2} - n$.

Now consider the effect of a random assignment of vertical dividers. It is easy to deduce that with high probability no divider will hit the vertical heavy block. Let B denote the cost of the heaviest column block C_B. Let b satisfy $b = B(p - b)/Z$. We round b to the nearest integer.

After this first sweep, the algorithm proceeds deterministically. The second sweep must use b horizontal dividers on the thick horizontal block and $p - b$ on the thick vertical block. The cost of each block in the former is $B/b \approx Z/(p - b)$, while the cost of the latter is clearly $Z/(p - b)$.

On the third sweep, the algorithm must similarly use b vertical dividers on the thick vertical block and $p - b$ on the thick horizontal block. The cost of each block is then about $Z/b(p - b)$. No significant changes can then be made to either the horizontal or the vertical dividers independently to decrease this value.

We have skipped over the detail of the "joint block", the only block that contains elements from both thick blocks. Its size may bias the assignment of dividers somewhat, resulting in small oscillations. None of them can make significant difference, and in fact, cannot change the number of dividers used to partition either heavy block.

To wrap up this analysis, compare the algorithm's solution to the solution that on each side uses one divider to separate the heavy blocks and $p/2 - 1$ dividers on each of the them. The cost of this solution is then Z/p^2, and the ratio between the two solutions $p^2/(p - b)b \approx p/b$. From Lemma 6, with high probability, the value of b is $O(\ln p)$. Hence, the performance ratio is at least $\Omega(p/\ln p)$, expected and with high probability.

The proof of the lower bound for the uniform partition is left as an exercise. ∎

We can also observe that for any number of repetitions of this random procedure, within any polynomial of p, yields a performance ratio of at least $\Omega(p/\log p)$. We also remark that the ratio can be shown to be $\theta(p/\log p)$.

Remark: Recall that our basic iterative improvement algorithm starts with an optimal vertical partition without any horizontal dividers. We might view this as starting with a totally degenerate initial partition of the rows. On a random initial partition the algorithm initially performs more like on a uniform partition than when started with no horizontal dividers.

3.5 Extensions

Other cost functions While the sum of the elements within a block is usually the most natural measure, other cost functions may be more appropriate in certain applications. For some examples, see [11]. The results of this paper can easily be extended to other reasonable cost functions, in particular the following class.

Corollary 2. *General Block Partitioning can be approximated within* $O(\sqrt{p})$ *for any sub-additive cost function, i.e. if* $\phi(B) \leq \phi(B_1) + \phi(B_2)$ *whenever* $B = B_1 \cup B_2$.

58

Higher dimensions Our analysis of the 2-D case extends straightforward to higher dimensional matrices.

Claim. The iterative refinement algorithm attains a performance ratio $\theta(p)$ on three-dimensional matrices.

This generalizes to a ratio of $p^{d/2}$ for d-dimensional matrices. Matching lower bounds are straightforward generalizations of the 2-D case.

While this bound may appear weak, it is a considerable improvement over the alternatives. An oblivious assignment, where we assign dividers in each dimension independently, only guarantees a W/p bound, or a p^{d-1} ratio. And the simplest partition – uniform assignment – can be as much as p^d away from optimal.

References

1. B. ASPVALL, M. M. HALLDÓRSSON, AND F. MANNE, Approximating the Generalized Block Distribution of a Matrix, Institutt for Informatikk, TR-141, University of Bergen, Dec. 1997.
2. B. CHAPMAN, P. MEHROTRA, AND H. ZIMA, *Extending HPF for advanced data parallel applications*, IEEE Trans. Par. Dist. Syst., (Fall 1994), pp. 59–70.
3. G. N. FREDERICKSON, *Optimal algorithms for partitioning trees and locating p-centers in in trees*, in Proceedings of second ACM-SIAM Symposium on Discrete Algorithms, 1991, pp. 168–177.
4. M. GRIGNI AND F. MANNE, *On the complexity of the generalized block distribution*, in Proceedings of Irregular'96, the third international workshop on parallel algorithms for irregularly structured problems, Lecture Notes in Computer Science 1117, Springer, 1996, pp. 319–326.
5. *High Performance Fortran Language Specification 2.0*, January 1997. Available from http://www.crpc.rice.edu/HPFF/home.html.
6. S. KHANNA, S. MUTHUKRISHNAN, AND S. SKIENA, *Efficient array partitioning*. Proceedings of the 24th International Colloquium on Para Automata, Languages, and Programming (ICALP), Lecture Notes in Computer Science 1256, Springer, 1997, pp. 616–626.
7. F. MANNE, *Load Balancing in Parallel Sparse Matrix Computations*, PhD thesis, University of Bergen, Norway, 1993.
8. F. MANNE AND T. SØREVIK, *Partitioning an array onto a mesh of processors*, in Proceedings of '96, Workshop on Applied Parallel Computing in Industrial Problems and Optimization, Lecture Notes in Computer Science 1184, Springer, 1996, pp. 467–477.
9. A. MINGOZZI, S. RICCIARDELLI, AND M. SPADONI, *Partitioning a matrix to minimize the maximum cost*, Disc. Appl. Math., 62 (1995), pp. 221–248.
10. D. M. NICOL, *Rectilinear partitioning of irregular data parallel computations*, J. Par. Dist. Comp., (1994), pp. 119–134.
11. B. OLSTAD AND F. MANNE, *Efficient partitioning of sequences*, IEEE Trans. Comput., 44 (1995), pp. 1322–1326.

An Optimal Algorithm for Computing Visible Nearest Foreign Neighbors Among Colored Line Segments

Thorsten Graf[1] and Kamakoti Veezhinathan[2]

[1] Department ICG-4, Research Center Jülich, 52425 Jülich, Germany,
email:t.graf@fz-juelich.de
[2] Institute of Mathematical Sciences, CIT Campus, Chennai - 600 113, India
email:kama@imsc.ernet.in

Abstract. Given a set S of n colored line segments in $\mathbb{R}^2$ that may intersect only in endpoints. Let $c(u)$ denote the color of a line segment $u \in S$ chosen from $\chi \leq n$ different colors. A line segment $v \in S$ is a *visible nearest foreign neighbor* of $u \in S$ if v is a nearest foreign neighbor of u in S, i.e. $c(u) \neq c(v)$ and no segment with a color different from $c(u)$ is closer to u than v, and if there exist points $u' \in u$ and $v' \in v$ realizing the distance between u and v that are *visible* for each other, i.e. the open segment connecting u' and v' is not intersected by an open line segment in S. We present the first optimal $\Theta(n \log n)$ algorithm that computes for each line segment $u \in S$ all its visible nearest foreign neighbors. The algorithm finds applications in polygon arrangement analysis, VLSI design rule checking and GIS.

1 Introduction

1.1 Problem definition

Let S be a set of n line segments in $\mathbb{R}^2$ which may intersect only in their endpoints. Throughout the paper we assume that the line segments in S are in general position, i.e. not two of them are parallel, which can be simulated by actual or conceptual perturbation of the input. The Euclidean distance between two line segments $u, v \in S$ is defined by $d(u, v) := \min\{d(u', v') | u' \in u, v' \in v\}$ where $d(u', v')$ denotes the Euclidean distance between the points u' and v'. We assign each line segment $u \in S$ a color $c(u) \in \mathbb{N}$; w.l.o.g. we assume that $\cup_{u \in S} c(u) = \{1, \ldots, \chi\}$ with $\chi \leq n$. We define the subsets $S_i := \{u \in S | c(u) = i\}$, $1 \leq i \leq \chi$. The Euclidean *foreign distance* $d_f(u, v)$ between two line segments $u, v \in S$, $u \neq v$, is defined as follows:

$$d_f(u, v) := \begin{cases} d(u, v) & \text{if } u \notin S_{c(v)} \\ \infty & \text{otherwise} \end{cases}$$

Definition 1. *A line segment* $v \in S$ *is a visible nearest foreign neighbor of* $u \in S$ *if* v *is a nearest foreign neighbor of* u, *(i.e.* $d_f(u, v) = \min\{d_f(u, r) | r \in$

$S \setminus S_{c(u)}\} < \infty$), and if there exist points $u' \in u$ and $v' \in v$ such that $d_f(u, v) = d(u', v')$, and v' and u' are visible for each other (i.e. the open line segment between u' and v' is not intersected by an open line segment in S).

Definition 2. The *Visible-Nearest-Foreign-Neighbors Problem (VNFNP)* is defined as follows: Out of all nearest foreign neighbors of each line segment $u \in S$, find those that are visible for u according to Definition 1.

In Voronoi diagram nomenclature, the *VNFNP* is to perform a post office query for each line segment $u \in S$ inside the Voronoi diagram of the set $S \setminus S_{c(u)}$ where the queries are constrained by the visibility condition. Notice, that a visible nearest foreign neighbor $v \in S$ of $u \in S$ is also a nearest foreign neighbor of u. Due to our assumption of general position we see that the points $u' \in u$ and $v' \in v$ realizing the distance between u and v can be computed for given u and visible nearest foreign neighbor v of u in $O(1)$ time.

1.2 Related work and lower bounds

Two fundamental problems in Computational Geometry are the *Closest-Pair Problem (CPP)* and the *All-Nearest-Neighbors Problem (ANNP)* for a set of n objects ([15]). The *Closest-Foreign-Pair Problem (CFPP)* and the *All-Nearest-Foreign-Neighbors Problem (ANFNP)* are generalizations of the *CPP* and the *ANNP* where the input is a set of n *colored* objects. $\Omega(n \log n)$ is a lower bound for the time taken by the *CPP* ([15]) and hence for the *CFPP* in the algebraic decision tree model of computation ([8]). This implies an $\Omega(n \log n)$ lower bound for the time complexity of the *VNFNP* in the algebraic decision tree model of computation. $\Omega(n)$ is a trivial lower bound for the space complexity of the *VNFNP*. The optimal $\Theta(n \log n)$ algorithms known so far for the *CFPP* (e.g. [8]) and the *ANFNP* (e.g. [1, 10]) can only be run for colored points but not for colored line segments. The algorithms given in [2,9] solve the *CPP* and the *ANNP* for n line segments, respectively, in optimal $\Theta(n \log n)$ time, but cannot solve the *CFPP* and the *ANFNP* for line segments. So far, there exists no optimal algorithm for the *VNFNP* or the *ANFNP* of line segments.

1.3 Main result and discussion

Notice that, to solve the *VNFNP* it is enough to locate for every line segment $u \in S$ the Voronoi cells in the Voronoi diagram $\mathcal{VD}(S \setminus S_{c(u)})$ that are intersected by u; unfortunately this takes total $\Omega(n^2)$ time in the worst case. Furthermore constructing all χ Voronoi diagrams $\mathcal{VD}(S \setminus S_i)$, $i \in \{1, \dots, \chi\}$, takes much more time than $O(n \log n)$ unless χ is very small. The main result of this paper is the following:

Theorem 1. *The VNFNP can be solved in optimal $O(n \log n)$ time using $O(n)$ space.*

In particular, we will show in the proof of Theorem 1 that the total number of visible nearest foreign neighbors in S is $O(n)$. It is worthwhile to notice that if (u, v) is a closest foreign pair in the line-segment set S, then u and v are visible nearest foreign neighbors of each other. If $v \in S$ is a visible nearest foreign neighbor of $u \in S$, then by definition v is also a nearest foreign neighbor of u; the inverse implications are not true in general. Hence, the *VNFNP* is somewhere "between" the *CFPP* and the *ANFNP*.

A simple approach based on the Voronoi diagram $\mathcal{VD}(S)$ for computing all pairs of *visible reciprocal foreign neighbors*, i.e. pairs of line segments that are visible nearest foreign neighbors of each other, which in turn solves the *CFPP*, fails for the *VNFNP*:

Let $u, v \in S$ be visible reciprocal foreign neighbors. Denote by $u' \in u$ and $v' \in v$ two points realizing the foreign distance between u and v. It can be shown easily that the open circle with center $m = ((u'.x + v'.x)/2, (u'.y + v'.y)/2)$ and radius $d(u', v')/2$ cannot be intersected by any line segment in S. Hence m is part of the Voronoi diagram $\mathcal{VD}(S)$ and lies on the boundaries of u's and v's Voronoi regions in $\mathcal{VD}(S)$ which implies that u and v are Voronoi neighbors in $\mathcal{VD}(S)$. Since $\mathcal{VD}(S)$ can be computed in $O(n \log n)$ time ([7]), by Euler's formula on planar graphs ([15]) all visible reciprocal foreign-neighbor pairs can be computed in total $O(n \log n)$ time using $O(n)$ space. It is easy to see that in $O(n)$ time the solution for the *CFPP* on the set S of colored line segments can be obtained from the output of this algorithm.

A line segment and its visible nearest foreign neighbors need not be neighbors in $\mathcal{VD}(S)$ as the counter example in Figure 1 shows for two colors in S (indicated by different line widths):

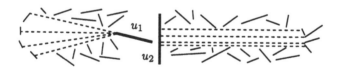

Fig. 1. Counter example for two colors in S

The line segments u_1 and u_2 are visible nearest foreign neighbors of the line segments connected to them by dashed lines, but none of these line segments is a Voronoi neighbor of u_1 or u_2 in $\mathcal{VD}(S)$.

1.4 Numerical considerations

Computing the Voronoi diagram for a set of line segments is studied in ([4]). In practice, the computation of $\mathcal{VD}(S)$ for a set S of line segments raises crucial numerical problems: The edges of $\mathcal{VD}(S)$ are either straight segments or parabolic segments ([17]), and computing Voronoi vertices involves intersecting bisectors with each other. Although in the divide-and-conquer algorithm ([17]) there is never a need to compute Voronoi vertices by intersecting pairs of parabolas (see [17] for details) the numerical problem of reliably intersecting a straight segment and a parabolic segment remains. The numerical and geometric details of the plane-sweep algorithm ([7]) are even more difficult to handle since line segments, parabolas and in addition hyperbolas have to be intersected with each other. Due to these numerical problems in the computation of $VD(S)$ the algorithm for computing visible reciprocal foreign neighbors as given in section 1.3 is not of practical relevance.

As we will see in section 3, although very much related to the line segment Voronoi diagram, our algorithm solves the *VNFNP* avoiding the numerically unreliable computation of the latter but uses piecewise linear approximations of $VD(S)$ ([13]), point Voronoi diagrams ([15]), trapezoidal decompositions ([15]), and generalized Bentley-Ottmann sweeps ([3]) instead, all for which reliable implementations are known ([5, 6, 14]). The same approach can be applied to the algorithm given in section 1.3 for computing visible reciprocal foreign neighbors to make it more practical.

1.5 Applications

Arrangement analysis for colored polygonal objects. Our algorithm for the *VNFNP* can be used to analyze the arrangement given by a set S of colored polygonal objects: The set S can be viewed as a collection of n colored edges.

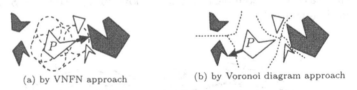

(a) by VNFN approach   (b) by Voronoi diagram approach

Fig. 2. Foreign neighbor computed for polygon P

The color of an edge is the color of the object to which it belongs. Using our algorithm for the *VNFNP* we can find in $O(n \log n)$ time for each object the visible nearest foreign neighbors (Fig. 2 (a)) which gives much better information about the foreign neighborhoods in S than e.g. computing foreign Voronoi neighbors in $VD(S)$ using the algorithm [13] (Fig. 2 (a)).

Design rule checking of VLSI circuits. The design rules for VLSI circuits, like the Mead-Conway rules ([16]) state that any two objects in a VLSI circuit that lie *adjacent* to each other, should be a minimum distance apart and this distance depends upon the class of objects making the pair. For example, a polysilicon wire should be separated from a diffusion wire by a distance of at least λ, while, two polysilicon wires should be separated from each other by a distance of 2λ. These are called the λ-parameters where the value of λ is fixed by the technology. Our algorithm for the *VNFNP* finds direct applications in the design-rule-checking problem.

GIS (geo information systems). Algorithms for computing valid pairs of geometric objects regarding specific criteria are fundamental in GIS ([12]). To mention one such GIS problem, our algorithm for the *VNFNP* can be used e.g. to compute for a polygonal coastline and polygonal islands with total complexity n, locally shortest straight-line connections between the islands and the coastline in $O(n \log n)$ time.

2 Reduction to bichromatic VNFNPs

The *bichromatic VNFNP* (=*BVNFNP*) is the *VNFNP* for a line-segment set S containing only two different colors. In this section we show how to reduce the *VNFNP* in $O(n \log n)$ time to at most χ instances of the *BVNFNP* with total $O(n)$ line segments: Each set S_i, $i \in \{1, \ldots, \chi\}$, is assigned an initially empty *candidate set* C_i. For each $i \in \{1, \ldots, \chi\}$ we insert those line segments $u \in S \setminus S_i$ into C_i which are Voronoi neighbor of a line segment in S_i inside the Voronoi diagram $\mathcal{VD}(S)$ of S, where $\mathcal{VD}(S)$ can be computed in $O(n \log n)$ time using $O(n)$ space ([7]). To get the main idea, we tentatively assume the absence of numerical problems in the computation of $\mathcal{VD}(S)$. The line segments are inserted into the candidate sets during a traversal of $\mathcal{VD}(S)$. By Euler's formula ([15]) this can be done in $O(n)$ time and the number $\sum_{i=1}^{\chi} |C_i|$ of line segments thus inserted into the candidate sets is at most $O(n)$ ([15]). By Theorem 2, in which visibility is not needed, a solution of the *VNFNP* can be computed in $O(n)$ time from the solutions of the *BVNFNP*s for the sets S_i and C_i (where the line segments in C_i tentatively lose their original color). The visibility condition is used in our algorithm solving the *BVNFNP*s; based on the visibility we define "kernels" of Voronoi regions which our algorithm can process efficiently (sections 3.1-3.3).

Theorem 2. *If $v \in S$ is a nearest foreign neighbor of $u \in S$ then $v \in C_{c(u)}$.*

Proof. Let K denote a circle that touches v and u such that only segments in $S_{c(u)}$ may intersect K in inner points; K exists since v is a nearest foreign neighbor of u. We shrink K such that at any time during the shrinking process the circle touches v in the same point as the original circle K does. The shrinking process stops as soon as the current circle is no longer intersected in inner points by segments in S. The final circle K' touches v and a segment in $S_{c(u)}$ (which is not necessarily the segment u). The centerpoint of K' is part of a Voronoi edge in $\mathcal{VD}(S)$ and the theorem follows. $\qquad\square$

We work around the crucial numerical problems that are present in the computation of $\mathcal{VD}(S)$ by using a piecewise linear approximation of $\mathcal{VD}(S)$ instead ([13]). This diagram decomposes the plane into total $O(n)$ "core regions" around the segments, and "spoke regions" for which the closest line segment is found among two candidates. Hence, all line-segment Voronoi neighbor pairs in $\mathcal{VD}(S)$ can be computed from the piecewise linear approximating diagram in $O(n)$ time.

Since candidate sets C_i may remain empty, i.e. $|C_i| = 0$, the number of BVNFNPs to be solved can be less than χ. Assuming that solving a BVNFNP with total m line segments takes $O(T(m))$ time, the VNFNP can be solved in $O(\sum_{i=1}^{\chi} T(|C_i| + |S_i|))$ time. We show in section 3 that $T(m) = O(m \log m)$. Since $\sum_{i=1}^{\chi}(|C_i| + |S_i|) = O(n)$ our algorithm for solving the VNFNP runs in total $O(n \log n)$ time. The following pseudo-code gives a rough idea of the entire VNFNP algorithm:

Algorithm VNFNP

1. **for** all S_i, $1 \leq i \leq \chi$, **do** { *compute the candidate set C_i;* } (section 2)
2. **for** all BVNFNPs (S_i, C_i), $1 \leq i \leq \chi$, **do**
   (a) *tentatively color all segments in C_i with a unique color $\neq i$;* (section 2)
   (b) *preprocess $S := S_i \cup C_i$* (section 2)
   (c) **for** all segments u in S **do**
      i. *compute visible nearest foreign neighbors v of u for which $d_f(u, v)$ can be realized in an endpoint of u; use approximating Voronoi diagram $\mathcal{VD}'(S \setminus S_{c(u)})$ and point location structure on $\mathcal{VD}'(S \setminus S_{c(u)})$* (Theorem 3 in section 3)
      ii. *compute visible nearest foreign neighbors v of u for which $d_f(u, v)$ can only be realized in an inner point of u; use Voronoi diagram of endpoints in $S \setminus S_{c(u)}$, trapezoidal decompositions and generalized Bentley-Ottmann sweeps (sections 3.1-3.3)*

---

3   The algorithm for the BVNFNP

Let B be a set of n_B blue line segments and R be a set of n_R red line segments in the plane; define $N := n_B + n_R$. The line segments in $S := B \cup R$ may only intersect in endpoints. Let B' denote the set of line-segment endpoints in B and let R' denote the set of line-segment endpoints in R. Due to our assumption of general position, for all $u \in S$ the distance $d_f(u, v)$ between u and a visible nearest foreign neighbor $v \in S$ of u can be realized by at least one endpoint. Theorem 3, in which visibility is not needed, presents a method to find the (visible) nearest foreign neighbors of $u \in S$ that are related to its endpoints:

Theorem 3. *Those nearest foreign neighbors $v \in S$ of $u \in S$ for which $d_f(u, v)$ can be realized in an endpoint u' of u can be computed in total $O(N \log N)$ time using $O(N)$ space.*

Proof. We first consider the points $u' \in B'$. We compute the piecewise linear approximation $\mathcal{VD}'(R)$ of $\mathcal{VD}(R)$ in $O(n_R \log n_R)$ time ([13]). Since $\mathcal{VD}'(R)$ is a planar subdivision, after a preprocessing that requires $O(N \log N)$ time, the $n(u')$ candidates for the nearest foreign neighbor segment for each query point

$u' \in B'$ can be computed in $O(n(u') + \log n_R)$ time ([11]); both $VD'(R)$ and the additional data structures generated by the preprocessing require $O(n)$ space ([13, 11]). Since by Euler's formula ([15]) $\sum_{u' \in B'} n(u') = O(|R|)$ the Theorem follows applying a symmetric argument for the points $u' \in R'$. $\qquad \square$

It therefore remains to consider the case that the foreign distance of $u \in S$ to a visible nearest foreign neighbor $v \in S$ can only be realized by an inner point $\tilde{u}$ of u and an endpoint $v' \in v$, i.e. $d_f(u, v) = d(\tilde{u}, v')$ (Figure 1 illustrates such a case). Denote by $\tilde{R} \cup \tilde{B}$ the discrete set of these inner red and blue points $\tilde{u}$. In sections 3.1-3.3 we show how to compute the point set $\tilde{R}$, symmetric arguments can be applied to compute $\tilde{B}$.

We repeat well-known definitions of the Voronoi region $VR(p) := \{t \in \mathbb{R}^2 : d(p, t) = \min_{b \in B'} d(b, t)\}$, $p \in B'$, the Voronoi polygon $VP(p) := \partial VR(p)$, $p \in B'$, and the Voronoi diagram $VD(B') := \bigcup_{p \in B'} VP(p)$. Denote by $\sigma_p(r)$ the subsegment of $r \in R$ which we obtain by clipping r against $VR(p)$, $p \in B'$, more precisely

$$\sigma_p(r) := \begin{cases} VR(p) \cap r & : \quad VR(p) \cap r \neq \emptyset \\ nil & : \quad \text{otherwise} \end{cases}$$

We define the line-segment sets

$$R^*(p) := \{\sigma_p(r) | r \in R, \sigma_p(r) \neq nil\}, p \in B' \qquad \text{and} \qquad R^* := \bigcup_{p \in B'} R^*(p)$$

Each line segment $r \in R^*$ which is not a single point is contained in at most two of the line-segment sets $R^*(\cdot)$. Since any points $p, q \in B'$, $p \neq q$, and any segment $r \in R^*(p)$ satisfy the inequality $d(p, r) \leq d(q, r)$ we easily obtain:

Lemma 1. *For each line segment $r \in R$ the minimal distance between r and the points in B' is given by $\min_{p \in B'} d(r, p) = \min_{p \in B'} d(\sigma_p(r), p)$.*

3.1 Filtering the line segments in R^*

Computing the complete sets $R^*(p)$, $p \in B'$, is poor since there are up to $\Theta(N^2)$ line segments in R^*. In the description of the filtering process that reduces the number of line segments $r \in R^*$ computed by our algorithm to $O(N)$, which is the part of our algorithm in which visibility is crucial, we need the following:

Definition 3. *The inner points of $p \in B'$ are those points $q \in VR(p)$ for which there exists a path inside $VR(p)$ from p to q which is not intersected – except in its endpoints p and q – by line segments in R^*. The inner line segments of $p \in B'$ are those line segments in $R^*(p)$ which contain at least one inner point of p; the set of inner line segments is denoted by $R_I^*(p)$. The subset $\tilde{R}_I^*(p)$ of $R_I^*(p)$ contains those inner line segments of p for which both endpoints lie on $VP(p)$.*

Figure 3 gives an illustration: The set of all inner points of p appears as shaded portion of $\mathcal{VR}(p)$. The sets $\tilde{R}_I^\star(p)$ and $R_I^\star(p) \setminus \tilde{R}_I^\star(p)$ each contain three inner line segments of p.

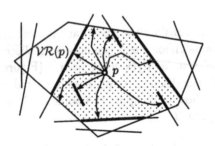

Fig. 3. Inner points and inner line segments of p

The arrows in Figure 3 represent paths from p to some of its inner points. From Lemma 1 we obtain easily the following:

Lemma 2. *For all points $q \in \tilde{R}$ there exists an endpoint $p \in B'$ and a subsegment $r \in R_I^\star(p)$ such that $q \in r$.*

By Lemma 2 it suffices for the computation of the point set $\tilde{R}$ to compute only the distances between the points $p \in B'$ and their inner line segments. Theorem 4 ensures that there are only total $O(N)$ inner line segments.

Theorem 4. *The total number of inner line segments is $O(N)$.*

Proof. By Euler's formula on planar graphs ([15]) at most $O(N)$ line segments in $R^\star$ lie on edges of $\mathcal{VD}(B')$. We may therefore assume in the following that such line segments do not exist in $R^\star$. Fix a point $p \in B'$. For each line segment in $\tilde{R}_I^\star(p)$ there exists a vertex of $\mathcal{VP}(p)$ that is not an inner point of p. It is easy to verify that no vertex of $\mathcal{VP}(p)$ is considered by more than one line segment in $\tilde{R}_I^\star(p)$. Together with Euler's formula we obtain that

$$\sum_{p \in B'} |\tilde{R}_I^\star(p)| + \sum_{p \in B'} |R_I^\star(p) \setminus \tilde{R}_I^\star(p)| = O(N)$$

which completes the proof. $\qquad\square$

Given all inner line segments of the points $p \in B'$ computing the set $\tilde{R}$, and thus the remaining visible nearest foreign neighbors of the red line segments that cannot be computed by the argument given in Theorem 3, can be done easily in $O(N \log N)$ time using $O(N)$ space: Denote by Q the set of line segments that we obtain by connecting each point $p \in B'$ with the closest points on its inner line segments. Apply the line-segment intersection algorithm given in [3] and remove an inner line segment of p from further consideration as soon as the first intersection point of the corresponding segment in Q and a red line segment is computed. From the remaining inner line segments of the points in B' the set $\tilde{R}$ can be obtained easily in $O(n)$ time.

It therefore remains to show how to compute all $O(N)$ inner line segments in $O(N \log N)$ time. By point location queries on the Voronoi diagram $\mathcal{VD}(B')$ we can compute all $O(N)$ inner line segments in $\cup_{p \in B'} [R_I^*(p) \setminus \tilde{R}_I^*(p)]$ in time $O(N \log N)$ using linear space ([15]). In the rest of the paper we show how to compute the line segments in $\cup_{p \in B'} \tilde{R}_I^*(p)$ in time $O(N \log N)$. For ease of notation we use the following definition:

Definition 4. *For $p \in B'$ the kernel $\kappa(p)$ of its Voronoi region $\mathcal{VR}(p)$ is the boundary of the closure of the inner points of p.*

The kernel $\kappa(p)$, $p \in B'$, consists of the line segments in $\tilde{R}_I^*(p)$ and segments of $\mathcal{VP}(p)$. In Figure 3 on page 66, the kernel $\kappa(p)$ of p is the boundary of the shaded region. By Theorem 4 and Euler's formula the total complexity of the kernels is $O(N)$. In the rest of the paper we show how to compute all kernels $\kappa(p)$, $p \in B'$, in total $O(N \log N)$ time using $O(N)$ space.

3.2 Computing single kernel points

The algorithm for computing the kernels will traverse each kernel only once. To ensure that no line segments in $R^* \setminus [\cup_{p \in B'} \tilde{R}_I^*(p)]$ will be considered we compute for each point $p \in B'$ an arbitrary point of the kernel $\kappa(p)$ first.

Theorem 5. *For all points $p \in B'$ a point $p' \in \kappa(p)$ can be computed in total time $O(N(\log n_B + \log n_R))$ using $O(N)$ space.*

Proof. We compute a trapezoidal decomposition $H(R)$ for the line segments in R by inserting vertical attachments at all points in R' (Fig. 4). This can be done in $O(n_R \log n_R)$ time using a sweep similar to the Bentley-Ottmann sweep ([3]). After a preprocessing time of $O(n_R \log n_R)$ we can determine for each point $p \in B'$ a trapezoid $t(p) \in H(R)$ for which $p \in t(p)$ in total $O(n_B \log n_R)$ time ([11]). Starting at p we walk up a path $P(p)$:

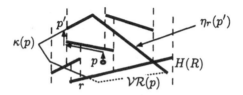

Fig. 4. Path $P(p)$ inside $H(R)$ of a point $p \in B'$

The first segment of $P(p)$ is the (possibly empty or infinite) vertical line segment from p to the top border of $t(p)$ (possibly at infinity). The remaining part of the path $P(p)$ is built by portions of line segments in $R_I^*(p)$ and vertical attachments in $H(R)$, or is empty. Line segments in $R_I^*(p)$ are traversed only non-decreasing in the y-direction and vertical attachments are traversed only starting from their corresponding point in R'. Notice that we do not traverse $H(R)$ cell by cell, i.e. we traverse a subsegment of $r \in R_I^*(p)$ in a single step skipping all vertical attachments ending on inner points of r (see Fig. 4). As the endpoint p' of the path $P(p)$, $p \in B'$, we choose the first intersection point with $\mathcal{VP}(p)$ if such a point exists. In particular this is the case if the Voronoi region $\mathcal{VR}(p)$ is bounded. If $\mathcal{VR}(p)$ is unbounded it may happen that $P(p)$ ends with an infinite vertical line segment which does not hit $\mathcal{VP}(p)$. In this case we choose as endpoint p' of $P(p)$ an arbitrary point on $\mathcal{VP}(p)$ at infinity. Obviously the paths $P(p)$, $p \in B'$, are disjoint. Since the total number of inner line segments is $O(N)$ by Theorem 4 and the number of segments in a path $P(p)$, $p \in B'$, along vertical attachments in $H(R)$ differs from the number of segments in the path $P(p)$ along segments in $R_I^*(p)$ by at most 2, the paths consist of total $O(N)$ line segments. Each line segment in a path $P(p)$, $p \in B'$, can be computed in $O(\log n_B)$ time using a point-in-polygon test on $\mathcal{VP}(p)$ ([15]). Hence all paths $P(p)$, $p \in B'$, can be computed in total $O(N \log n_B)$ time. Summing up we obtain that the points $p' \in \kappa(p)$, $p \in B'$, can be computed in total $O(N(\log n_B + \log n_R))$ time. With standard data structures the algorithm can be implemented using $O(N)$ space. □

3.3 Computing the kernels

By a *pseudo-line* we denote a polygonal chain which is monotone in the x-direction. For a point $q \in \kappa(p)$, $p \in B'$, denote by $\eta_r(q)$ (resp. $\eta_l(q)$) the pseudo-line along $\kappa(p)$ to the right (resp. to the left) of q with maximal length (Fig. 4); pseudo-lines may be empty, i.e. consist of a single point.

Theorem 6. *Given for each point $p \in B'$ a point $p' \in \kappa(p)$, the kernels $\kappa(p)$, $p \in B'$, can be computed in total $O(N \log N)$ time using $O(N)$ space.*

Proof. Our algorithm for computing the kernels performs alternately left-to-right and right-to-left sweeps which are similar to the Bentley-Ottmann sweep ([3]). We start with a left-to-right sweep. The *event-point schedule* is initialized with the left endpoints p' and the right endpoints of all non-empty pseudo-lines $\eta_r(p')$, $p \in B'$, and the points in R'. The pseudo-lines and the red line segments are inserted into the *sweep-line status* when the sweep line encounters their left endpoints in the event point schedule; they are removed from the sweep-line status when the sweep line encounters their right endpoint in the event-point schedule if they have not been removed before.

The sweep-line status stores pseudo-lines and red line segments in the order of their intersections with the sweep line. Since the red line segments do not intersect and the pseudo-lines need not change their ordering in the sweep-line status we can compute the x-minimal intersection point q of a pseudo-line and a red line segment by testing pairs of line segments and pseudo-lines which become neighbors in the sweep-line status. Assume that this intersection point q is realized by $\eta_r(p')$ and $r \in R^*$ (Fig. 4); since $p' \in \kappa(p)$ we obtain that $r \in R_l^*(p)$. The algorithm continues if r is not part of $\kappa(p)$ or if r lies on $\mathcal{VP}(p)$ which can be determined in $O(\log n_B)$ time using point location on $\mathcal{VD}(B')$; there are at most $O(N)$ such segments r. If r is part of $\kappa(p)$ and does not lie on $\mathcal{VP}(p)$ then we remove the pseudo-line $l := \eta_r(p')$ from the sweep-line status. There exists another intersection point $(\kappa(p) \cap r) \setminus \{p'\}$ which we again denote by p' in the following. We generate a new insertion event in the event-point schedule at p' for $\eta_r(p')$ if $p' \in l$ and $\eta_r(p')$ is not empty. After processing the intersection point q the sweep continues. The sweep is complete after all pseudo-lines $\eta_r(p')$, $p \in B'$, and all line segments in R have been swept over and the sweep-line status is again empty. Finally we replace each point p', $p \in B'$, which lies on the original pseudo-line l of the original point $p' \in \kappa(p)$, by the right endpoint of $\eta_r(p')$ since this pseudo-line contributes all its edges to the kernel $\kappa(p)$. Figure 5 shows three pseudo-lines $\eta_r(p_1'), \eta_r(p_2')$ and $\eta_r(p_3')$ and six red line segments. The intersection points that are computed during the left-to-right sweep are surrounded by squares.

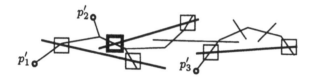

Fig. 5. Intersection points found during the left-to-right sweep

Note that the intersection point surrounded by the fat square is only computed for the pseudo-line $\eta_r(p_2')$ but not for $\eta_r(p_1')$. In the second sweep from right to left we repeat the process with the pseudo-lines $\eta_l(p')$ for the current points p', $p \in B'$, and the red line segments. It is obvious that we need at most three sweeps for computing the kernel $\kappa(p)$ for all points $p \in B'$. Since the total complexity of the kernels is $O(N)$ by Theorem 4 each sweep can be performed in $O(N \log N)$ time. Using standard data structures the algorithm can be implemented using $O(N)$ space. The Theorem follows. $\qquad \square$

4   Conclusion and open problem

We have presented an optimal algorithm for the *VNFNP* of colored line segments with applications. The algorithm can be implemented using only data structures for which numerically reliable implementations are known. To the best of our knowledge, this is the first algorithm addressing a colored proximity problem for geometric non-point sites. Solving the *ANFNP* for n colored line segments in $O(n \log n)$ time remains an open problem.

References

1. A. Aggarwal, H. Edelsbrunner, P. Raghavan, and P. Tiwari. Optimal time bounds for some proximity problems in the plane. *Inform. Process. Lett.*, 42(1):55–60, 1992.
2. F. Bartling and K. Hinrichs. A plane-sweep algorithm for finding a closest pair among convex planar objects. In *Proc. 9th Sympos. Theoret. Aspects Comput. Sci.*, volume 577 of *Lecture Notes in Computer Science*, pages 221–232. Springer-Verlag, 1992.
3. J. L. Bentley and T. A. Ottmann. Algorithms for reporting and counting geometric intersections. *IEEE Trans. Comput.*, C-28:643–647, 1979.
4. C. Brunikel, K. Mehlhorn, and S. Schirra. How to compute the voronoi diagram of line segments: Theoretical and experimental results. In *Proc. European Symposium on Algorithms*, volume 855 of *Lecture Notes in Computer Science*, pages 227–239. Springer-Verlag, 1994.
5. S. Fortune. Numerical stability of algorithms for 2-d Delaunay triangulations and Voronoi diagrams. In *Proc. 8th Annu. ACM Sympos. Comput. Geom.*, pages 83–92, 1992.
6. S. Fortune and V. Milenkovic. Numerical stability of algorithms for line arrangements. In *Proc. 7th Annu. ACM Sympos. Comput. Geom.*, pages 334–341, 1991.
7. S. J. Fortune. A sweepline algorithm for Voronoi diagrams. *Algorithmica*, 2:153–174, 1987.
8. T. Graf and K. Hinrichs. Algorithms for proximity problems on colored point sets. In *Proc. 5th Canad. Conf. Comput. Geom.*, pages 420–425, Waterloo, Canada, 1993.
9. T. Graf and K. Hinrichs. A plane-sweep algorithm for the all-nearest-neighbors problem for a set of convex planar objects. In *Proc. 3rd Workshop Algorithms Data Struct.*, volume 709 of *Lecture Notes in Computer Science*, pages 349–360. Springer-Verlag, 1993.
10. T. Graf and K. Hinrichs. Distribution algorithms for the all-nearest-foreign-neighbors problem in arbitrary L^t metrics. In *Proc. 6th Canad. Conf. Comput. Geom.*, pages 69–74, 1994.
11. D. G. Kirkpatrick. Optimal search in planar subdivisions. *SIAM J. Comput.*, 12:28–35, 1983.
12. I. Masser and M. Blakemore, editors. *Handling geographical information: methodology and potential applications.* Long Scientific & Technical, 1991.
13. M. McAllister, D. Kirkpatrick, and J. Snoeyink. A compact piecewise-linear Voronoi diagram for convex sites in the plane. *Discrete Comput. Geom.*, 15:73–105, 1996.
14. K. Mehlhorn and S. Näher. Implementation of a sweep line algorithm for the straight line segment intersection problem. Report MPI-I-94-160, Max-Planck-Institut Inform., Saarbrücken, Germany, 1994.
15. F. P. Preparata and M. I. Shamos. *Computational Geometry: An Introduction.* Springer-Verlag, New York, NY, 1985.
16. J. D. Ullman. *Computational Aspects of VLSI Design.* Computer Science Press Inc., Maryland, 1984.
17. C. K. Yap. An $O(n \log n)$ algorithm for the Voronoi diagram of a set of simple curve segments. *Discrete Comput. Geom.*, 2:365–393, 1987.

Moving an Angle Around a Region

Frank Hoffmann[1], Christian Icking[2], Rolf Klein[2], and Klaus Kriegel[1]

[1] Freie Universität Berlin, Institut für Informatik, Takustr. 9, D-14195 Berlin.
[2] FernUniversität Hagen, Praktische Informatik VI, Feithstr. 142, D-58084 Hagen.***

Abstract. Let D be a connected region inside a simple polygon, P. We define the *angle hull*, $\mathcal{AH}(D)$, of D to be the set of all points in P that can see two points of D at a right angle. We show that the perimeter of $\mathcal{AH}(D)$ cannot exceed the perimeter of the relative convex hull of D by more than a factor of 2. A special case occurs when P equals the full plane. Here we prove a bound of $\frac{\pi}{2}$. Both bounds are tight, and corresponding results are obtained for any other angle.

Key words: Angle hull, arrangements of circles, computational geometry, convex hull, curve length, envelopes of circles, motion planning, polygon.

1 Introduction

In the analysis of on-line navigation algorithms for autonomous robots in unknown environments very often structural properties of planar curves have proved very useful for estimating their lengths; see [6, 8, 9]. In this paper we provide a new result of this type that helps to significantly improve the best upper bound known so far [6] for exploring an unknown simple polygon; see Sect. 2. The types of curves studied here seem to be interesting in their own right, and the proofs reveal some interesting properties of circles.

Suppose we are given a bounded, connected region D in the plane. For convenience, we shall assume that D is a simple polygon; our results can be easily generalized by approximation. Now we take a right angle and move it around D in such a way that D is contained between, and touched by, the two sides of the angle; see Fig. 1. During this motion the angle's apex describes a certain closed curve that is of interest to us. All points enclosed by this curve, and no other, can see two points of D at a 90° angle; we call this point set the *angle hull* of D, denoted by $\mathcal{AH}(D)$.

The structure of $\mathcal{AH}(D)$ can be described quite easily. Let $\mathcal{CH}(D)$ denote the convex hull of D. As the right angle moves around D, its two sides can touch only vertices of $\mathcal{CH}(D)$. While vertices v, w are being touched the angle's apex describes a boundary segment of the circle spanned[1] by v and w, as follows from Thales' theorem. Consequently, $\mathcal{AH}(D)$ contains $\mathcal{CH}(D)$, and the boundary of $\mathcal{AH}(D)$ consists of circular arcs. It is the outer envelope of the set of circles

---

*** This work was supported by the Deutsche Forschungsgemeinschaft, grant Kl 655/8-2.

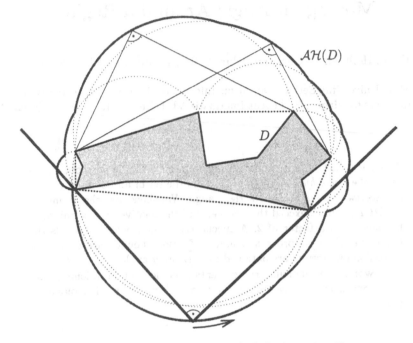

Fig. 1. Drawing the angle hull $\mathcal{AH}(D)$ of a region D.

spanned by all pairs of vertices of $\mathcal{CH}(D)$, and it can be constructed in time linear in the number of vertices of $\mathcal{CH}(D)$ once the convex hull of D is given.

We are interested in how much the perimeter of $\mathcal{AH}(D)$ can exceed, in length, the perimeter of the convex hull of D. It is not too hard to see that a tight upper factor of $\frac{\pi}{2}$ applies, as will be shown in Sect. 3.

In our application, that will be briefly sketched in Sect. 2, we have to deal with a more complex situation. Here D is contained in a simple polygon P. Again, $\mathcal{AH}(D)$ is defined to be the set of all points of P that can see two points of D at a right angle, but now the edges of P give rise to visibility constraints.

The boundary of $\mathcal{AH}(D)$ may contain segments of edges of P as well as circular segments. But now the corresponding circles may be spanned by vertices of D and vertices of P. Indeed, if a is a boundary point of $\mathcal{AH}(D)$ not situated on the boundary of P then there exists a right angle with apex a such that either of its sides touches a convex vertex of D that is included in the angle, or it touches a reflex vertex[2] of P that is excluded; see Fig. 2. Any combination of these cases is possible.

The angle hull $\mathcal{AH}(D)$ contains the relative convex hull[3] $\mathcal{RCH}(D)$, of D. Clearly, $\mathcal{AH}(D)$ only depends on $\mathcal{RCH}(D)$, rather than D. Therefore, we may

---

[1] The smallest circle containing two points, v and w, is called the circle *spanned* by v and w.

[2] A vertex of a polygon is called *reflex* if its internal angle exceeds 180°, as opposed to convex.

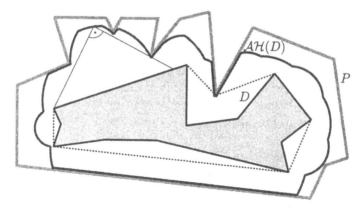

Fig. 2. The angle hull $\mathcal{AH}(D)$ inside a polygon P.

assume D itself to be relatively convex; see in Sect. 4 below. In this setting, estimating the perimeter of $\mathcal{AH}(D)$ turns out to be much harder a task. In Sect. 4 we prove that the ratio of the perimeters of $\mathcal{AH}(D)$ and of the relative convex hull of D is bounded by 2, and that this bound is tight.

2 The application: Exploring an unknown environment

The purpose of this section is to motivate our interest in the angle hull.

Suppose a mobile robot has to explore an unknown simple polygon. The robot starts walking at a given boundary point, s; it is equipped with a vision system that continuously provides the visibility polygon of the robot's current position. When each point of the polygon has been visible at least once, the robot returns to s.

The problem is in designing a *competitive* exploration strategy, i.e. one that guarantees the robot's path never to exceed in length a fixed constant times the length of the shortest watchman tour through s.

The *off-line* version of this problem, where the polygon is known, has been extensively studied. Chin and Ntafos [3] have shown that it is sufficient and necessary for the shortest watchman tour from s to visit the prolongations of some edges, the so-called essential cuts. Moreover, they have shown that the shortest watchman tour can be computed in time $O(n^4)$; their result has later been improved on by Tan and Hirata [11] to $O(n^2)$. Carlsson et al. [2] have proven that the shortest watchman tour without a specified point s can be computed in time $O(n^3)$. Furthermore, Carlsson and Jonsson [1] proposed an $O(n^6)$ algorithm for computing the shortest path inside a simple polygon from which each point of the boundary is visible, where start and end point are not specified.

---

[3] The *relative convex hull* of a subset D of a polygon P, $\mathcal{RCH}(D)$, is the smallest subset of P that contains D and, for any two points of D, the shortest path in P connecting them.

For the *on-line* problem, Deng et al. [4] were the first to claim that such a competitive strategy does in fact exist. In [5] they gave a proof for the rectilinear case, where a simple greedy strategy can be applied that performs surprisingly well.

For the general case, we have presented a strategy with a proven factor of 133 in [6]. Although this value compares favorably against the bound discussed in [4], which was in the thousands, it is still too large to be attractive to practitioners. The difficulty is in providing tight length estimates for curves of complex structure.

The concept, and analysis, of the *angle hull* presented in this paper has led to a breakthrough. In a forthcoming paper [7] we are presenting a modified strategy for exploring simple polygons with a competitive factor of 26.5.

To illustrate one relationship between the angle hull and the exploration problem we discuss the situation depicted in Fig. 3. For brevity, only a simple example is shown.

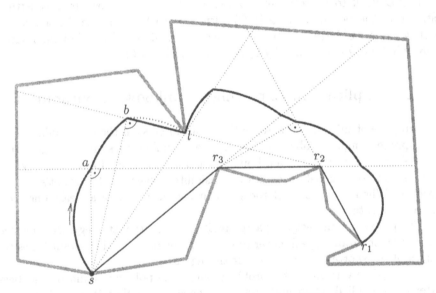

Fig. 3. Exploring an unknown polygon: a simple case.

At the start point, s, the robot discovers two reflex vertices that are obstructing its view. The adjacent edges of one of them, vertex l, are situated to the left of the shortest path from s to l, and so we call l a *left vertex*. Analogously, vertex r_3 is a *right vertex*. The *cut* of such a vertex means the prolongation of its invisible adjacent edge. We call a vertex *discovered* after it has once been visible, and we call it *fully explored* when its cut has been reached. Exploring a polygon is equivalent to visiting all cuts.

In a first phase, the robot explores right reflex vertices only, ignoring left ones. To this end, it proceeds by the following rules.

1. Always explore the clockwise first[4] among those right reflex vertices that have been discovered but not yet fully explored.
2. To explore a right reflex vertex, r, follow the clockwise oriented circle spanned by r and by the last vertex on the shortest path from s to the robot's current position.
3. When the view to the vertex r currently explored gets blocked (or when the boundary is hit) walk straight towards r (or follow the boundary) until motion according to (2) becomes possible again.

In Fig. 3, r_3 is the only right reflex vertex visible from the start point, s. The robot approaches r_3 along the circle spanned by s and r_3, with the intention of reaching the cut of r_1.

Before r_3's cut is reached, right reflex vertex r_2 is detected, at point a. Since r_2 comes before r_3 in clockwise order, the robot now has to switch to exploring r_2, by following the circle spanned by r_2 and by s. This circle is also passing through a, so that rule (2) can in fact be applied. While exploring r_2, after point b the view to r_2 would get blocked, so the robot walks straight to l, according to (3).

The robot proceeds according to these rules until it arrives at r_1.

Now the crucial observation is the following: The robot's path from s to r_1 is part of the boundary of the angle hull of the shortest path from s to r_1 (except for the straight piece from b to l that is replaced with the circular arc spanned by s and l.) Thus, we can apply Theorem 1 and conclude that the robot's path to r_1 is at most twice as long as the shortest path!

Actually, in [7] we apply Theorem 1 a second time to a higher level of the strategy, where the set D is the optimum watchman tour. So the application benefits even twice from the result provided in this paper.

3   The angle hull in the plane

Consider a circle with diameter d and a circular arc a on its boundary. Two lines passing through the endpoints of the arc and an arbitrary third point on the boundary always intersect in the same angle α, by the "generalized Thales' theorem". For the length of a, we have $|a| = \alpha d$. This equality will turn out to be quite useful, e. g. in the following lemma.

Lemma 1. *For a convex polygon, D, the boundary of its angle hull, $\mathcal{AH}(D)$, is at most $\frac{\pi}{2}$ times longer than the boundary of D.*

Proof. As described before, the boundary of $\mathcal{AH}(D)$ is the locus of the apex of a right angle which is rotated around D while the angle's two sides touch D from the outside. It consists of circular arcs, each one spanned by two vertices of D.

The length of one such circular arc, a, is the distance between the two spanning vertices, d, times the angle, α, by which the right angle rotates to generate a; see Fig. 4.

The distance, d, is clearly not greater than the length of the boundary of D between the two spanning vertices. The length, L, of the whole boundary of

---

[4] With repect to the order of vertices on the polygon's boundary, starting from s.

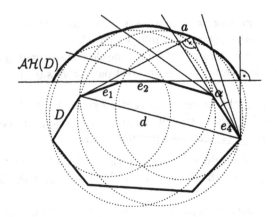

Fig. 4. The length of arc a equals $\alpha d \leq \alpha(|e_1| + \cdots + |e_4|)$; edge e_2 contributes only to the arcs above the horizontal line.

$\mathcal{AH}(D)$ is the sum over all such circular arcs a_i and we get a formula of the following form.

$$L = \sum_i a_i \leq \sum_i \alpha_i (e_{i_s} + e_{i_s+1} + \cdots + e_{i_t})$$

Here, α_i is the corresponding angle for arc a_i and $e_{i_s}, \ldots, e_{i_t}$ denote the edge lengths of D between the two spanning vertices of a_i. Each e_j contributes only to the part of the boundary of $\mathcal{AH}(D)$ that is separated from D by the line through e_j; see edge e_2 in Fig. 4. In order to generate this boundary part, the right angle rotates by exactly 90°. Thus, after re-ordering the sum on the right hand side we have

$$L \leq \sum_j e_j (\alpha_{j_s} + \cdots + \alpha_{j_t}),$$

where the angles $\alpha_{j_s}, \ldots, \alpha_{j_t}$ belonging to edge e_j add up to a right angle. Thus, we obtain

$$L \leq \frac{\pi}{2} \sum_j e_j$$

which concludes the proof. □

For D being a line segment, $\mathcal{AH}(D)$ is a circle with diameter D, so the upper bound of $\frac{\pi}{2} \approx 1.571$ is tight. Other examples with this property are triangles with only acute angles, or rectangles.

4 The angle hull inside a polygon

Let D be a relative convex polygon within a simple polygon, P. In this section we want to show that the perimeter of the angle hull of D is at most twice as long as the perimeter of D, and that this bound is tight.

In the following (Lemmata 2 through 5) we consider as region D only a line segment, before turning to the general case. It is clear that the angle hull of a line segment goes through its endpoints, so we can restrict ourselves to treat only the situation on one side of D.

A priori it is not obvious why the presence of the surrounding polygon, P, should result in an upper bound bigger than $\frac{\pi}{2}$. The following lemma shows that the factor can come arbitrarily close to 2.

Lemma 2. *Let $\varepsilon > 0$. There is a polygon, P, and a relative convex region, D, inside P, for which the boundary of the angle hull $\mathcal{AH}(D)$ with respect to P is longer than $2 - \varepsilon$ times the boundary of D.*

Proof. As our region D, we take a horizontal line segment of length 1. Let $p_0, \ldots, p_n$ be equidistant points on the halfcircle spanned by D, where p_0 and p_n are the endpoints of D; see Fig. 5. From each point p_i we draw the right angle to the endpoints of D. Let P be the upper envelope of these angles. Then we have $P = \mathcal{AH}(D)$ by construction.

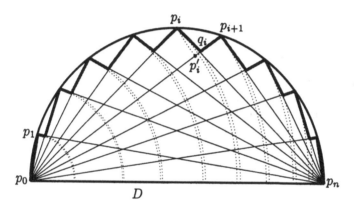

Fig. 5. The boundary of the upper envelope of the right angles is less than $2|D|$.

We want to show that the length of this jagged line is less than 2, but comes arbitrarily close to 2, as n increases. Let q_i be the intersection of the segments $p_0 \, p_{i+1}$ and $p_i \, p_n$. If we rotate, for all i, the ascending segments $q_i \, p_{i+1}$ about p_0 onto D; see the dotted arcs in Fig. 5, these segments cover disjoint pieces of D, so the total length of all ascending segments is always less than 1. By symmetry, the same bound holds for the descending segments. It remains to show that the ascending length can come arbitrarily close to 1.

Consider the triangle $p_i \, q_i \, p_i'$, where p_i' is the orthogonal projection of p_i onto $p_0 \, q_i$. Point p_0 is closer to p_i' than to p_i, so for the distances from p_0 to p_i and to q_i we have

$$|p_0 \, q_i| - |p_0 \, p_i| \le |p_0 \, q_i| - |p_0 \, p_i'| = |p_i' \, q_i| = |p_i \, q_i| \sin \frac{\pi}{2n}.$$

The total length of all ascending segments is therefore 1 minus the following rest.

$$\sum_i (|p_0\, q_i| - |p_0\, p_i|) \leq \sin\frac{\pi}{2n} \sum_i |p_i\, q_i| \leq \sin\frac{\pi}{2n}$$

For $n \to \infty$, this tends to 0. The last inequality holds because $\sum_i |p_i\, q_i| \leq 1$ is the length of all descending segments. □

The proof also works for non-equidistant points as long as the maximum distance between subsequent points tends to 0. We are obliged to R. Seidel [10] for this elegant proof of Lemma 2.

Interestingly, jagged lines like the one used in the proof of Lemma 2 continue to be very useful in the proof of the upper bound. For any circular arc a which is part of $\mathcal{AH}(D)$, we can construct a jagged line by distributing auxiliary points along a and by taking the upper envelope of the right angles at these points whose sides pass through the two spanning vertices of a. We denote with *jagged length of a* the limit of the lengths of these jagged lines as the maximum distance between subsequent points tends to 0. This limit is well-defined, i. e. it does not depend on how the points are chosen.

The jagged length of an arc is always greater than the arc length itself. From the proof of Lemma 2 we obtain the following.

Corollary 1. *The jagged length of a halfcircle of diameter 1 equals 2.*

Now we turn to proving that 2 is also an upper bound for the length of the angle hull divided by the perimeter of D. We start with the simple case where the polygon's boundary contributes only one tooth as a relevant obstacle for the angle hull; see Fig. 6. This tooth is supposed to be nearly vertical and very thin such that only vertex p really matters. Therefore, in the sequel we shall simply refer to the *obstacle points* p in this sense.

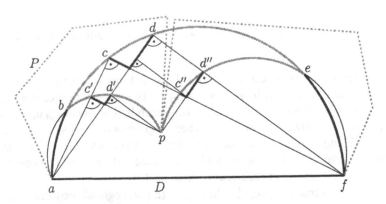

Fig. 6. In case of a line segment D and a single obstacle point p the angle hull is the inner envelope of the circles. The upper jag from c to d equals in length the sum of the lengths of the two lower jags.

In Fig. 6, the upper part of the angle hull consists of the following parts. From a to b and from e to f, it follows the big halfcircle spanned by D. But from b to e, the hull uses the smaller halfcircles via p. To find an upper bound, one would like to bound the lengths of the arc from b to p and from p to e, which are the new parts of the angle hull, in terms of the length of the arc from b to e, but it is not clear how to do this. For the jagged length, however, a suprisingly elegant equality holds.

Lemma 3. *Assume a situation as in Fig. 6, i. e. we have a line segment D of length 1 and a single obstacle point p. Then the jagged length of $\mathcal{AH}(D)$ with respect to p taken from one endpoint of D to the other is the same as the jagged length of the halfcircle spanned by D, namely 2.*

Proof. Let c and d be two points on the arc from b to e; see Fig. 6. We draw the right angles from these points to the endpoints of D and consider the jag between c and d.

The sides of the two right angles intersect the smaller arc from b to p in c' and d' and the arc from p to e in c'' and d''. By Thales' theorem we can also inscribe two right angles into each of these halfcircles, such that we get two small jags in each of them.

Surprisingly, the jag between c and d is exactly of the same length as the two other jags combined: consider the ascending part of the jag from c'' to d''. It is parallel to the line from a to d, so if we translate it from d'' to d it covers part of the jag there. But the rest of the ascending part of this jag has the same length as the ascending part of the jag from c' to d' since they are corresponding sides of congruent triangles; see Fig. 6. By symmetry, the same also holds for the descending parts.

Since this argument can be applied to any set of jags, the claim follows. □

As a consequence, we can argue as follows. The arc length of the angle hull shown in Fig. 6 is smaller than its jagged length which has just been shown to be equal to the jagged length of the halfcircle, which equals 2 by Lemma 2. The next step generalizes this result to more than one obstacle points.

Lemma 4. *Let D be a line segment of length 1, and let P be a surrounding polygon contributing only a finite set of obstacle points. Then the arc length of $\mathcal{AH}(D)$ with respect to P from one endpoint of D to the other is less than 2.*

Proof. The angle hull is composed of circular arcs, each arc is spanned by two points, either obstacle points or endpoints of the line segment. In Fig. 7, notation be, for example, denotes an arc spanned by obstacle points b and e, while b is to the left of e.

The number of arcs is proportional to the number of points, but there can be several arcs which are spanned by the same pair of points, for example af appears twice in Fig. 7. Furthermore, one point can contribute to many arcs which may not be connected. Point b, for example, contributes to arcs ab, bf, and bc, but also to bd, which is far away from the others.

Now we proceed by induction. If there is only one obstacle point, we have seen in Lemma 3 that we can remove it without altering the jagged length of the angle hull.

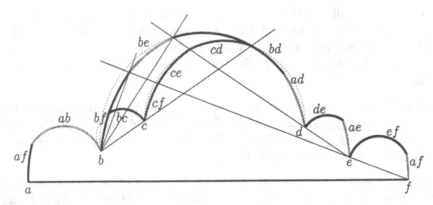

Fig. 7. The angle hull below 4 obstacle points b, c, d, e, and the situation after removing point c.

For more than one obstacle points, we would like to apply the same technique. However, it seems not clear how to show that the jagged length is preserved under removal of an arbitrary point.

We choose to remove the last vertex of the upwards convex chain of obstacle points, starting with the left endpoint of D. In Fig. 7, this is point c. It has the following properties. It co-spans only one arc (namely bc) to its left and one or more arcs to its right (cf, ce, cd). All these arcs co-spanned by c must be connected.

Fig. 7 shows how c is eliminated by applying three times the technique of Fig. 6. Segment bc is divided into three parts. The left part of bc can be combined with cf to a prolongation of bf with equal jagged length, the middle part together with ce gives a new segment be, and the right part plus cd equals in jagged length the prolongation of bd. As a result, we get the angle hull with respect to the obstacle set without point c, which is still of the same jagged length. □

Now let us turn to an arbitrary surrounding polygon P that influences the angle hull not only with acute reflex vertices but also with its edges.

Lemma 5. *Let D be a line segment of length 1, and let P be a simple polygon surrounding D. The arc length of $AH(D)$ with respect to P from one endpoint of D to the other is less than 2.*

Proof. First, we consider the angle hull $\mathcal{AH}_1(D)$ with respect to only the vertices of P as obstacle points. Its arc length is less than 2, by Lemma 4.

Now also the edges come into play. The angle hull $\mathcal{AH}_2(D)$ with respect to the whole of P contains circular arcs and some pieces of P's edges, for an example see Fig. 2. The circular arcs of $\mathcal{AH}_2(D)$ are also part of $\mathcal{AH}_1(D)$.

For every piece of an edge which contributes to $\mathcal{AH}_2(D)$, the piece's two endpoints are also on the boundary $\mathcal{AH}_1(D)$. Therefore, $\mathcal{AH}_2(D)$ can only be shorter than $\mathcal{AH}_1(D)$. □

So far, the region D was assumed to be a line segment. Now we turn to the general case where D is a simple polygon which is relatively convex with respect to P.

All reflex vertices of D, if they exist, are also vertices of P, due to the relative convexity of D. Also the boundary of $\mathcal{AH}(D)$ passes through these points. Therefore, we can subdivide at these points, and we consider only the case where D is a convex chain with endpoints on the boundary of P or a closed convex polygon.

For dealing with the edges of P, the technique of Lemma 5 can be applied without modification. Furthermore, we can remove the vertices of P (obstacle points) one by one while maintaining the jagged length with an approach analogous to the one used in Lemmata 3 and 4. It remains to show the following generalization of Corollary 1.

Lemma 6. *Let D be a convex chain (without surrounding polygon). The jagged length of the angle hull $\mathcal{AH}(D)$ between the endpoints of D equals twice the length of D.*

Proof. We proceed by induction on the number of edges of D. For a single line segment the claim follows from Corollary 1. We demonstrate how the inductive step works only for a two-edge chain D; see Fig. 8. Remark the unexpected similarity to Lemma 3 (and to Fig. 6).

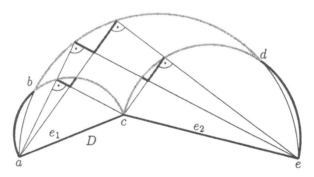

Fig. 8. Angle hull $\mathcal{AH}(D)$ is the outer envelope of the circles. The jagged length of $\mathcal{AH}(D)$ equals $2(|e_1| + |e_2|)$.

The angle hulls of e_1 and e_2 are the halfcircles spanned. Their jagged lengths are $2|e_1|$ and $2|e_2|$, resp., by Corollary 1. The boundary of $\mathcal{AH}(D)$ is the outer envelope of the three circles, the small arcs from b to c and from c to d are replaced by the big arc from b to d. But in the proof of Lemma 3 we have seen that the sum of the two smaller jagged lengths equals the jagged length of the big arc.

So the jagged length of $\mathcal{AH}(D)$ equals $2(|e_1| + |e_2|)$. These arguments also apply for more than two edges and for closed convex chains. $\square$

With all reduction steps done, we now formulate the main result.

Theorem 1. *Let P be a simple polygon containing a relative convex polygon D. The arc length of the boundary of the angle hull, $\mathcal{AH}(D)$, with respect to P is less than 2 times the length of D's boundary. This bound is tight.*

5 Generalizations and open questions

We have introduced a new type of hull operator that suits well our needs in analysis of online navigation strategies. But clearly the angle hull is interesting in its own right with many possible generalizations.

For example, what happens if the prescribed angle α is different of 90°? Angle hulls generalized in that way can be analyzed much in the same way as we have seen in Sections 3 and 4. The resulting bounds are $(\pi - \alpha)/\sin\alpha$ in the unrestricted and $2(\cos\alpha + 1)/\sin^2\alpha$ in the restricted case.

Still open questions are: Does each subset of A have a shorter angle hull than A? Does the iterated construction of the angle hull approximate a circle? How can angle hulls be analyzed in three dimensions?

References

1. S. Carlsson and H. Jonsson. Computing a shortest watchman path in a simple polygon in polynomial-time. In *Proc. 4th Workshop Algorithms Data Struct.*, volume 955 of *Lecture Notes Comput. Sci.*, pages 122–134. Springer-Verlag, 1995.
2. S. Carlsson, H. Jonsson, and B. J. Nilsson. Finding the shortest watchman route in a simple polygon. In *Proc. 4th Annu. Internat. Sympos. Algorithms Comput.*, volume 762 of *Lecture Notes Comput. Sci.*, pages 58–67. Springer-Verlag, 1993.
3. W.-P. Chin and S. Ntafos. Watchman routes in simple polygons. *Discrete Comput. Geom.*, 6(1):9–31, 1991.
4. X. Deng, T. Kameda, and C. Papadimitriou. How to learn an unknown environment. In *Proc. 32nd Annu. IEEE Sympos. Found. Comput. Sci.*, pages 298–303, 1991.
5. X. Deng, T. Kameda, and C. H. Papadimitriou. How to learn an unknown environment I: the rectilinear case. Technical Report CS-93-04, Department of Computer Science, York University, Canada, 1993.
6. F. Hoffmann, C. Icking, R. Klein, and K. Kriegel. A competitive strategy for learning a polygon. In *Proc. 8th ACM-SIAM Sympos. Discrete Algorithms*, pages 166–174, 1997.
7. F. Hoffmann, C. Icking, R. Klein, and K. Kriegel. The polygon exploration problem: A new strategy and a new analysis technique. In *Proc. 3rd International Workshop on Algorithmic Foundations of Robotics*, 1998.
8. C. Icking and R. Klein. Searching for the kernel of a polygon: A competitive strategy. In *Proc. 11th Annu. ACM Sympos. Comput. Geom.*, pages 258–266, 1995.
9. C. Icking, R. Klein, and L. Ma. How to look around a corner. In *Proc. 5th Canad. Conf. Comput. Geom.*, pages 443–448, 1993.
10. R. Seidel. Personal communication, 1997.
11. X. Tan and T. Hirata. Constructing shortest watchman routes by divide-and-conquer. In *Proc. 4th Annu. Internat. Sympos. Algorithms Comput.*, volume 762 of *Lecture Notes Comput. Sci.*, pages 68–77. Springer-Verlag, 1993.

Models and Motion Planning[*]

Mark de Berg[1], Matthew J. Katz[2][**], Mark Overmars[1],
A. Frank van der Stappen[1], and Jules Vleugels[1]

[1] Department of Computer Science, Utrecht University,
P.O.Box 80.089, 3508 TB Utrecht, the Netherlands.
Email: {markdb,markov,frankst,jules}@cs.uu.nl
[2] Department of Mathematics and Computer Science,
Ben-Gurion University of the Negev, Beer-Sheva 84105, Israel.
Email: matya@cs.bgu.ac.il

Abstract. We study the consequences of two of the realistic input models proposed in the literature, namely unclutteredness and small simple-cover complexity, for the motion planning problem. We show that the complexity of the free space of a bounded-reach robot with f degrees of freedom is $O(n^{f/2})$ in the plane, and $O(n^{2f/3})$ in three dimensions, for both uncluttered environments and environments of small simple-cover complexity. We also give an example showing that these bounds are tight in the worst case. Our bounds fit nicely between the $\Theta(n^f)$ bound for the maximum free-space complexity in the general case, and the $\Theta(n)$ bound for low-density environments.

1 Introduction

It is well known that the maximum complexity of the free space of a robot with f degrees of freedom moving in a scene consisting of n disjoint obstacles can be $\Omega(n^f)$. Consequently, exact motion-planning algorithms often have a worst-case running time of at least the same order of magnitude. This is probably one of the reasons that most of the exact algorithms were never implemented. One exception is Bañon's implementation [3] of the $O(n^5)$ algorithm of Schwartz and Sharir [14] for a ladder moving in a two-dimensional workspace, which performs surprisingly well, and much better than the worst-case theoretical analysis predicts. The reason is that the running time of the algorithm is sensitive to the actual complexity of the free space, and this is in practice far less than the $\Theta(n^f)$ worst-case bound.

These observations inspired research [1, 2, 5–9, 11, 13, 15–19] where geometric problems are studied under certain (hopefully realistic) assumptions on the input—in the case of motion planning: the environment in which the robot is

---

[*] This research was partially supported by the Netherlands' Organization for Scientific Research (NWO).
[**] Supported by the Israel Science Foundation founded by the Israel Academy of Sciences and Humanities.

moving. The goal of this line of research is to be able to predict better the practical performance of algorithms. For instance, van der Stappen et al. [16] studied the free-space complexity for a *bounded-reach robot* moving in environments consisting of *fat obstacles*. They showed that in this restricted type of environments the worst-case free-space complexity is only $\Theta(n)$, which is more in line with the experimental results of Bañon. Van der Stappen and Overmars [17] used this result to obtain an efficient algorithm for robot motion planning amidst fat obstacles. These results were extended to the more general setting of low-density environments by van der Stappen et al. [18].

Recently, de Berg et al. [4] brought together various of the *realistic input models* that were proposed in the literature, namely *fatness, low density, unclutteredness,* and *small simple-cover complexity*—see Sect. 2 for formal definitions of these models. They showed that these models form a strict hierarchy in the sense that fatness implies low density, which in turn implies unclutteredness, which implies small simple-cover complexity, and that no other implications exist between the models. A natural question that arises is whether the results of van der Stappen et al. [18] remain valid when, instead of a low-density scene, we assume a more general setting, like an uncluttered scene or a scene with small simple-cover complexity. In other words, does the complexity of the free space of a bounded-reach robot with f degrees of freedom moving in an uncluttered scene (alternatively, in a scene with small simple-cover complexity) remain $O(n)$?

The main result of this paper, which we obtain in Sect. 4, is a negative answer to this question. We prove that the maximum complexity of the free space of a bounded-reach robot moving in either an uncluttered scene or a scene with small simple-cover complexity is $\Theta(n^{f/2})$ when the workspace is two-dimensional, and $\Theta(n^{2f/3})$ when the workspace is three-dimensional. These bounds fit nicely between the $\Theta(n)$ bound for low-density scenes and the $\Theta(n^f)$ bound for general scenes.

In our upper-bound proofs we use so-called *guarding sets* for the set of obstacles in the environment. A κ-*guarding set* for a set of objects against a family $\mathcal{R}$ of ranges is a set G of points such that any range from $\mathcal{R}$ without a point from G in its interior intersects at most κ objects. Guarding sets are of independent interest for several reasons. First of all, the concept of guarding sets provides a unifying framework for the unclutteredness model and the small simple-cover complexity model in the plane: a scene is κ-cluttered if and only if the bounding-box vertices of the objects form a κ-guarding set (this is the definition of unclutteredness [5]), and we show that a planar scene has small simple-cover complexity if and only if it has a κ-guarding set of linear size (for some constant κ). De Berg's algorithm [5] for computing linear-size binary space partitions (BSPs) in fact works for scenes admitting a linear size guarding set. This implies that scenes with small simple-cover complexity also admit a linear-size BSP. A second aspect that makes guarding sets interesting is their connection with ε-nets, which we discuss briefly in Sect. 3. (For an introduction to ε-nets see, for example, the survey paper by Matoušek [10].)

In this paper we introduce the concept of guarding sets, and we prove a number of basic results about them. In particular, we show in Sect. 3.1 that asymptotically the type of the ranges in the family $\mathcal{R}$—hypercubes, balls, and so on—does not matter, as long as the ranges are convex and fat. We also study the relation between the distribution of the objects and the distribution of guards. Due to space limitations, all proofs have been omitted from this extended abstract.

2 Preliminaries

In this paper we shall be dealing a lot with squares, cubes, rectangles, and so on. These are always assumed to be axis-aligned. All geometric objects we consider are open; in particular, if we talk about a point lying in a square or cube, we mean that the point lies in the interior of the square or cube. Furthermore, all objects we consider are assumed to have constant complexity. The dimension of the space is denoted as d.

The *size* of a square (more generally, of a hypercube) is defined to be its edge length, and the size of an object is the size of a smallest enclosing hypercube for the object. An *L-shape* is the geometric difference of a hypercube σ with a hypercube $\sigma' \subseteq \sigma$ of less than half its size and sharing a vertex with it. An L-shape can be covered by $2^d - 1$ hypercubes contained in it: for each vertex v of σ not shared with σ', take the hypercube of maximal size with v as a vertex and contained in $\sigma \setminus \sigma'$.

We will often require a notion of an object or point being close to a given hypercube. The *vicinity* of a hypercube σ is the hypercube obtained by scaling σ with a factor of 5/3 with respect to its center. Thus, if we partition the vicinity of σ into 5^d equal-sized subhypercubes, then σ consists of the 3^d middle subhypercubes.

We now briefly describe the four input models that play a role in this paper.

Intuitively, an object is called *fat* if it contains no long and skinny parts. The definition we use in this paper is the following. Let $\mathcal{P}$ be an object in $\mathbb{R}^d$, and define $U(\mathcal{P})$ as the set of all balls centered inside $\mathcal{P}$ whose boundary intersects $\mathcal{P}$. We say that the object $\mathcal{P}$ is β-fat if for all balls $B \in U(\mathcal{P})$, $\text{vol}(\mathcal{P} \cap B) \geq \beta \cdot \text{vol}(B)$. The *fatness* of $\mathcal{P}$ is defined as the maximal β for which $\mathcal{P}$ is β-fat.

The model of low density was introduced by van der Stappen et al. [18] and refined by Schwarzkopf and Vleugels [15]. It forbids any ball B to be intersected by many objects whose minimal-enclosing-ball radius is at least as large as the radius of B. In the definition, $\rho_{\text{meb}}(\mathcal{P})$ denotes the radius of the minimal enclosing ball of an object $\mathcal{P}$. Let $\mathcal{O}$ be a set of objects in $\mathbb{R}^d$. We say that $\mathcal{O}$ has λ-*low-density* if for any ball B, the number of objects $\mathcal{P}_i \in \mathcal{O}$ with $\rho_{\text{meb}}(\mathcal{P}_i) \geq \text{radius}(B)$ that intersect B is at most λ. The *density* of $\mathcal{O}$ is defined to be the smallest λ for which $\mathcal{O}$ is a λ-low-density scene. We say that a scene has *low density* if its density is a small constant.

Unclutteredness was introduced by de Berg [5] under the name *bounding-box-fitness* condition. The model is defined as follows. Let $\mathcal{O}$ be a set of objects in $\mathbb{R}^d$.

We say that $\mathcal{O}$ is κ-*cluttered* if any hypercube whose interior does not contain a vertex of one of the bounding boxes of the objects in $\mathcal{O}$ is intersected by at most κ objects in $\mathcal{O}$. We sometimes call a scene *uncluttered* if it is κ-cluttered for a small constant κ.

The following definition of simple-cover complexity is a slight adaptation of the original definition by Mitchell et al. [12], as proposed by de Berg et al. [4]. Given a scene $\mathcal{O}$, we call a ball δ-*simple* if it intersects at most δ objects in $\mathcal{O}$. Let $\mathcal{O}$ be a set of objects in $\mathbb{R}^d$, and let δ > 0 be a parameter. A δ-*simple cover* for $\mathcal{O}$ is a collection of δ-simple balls whose union covers the bounding box of $\mathcal{O}$. We say that $\mathcal{O}$ has (s, δ)-*simple-cover complexity* if there is a δ-simple cover for $\mathcal{O}$ of cardinality sn. We will say that a scene has *small simple-cover complexity* if there are small constants s and δ such that it has (s, δ)-simple-cover complexity.

We close this preliminary section by stating a result obtained by de Berg [5].

Lemma 1. (de Berg [5]) *Let $\mathcal{P}$ be a set of points in $\mathbb{R}^d$, and let $\sigma_{\mathcal{P}}$ be a hypercube containing all points from $\mathcal{P}$. Then there exists a partitioning of $\sigma_{\mathcal{P}}$ into $O(|\mathcal{P}|)$ hypercubes and L-shapes without points from $\mathcal{P}$ in their interior.*

3 Guarding sets

A guarding set for a collection of objects is, loosely speaking, a set of points that approximates the distribution of the objects. More precisely, guarding sets are defined as follows.

Definition 1. *Let $\mathcal{O}$ be a set of objects in $\mathbb{R}^d$, let $\mathcal{R}$ be a family of subsets of $\mathbb{R}^d$ called* ranges, *and let κ be a positive integer. A set G of points is called a* κ-*guarding set for $\mathcal{O}$ against $\mathcal{R}$ if any range from $\mathcal{R}$ not containing a point from G intersects at most κ objects from $\mathcal{O}$.*

We will often call the points in G *guards*, and we will refer to ranges not containing guards as *empty ranges*. Ranges that intersect more than κ objects are called *crowded*. Clearly, a crowded range cannot be empty.

Let's give an example. Suppose the set $\mathcal{O}$ is a set of n pairwise disjoint discs in the plane and that the family of ranges is the family of all squares. For a disc D, define G_D to be the set of the following five points: the center of D plus the topmost, bottommost, leftmost, and rightmost point of D. When a square σ intersects D, and σ does not contain a point from G_D in its interior, then D contains a vertex of σ. (Recall our convention that all geometric objects are open.) Hence, the set $G := \{G_D \mid D \in \mathcal{O}\}$ is a 4-guarding set of size $5n$ for $\mathcal{O}$ against the family of squares.

Let's consider another example. Again $\mathcal{O}$ is a set of n disjoint discs in the plane, but this time $\mathcal{R}$ is the family of all lines. Then it can easily happen that no finite κ-guarding set exists for κ < n: if, for instance, the centers of the discs are (almost) collinear, then there are infinitely many lines stabbing all the discs and no finite guarding set can capture all these lines.

Later, when we study the relation between guarding sets and some of the realistic input models proposed in the literature we shall see that in many settings there are finite-size—in fact, linear-size—guarding sets. But first we discuss the relation with a well-known concept from the literature, namely (weak) ε-nets, and we prove some basic facts about guarding sets.

A subset N of a given set $\mathcal{O}$ of n objects in $\mathbb{R}^d$ is called an ε-net with respect to a family $\mathcal{R}$ of ranges, if any empty range—any range not intersecting an object from N—intersects at most εn objects from $\mathcal{O}$. (The actual definition of ε-nets is more general than the definition we just gave: it is purely combinatorial.) A weak ε-net is defined similarly, except that N is not restricted to a subset of $\mathcal{O}$ but can be any object from the same class as those in $\mathcal{O}$.

If we take $\kappa = \varepsilon n$ then the notions of κ-guarding set and ε-net become very similar. The major difference is that a guarding set consists of points, whereas an ε-net is either a subset of the given set $\mathcal{O}$, or (in the case of a weak ε-net) an arbitrary set of objects from this class. This means that sometimes a finite-size guarding set does not exist, whereas an ε-net—and, hence, a weak ε-net—does exist. On the other hand, the fact that guarding sets consist of points makes further processing easier from an algorithmic point of view. Note that the notions of εn-guarding set and weak ε-net are identical if the objects in $\mathcal{O}$ are points.

3.1 Basic properties of guarding sets

From the second example given above it appears to be wise to restrict ourselves to fat ranges, since long and skinny ranges such as lines are likely not to admit finite-size guarding sets. Therefore we mainly consider two types of ranges, namely cubes (or squares) and balls (or discs). We start by considering the relation between these two types of ranges.

Theorem 1. *If there is a κ-guarding set of size m for $\mathcal{O}$ against squares in $\mathbb{R}^2$ or against cubes in $\mathbb{R}^3$, then there is a $(c\kappa)$-guarding set of size $O(\beta^{-1}m)$ for $\mathcal{O}$ against convex ranges that are at least β-fat, for some constants c and β with $0 < \beta \leqslant 1$.*

Theorem 1 implies that, from an asymptotic point of view, it doesn't matter whether we study guarding against hypercubes or against any other convex fat shape, such as balls. Hence, from now on we restrict our attention to squares (in $\mathbb{R}^2$) and cubes (in $\mathbb{R}^3$).

3.2 Guards in hypercubes

In this section we study the relation between the distribution of objects and the distribution of guards. We give a simple theorem stating that the number of guards in a hypercube is at least linear in the number of objects intersecting the hypercube. Observe that the reverse of the theorem is not necessarily true: a hypercube that is intersected by only few objects may contain many guards.

Theorem 2. *Let G be a κ-guarding set against hypercubes for a set $\mathcal{O}$ of objects. Any hypercube with exactly g guards in its interior is intersected by $O(\kappa g)$ objects.*

Remark 1. Although the theorem is not surprising and easy to prove, it has an interesting consequence. Let $\mathcal{O}$ be a set of n objects, and suppose we have a κ-guarding set G of size m, with $m = cn$ for some constant c, against hypercubes for $\mathcal{O}$. Now suppose we wish to have a κ'-guarding set for $\mathcal{O}$ for some $\kappa' > \kappa$, say for $\kappa' = \delta n$. Then we can obtain such a guarding set by taking an ε-net N for G with respect to the family of (open) hypercubes, for a suitable $\varepsilon = \Theta(\delta/\kappa)$: a hypercube not containing a point from N contains less than $\varepsilon \cdot m = \varepsilon \cdot cn$ guards from G, so by Theorem 2 it intersects $O(\varepsilon \kappa n)$ objects. Because the underlying range space has finite VC-dimension, there is such an ε-net of size $O((\kappa/\delta) \log(\kappa/\delta))$ [10]. Combining this with the result of de Berg [5] that any collection of disjoint fat objects has a linear size guarding set—see also Sect. 4—this implies the following ε-net-type result. Let $\mathcal{O}$ be a set of n disjoint fat objects in $\mathbb{R}^d$. There exists a guarding set of $O((\kappa/\delta) \log(\kappa/\delta))$ points such that any hypercube intersecting more than εn objects contains at least one guard.

3.3 Guards in vicinities

Theorem 3 below states another, more surprising, relation between the distribution of the objects and the distribution of the guards. Again we look at empty hypercubes, but this time we only look at objects that are at least as large as the hypercube, and not only consider the guards inside the hypercube but also the ones in its vicinity. (Recall that the vicinity of a hypercube σ is obtained by scaling σ with a factor of 5/3 with respect to its center.) We will show that the number of relatively large objects intersecting a hypercube σ in $\mathbb{R}^d$ cannot be more than (roughly) $\sqrt[d]{g}$, where g is the number of guards in the vicinity of σ.

We first reduce the problem to a simpler problem on so-called 3-blocks. Define a *3-block* to be an axis-parallel hypercube scaled by a factor of 1/3 along one of the coordinate axes. For a given hyperrectangle we define a *crossing object curve* to be a curve inside the hyperrectangle that is contained in one of the objects and that connects the two largest (and opposite) faces of the hyperrectangle. From now on, whenever we speak of a collection of object curves, we implicitly assume that the curves are contained in distinct objects. First we prove that if a hypercube is intersected by many larger objects, then there must be a 3-block in its vicinity that is crossed by many object curves.

Lemma 2. *Let σ be a hypercube that is intersected by a collection $\mathcal{O}$ of m objects that are at least as large as σ. Then there is a 3-block contained in the vicinity of σ that has $m/(2d3^d)$ crossing object curves.*

The next step is to establisch a relation between the number of crossing object curves and the number of guards in a 3-block B.

Lemma 3. *Let G be a κ-guarding set for a collection $\mathcal{O}$ of objects in $\mathbb{R}^d$. Let B be a 3-block with m crossing object curves, each contained in a distinct object of $\mathcal{O}$. Then there must be at least $3\lfloor m/6(\kappa+1) \rfloor^2$ guards inside B if $d = 2$, and $\lfloor \sqrt{m}/5(\kappa+1) \rfloor^3$ guards if $d = 3$.*

Combining the two lemmas above, we can prove that the number of relatively large objects intersecting a square (or cube) cannot exceed (roughly) the square (or cubic) root of the number of guards in its vicinity.

Theorem 3. *Let G be a κ-guarding set for a set $\mathcal{O}$ of objects in $\mathbb{R}^d$, for $d = 2$ or $d = 3$. Any square (or cube) σ whose vicinity contains exactly g guards from G is intersected by $O(\kappa^{d-1}(1 + \sqrt[d]{g}))$ objects from $\mathcal{O}$ that are at least as large as σ.*

4   Realistic input models and motion planning

In this section we study the complexity of the motion planning problem in two of the recently proposed *realistic input models* described in Sect. 2: unclutteredness and small simple-cover complexity. But first we discuss the relation between guarding sets and these models.

The relation between unclutteredness and guarding sets is immediate: a scene is κ-cluttered if and only if the bounding-box vertices form a κ-guarding set for it against hypercubes. The relation between simple-cover complexity and guarding sets is as follows: a scene has small simple-cover complexity if it has a linear-size guarding set. For the planar case the reverse is also true. More precisely, we have the following theorem.

Theorem 4. *Let $\mathcal{O}$ be a set of n objects in $\mathbb{R}^d$.*

(i) If $\mathcal{O}$ has a κ-guarding set of size m against hypercubes, then it has (s, δ)-simple-cover complexity for $s = 2^{5d+3}d(m/n)$ and $\delta = 6^d \kappa$.

(ii) For $d = 2$, if $\mathcal{O}$ has (s, δ)-simple-cover complexity, then it has a 8δ-guarding set against squares of size $O(sn)$.

We now turn our attention to the complexity of motion planning in two-dimensional workspaces that are either uncluttered or have small simple-cover complexity, and then extend the obtained results to three dimensions in Sect. 4.2.

4.1   The complexity of motion planning in 2D workspaces

Let $\mathcal{R}$ be a robot with f degrees of freedom, moving in a two-dimensional workspace amidst a set $\mathcal{O}$ of n obstacles. The robot $\mathcal{R}$ can be of any type: it can be a free-flying robot, a robotic arm, and so on. The only restriction is that it must have *bounded reach* [16], which is defined as follows. Let $p_{\mathcal{R}}$ be an arbitrary reference point inside $\mathcal{R}$. Then the *reach* of $\mathcal{R}$, denoted by reach($\mathcal{R}$), is defined as the maximum distance that any point of $\mathcal{R}$ can be from $p_{\mathcal{R}}$, taken over all possible configurations of $\mathcal{R}$. For instance, if $\mathcal{R}$ consists of two links of length 1 that are both attached to the origin, and the reference point is the tip of one of the links, then the reach of $\mathcal{R}$ is 2. (If the reference point would be the origin then the reach would be 1. For any two reference points on this particular robot, however, the two values reach($\mathcal{R}$) can be at most a factor of two apart.) A bounded-reach robot is now defined as a robot $\mathcal{R}$ with reach($\mathcal{R}$) $\leqslant c \cdot \min_{C \in \mathcal{O}}\{\text{size}(C)\}$, where c is a (small) constant.

In this section we study the complexity of the free space of a bounded-reach robot $\mathcal{R}$ under the assumption that the set of obstacles satisfies one of the models defined above. We prove an $\Omega(\kappa^f n^{f/2})$ lower bound on the free-space complexity for the most restricted model, namely for κ-cluttered scenes. Because unclutteredness implies small simple-cover complexity in the hierarchy of input models [4], this bound carries over to scenes with small simple-cover complexity. Moreover, we prove an $O(\kappa^f n^{f/2})$ upper bound for scenes with a linear-size guarding set against squares. By definition of unclutteredness and by Theorem 4, the upper bound immediately carries over to uncluttered scenes and scenes with small simple-cover complexity. Hence, in both models we get a tight bound of $\Theta(n^{f/2})$.

A lower bound for uncluttered scenes The robot $\mathcal{R}$ in our lower bound example consists of f links, which are all attached to the origin. The links have length $1+\varepsilon$, for a sufficiently small $\varepsilon > 0$. Obviously $\mathcal{R}$ has f degrees of freedom.

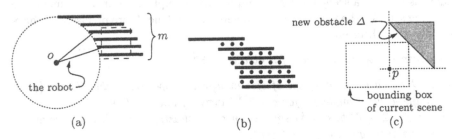

Fig. 1. (a) Part of the lower bound construction. (b,c) Adding bounding-box vertices to make the scene uncluttered.

The set of n obstacles for the case of a 2-cluttered planar scene is defined as follows. (Later we adapt the construction to get the bound for κ-cluttered scenes for larger κ.) Fix an integer parameter m; it will turn out later that the appropriate value for m is roughly $\sqrt{n}$. For a given integer i, let C_i be the horizontal segment of length 1 whose y-coordinate is i/m and whose left endpoint lies on the unit circle—see Fig. 1(a) for an example. Let $\mathcal{O}_1 := \{C_i \mid 1 \leqslant i \leqslant m\}$; this set forms a subset of the set of all obstacles. The remaining obstacles, which we describe later, are only needed to turn the environment into an uncluttered environment.

Consider any subset of f segments from $\mathcal{O}_1$. It is easy to see that there is a (semi-)free placement of $\mathcal{R}$ such that each segment in the subset is touched by a link of $\mathcal{R}$. Hence, the free-space complexity is $\Omega(m^f)$. When m is large, however, the set $\mathcal{O}_1$ does not form an uncluttered enviroment: the dashed square in Fig. 1(a) for instance, intersects $\Omega(m)$ obstacles without having a bounding-box vertex of one of the obstacles in its interior. This problem would disappear if between every pair of adjacent horizontal segments there would be a collection of $O(m)$ equal-spaced bounding-box vertices, as in Fig. 1(b). Notice that in total we need $\Theta(m^2)$ bounding-box vertices for this. We cannot add tiny obstacles between the segments to achieve this, because such obstacles would be much smaller than the robot, so the robot would no longer have bounded reach. There is no need, however, to add obstacles between the segments; we can also create bounding-box vertices there by adding obstacles outside the current scene. Suppose that we wish to have a bounding-box vertex at a given point $p = (p_x, p_y)$, and suppose that the current set of obstacles is contained in the rectangle $[x_{\min}, x_{\max}] \times [y_{\min}, y_{\max}]$. Then we add the right triangle Δ with vertices $(p_x, y_{\max} + x_{\max} - p_x)$, $(x_{\max} + y_{\max} - p_y, p_y)$, and $(x_{\max} + y_{\max} - p_y, y_{\max} + x_{\max} - p_x)$ as an obstacle—see Fig. 1(c). The point p is a bounding-box vertex of Δ, and Δ is disjoint from the current set of obstacles. By iteratively adding obstacles that generate the necessary bounding-box vertices between the segments in $\mathcal{O}_1$ we transform the cluttered environment into an uncluttered one. It is not difficult to see that we can even obtain a 2-cluttered environment in this manner, if we choose the distance between adjacent bounding-box vertices to be $1/m$.

We now have a collection of $\Theta(m^2)$ obstacles forming a 2-cluttered scene such that the free-space complexity is $\Omega(m^f)$. By choosing a suitable value for m (in the order of $\sqrt{n}$), we obtain a collection of n obstacles such that the free-space complexity is $\Omega(n^{f/2})$.

To get the general bound we replace each of the m segments in the set $\mathcal{O}_1$ by $\lfloor \kappa/2 \rfloor$ segments of length 1 that are quite close together. The left endpoints of these segments still lie on the unit circle. Since the original scene was 2-cluttered, the new scene is κ-cluttered. The number of f-fold contacts has increased to $\Omega(\kappa^f m^f)$. (Note that it is still possible to choose the value ε, which determines the length of the links of $\mathcal{R}$, small enough such that any f-tuple of segments in the new set $\mathcal{O}_1$ can be touched by a semi-free placement.) By again choosing m to be roughly $\sqrt{n}$ we get the final result.

Theorem 5. *The free-space complexity of a bounded-reach robot with f degrees of freedom moving in a two-dimensional κ-cluttered scene consisting of n obstacles can be $\Omega(\kappa^f n^{f/2})$.*

An upper bound for scenes with linear-size guarding sets We want to prove an upper bound on the complexity of the free space of a bounded-reach robot with f degrees of freedom moving in a scene with a linear-size guarding set. The global structure of our proof will be as follows. We construct a decomposition of the workspace into cells that are not much smaller than the robot. The decomposition will have the property that none of its cells can have too many obstacles close to it. This means that the robot cannot have too many f-fold contacts when its reference point lies inside any given cell. Summing the number of f-fold contacts over all the cells using Theorem 3 yields the desired bound on the number of features of the free space.

The decomposition we use is obtained by adapting the partitioning scheme described in Sect. 2. First we describe the exact properties that we require, and then show how to adapt the partitioning strategy to obtain a decomposition with the desired properties.

Let $\rho := 2 \cdot \text{reach}(\mathcal{R})$. Define the *expansion* $\hat{o}$ of an object o to be the Minkowski sum of o with a square of size 2ρ centered at the origin. Hence, $\hat{o}$ contains exactly those points that are at a L_∞-distance of less than ρ from o. Note that the expansion of a square σ is another square, whose edge length is 2ρ more than the edge length of σ. Let $\hat{\mathcal{O}} := \{\hat{C} \mid C \in \mathcal{O}\}$ denote the set of expanded obstacles.

Lemma 4. *Let $\mathcal{O}$ be a set of obstacles in $\mathbb{R}^2$ or $\mathbb{R}^3$, and let G be a κ-guarding set for $\mathcal{O}$. Let $\hat{\Sigma}$ be a bounding hypercube for the set $\hat{\mathcal{O}}$ of expanded obstacles. Then there exists a set S of cells that are either hypercubes or L-shapes with the following properties:*

(P1) the cells in S form a decomposition of $\hat{\Sigma}$;

(P2) the number of cells in S is $O(|G|)$;

(P3) every cell in S whose size is greater than 2ρ is intersected by $O(\kappa)$ expanded obstacles;

(P4) every cell $\sigma \in S$ whose size is less than or equal to 2ρ is a hypercube of size at least ρ.

Now that we have a suitable decomposition of the workspace, we can use Theorem 3 to prove our main result.

Theorem 6. *Let $\mathcal{R}$ be a bounded-reach robot with f degrees of freedom, with $f \geqslant 2$ being a constant, moving in a two-dimensional workspace containing a set $\mathcal{O}$ of n obstacles. If the set of obstacles has a κ-guarding set against squares, then the complexity of the free space is $O(\kappa^f n^{f/2})$.*

4.2 The complexity of motion planning in 3D workspaces

Having described the two-dimensional setting in the previous section, we now turn our attention to a robot R moving in a three-dimensional workspace amidst a set $\mathcal{O}$ of n obstacles. As in the two-dimensional case, the robot is allowed to be of any type—we only require that its reach is bounded. We prove an $\Omega(\kappa^{2f} n^{2f/3})$ lower bound on the complexity of the free space for κ-cluttered scenes, and an identical upper bound for scenes with a linear-size guarding set against squares. As before, this results in a tight bound of $\Theta(n^{2f/3})$ for both un-cluttered scenes; for scenes with small simple-cover complexity, the $\Omega(n^{2f/3})$ lower bound still holds, but the $O(n^{2f/3})$ upper bound is not implied since Theorem 4(ii) only holds for $d = 2$.

Theorem 7. *The free-space complexity of a bounded-reach robot with f degrees of freedom moving in a three-dimensional κ-cluttered scene of n obstacles is $\Omega(\kappa^{2f} n^{2f/3})$.*

Theorem 8. *Let $\mathcal{R}$ be a bounded-reach robot with f degrees of freedom, with $f \geqslant 2$ a constant, moving in a three-dimensional workspace that contains a set $\mathcal{O}$ of n obstacles. If the set of obstacles has a κ-guarding set against cubes, then the complexity of the free space is $O(\kappa^{2f} n^{2f/3})$.*

5 Concluding remarks

We have introduced the notion of a guarding set for a set of objects against a family of ranges, and we have proven that asymptotically the type of the ranges (such as hypercubes, balls, and so on) does not matter, as long as the ranges are convex and fat. We have also proven several interesting claims that form connections between the distribution of the guards and the distribution of the objects.

Scenes of objects admitting linear-size guarding sets constitute a new input model which is a natural generalization of de Berg's unclutteredness model. And indeed de Berg's algorithm for computing linear-size BSPs for uncluttered scenes applies also to scenes with linear-size guarding sets (the set of bounding box vertices is simply replaced by the guarding set). However, the new model is strictly more general than unclutteredness, and it is shown that in the plane it is equivalent to a model that has been proposed in the literature: small simple-cover-complexity.

Guarding sets were used to establish the main result of this paper: the complexity of the free-space of a bounded-reach robot with f degrees of freedom moving in an uncluttered scene is $\Theta(n^{f/2})$ in $\mathbb{R}^2$ and $\Theta(n^{2f/3})$ in $\mathbb{R}^3$; the planar upper and lower bounds, and the 3d lower bound, also hold for scenes with small simple-cover complexity. These bounds fits nicely between the $\Theta(n)$ bound for low-density scenes—which are more restrictive—, and the $\Theta(n^f)$ bound for general scenes.

We are currently developing several algorithms for computing a *small* guarding set for a given set of objects. These algorithms are also being evaluated experimentally.

Another useful direction of research would be to devise efficient algorithms for the new model; that is, algorithms that are more efficient than their more general counterparts. One such example is de Berg's logarithmic-time point-location structure in $\mathbb{R}^d$, which is of linear size and is easily constructed from the linear-size BSP tree that is obtained by his BSP algorithm.

References

[1] Pankaj K. Agarwal, M. J. Katz, and M. Sharir. Computing depth orders and related problems. *Comput. Geom. Theory Appl.*, 5:187–206, 1995.

[2] H. Alt, R. Fleischer, M. Kaufmann, K. Mehlhorn, S. Näher, S. Schirra, and C. Uhrig. Approximate motion planning and the complexity of the boundary of the union of simple geometric figures. *Algorithmica*, 8:391–406, 1992.

[3] J. M. Bañon. Implementation and extension of the ladder algorithm. In *Proc. IEEE Internat. Conf. Robot. Autom.*, pages 1548–1553, 1990.

[4] M. de Berg, M. J. Katz, A. F. van der Stappen, and J. Vleugels. Realistic input models for geometric algorithms. To appear; a preliminary version appeared in *Proc. 13th Annu. ACM Sympos. Comput. Geom.*, pages 294–303, 1997.

[5] Mark de Berg. Linear size binary space partitions for fat objects. In *Proc. 3rd Annu. European Sympos. Algorithms*, volume 979 of *Lecture Notes Comput. Sci.*, pages 252–263. Springer-Verlag, 1995.

[6] A. Efrat and M. J. Katz. On the union of κ-curved objects. In *Proc. 14th Annu. ACM Sympos. Comput. Geom.*, 1998. to appear.

[7] A. Efrat, M. J. Katz, F. Nielsen, and M. Sharir. Dynamic data structures for fat objects and their applications. In *Proc. 5th Workshop Algorithms Data Struct.*, pages 297–306, 1997.

[8] M. J. Katz. 3-D vertical ray shooting and 2-D point enclosure, range searching, and arc shooting amidst convex fat objects. *Comput. Geom. Theory Appl.*, 8:299–316, 1997.

[9] M. J. Katz, M. H. Overmars, and M. Sharir. Efficient hidden surface removal for objects with small union size. *Comput. Geom. Theory Appl.*, 2:223–234, 1992.

[10] J. Matoušek. Epsilon-nets and computational geometry. In J. Pach, editor, *New Trends in Discrete and Computational Geometry*, volume 10 of *Algorithms and Combinatorics*, pages 69–89. Springer-Verlag, 1993.

[11] J. Matoušek, J. Pach, M. Sharir, S. Sifrony, and E. Welzl. Fat triangles determine linearly many holes. *SIAM J. Comput.*, 23:154–169, 1994.

[12] J. S. B. Mitchell, D. M. Mount, and S. Suri. Query-sensitive ray shooting. In *Proc. 10th Annu. ACM Sympos. Comput. Geom.*, pages 359–368, 1994.

[13] M. H. Overmars and A. F. van der Stappen. Range searching and point location among fat objects. *J. Algorithms*, 21:629–656, 1996.

[14] J. T. Schwartz and M. Sharir. On the "piano movers" problem I: the case of a two-dimensional rigid polygonal body moving amidst polygonal barriers. *Commun. Pure Appl. Math.*, 36:345–398, 1983.

[15] Otfried Schwarzkopf and Jules Vleugels. Range searching in low-density environments. *Inform. Process. Lett.*, 60:121–127, 1996.

[16] A. F. van der Stappen, D. Halperin, and M. H. Overmars. The complexity of the free space for a robot moving amidst fat obstacles. *Comput. Geom. Theory Appl.*, 3:353–373, 1993.

[17] A. F. van der Stappen and M. H. Overmars. Motion planning amidst fat obstacles. In *Proc. 10th Annu. ACM Sympos. Comput. Geom.*, pages 31–40, 1994.

[18] A. F. van der Stappen, M. H. Overmars, M. de Berg, and J. Vleugels. Motion planning in environments with low obstacle density. Report UU-CS-1997-19, Dept. Comput. Sci., Utrecht Univ., Utrecht, Netherlands, 1997.

[19] Jules Vleugels. *On Fatness and Fitness—Realistic Input Models for Geometric Algorithms*. Ph.d. thesis, Dept. Comput. Sci., Utrecht Univ., Utrecht, the Netherlands, March 1997.

Constrained Square-Center Problems[*]

Matthew J. Katz[1]   Klara Kedem[1,2]   Michael Segal[1]

[1] Department of Mathematics and Computer Science, Ben-Gurion University of the Negev, Beer-Sheva 84105, Israel
[2] Computer Science Department, Cornell University, Upson Hall, Cornell University, Ithaca, NY 14853

Abstract. Given a set P of n points in the plane, we seek two squares whose center points belong to P, their union contains P, and the area of the larger square is minimal. We present efficient algorithms for three variants of this problem: In the first the squares are axis parallel, in the second they are free to rotate but must remain parallel to each other, and in the third they are free to rotate independently.

1   Introduction

In this paper we consider the problems of covering a given set P of n points in the plane by two constrained (discrete) squares, under different conditions. We call a square *constrained* if its center lies on one of the input points. In particular, we solve the following problems:

1. Find two constrained axis-parallel squares whose union covers P, so as to minimize the size of the larger square. We present an $O(n \log^2 n)$-time algorithm; its space requirement is $O(n \log n)$.
2. Find two constrained parallel squares whose union covers P, so as to minimize the size of the larger square. The squares are allowed to rotate but must remain parallel to each other. Our algorithm runs in $O(n^2 \log^4 n)$ time and uses $O(n^2)$ space.
3. Find two constrained squares whose union covers P, so as to minimize the size of the larger square, where each square is allowed to rotate independently. We present an $O(n^3 \log^2 n)$-time and $O(n^2)$-space algorithm for this problem.

The problems above continue a list of optimization problems that deal with covering a set of points in the plane by two geometric objects of the same type. We mention some of them: The two center problem, solved in time $O(n \log^9 n)$ by Sharir [18], and recently in time $O(n \log^2 n)$ by Eppstein [7] (by a randomized algorithm); the constrained two center problem, solved in time $O(n^{\frac{4}{3}} \log^5 n)$ by

[*] Work by M. Katz and K. Kedem has been supported by the Israel Science Foundation founded by the Israel Academy of Sciences and Humanities. K. Kedem has also been supported by the U.S.-Israeli Binational Science Foundation, and by the Mary Upson Award, College of Engineering, Cornell University.

Agarwal et al. [2]; the two line-center problem, solved in time $O(n^2 \log^2 n)$ by Jaromczyk and Kowaluk [11] (see also [9, 13]); the two square-center problem, where the squares are with mutually parallel sides (the unconstrained version of Problem 2 above), solved in time $O(n^2)$ by Jaromczyk and Kowaluk [10].

We employ a variety of techniques to solve these optimization problems. The decision algorithm of Problem 1 searches for the centers of a solution pair (of squares) in an implicit special matrix, using a technique that has recently been used in [6, 18]. For the optimization, a search in a collection of sorted matrices [8] is performed.

The decision algorithm of Problem 2 involves maintenance of dynamically changing convex hulls, and maintenance of an orthogonal range search tree that must adapt to a rotating axes system. For the optimization, we apply Megiddo's [14] parametric search. However, since our decision algorithm is not parallelizable, we had to find an algorithm that solves a completely different problem, but is both parallelizable and enables the discovery of the optimal square size when the parametric search technique is applied to it.

In Problem 3 we describe the sizes of candidate solution squares as a collection of curves. For a dynamically changing set of such curves, we transform the problem of determining whether their upper envelope has a point below some horizontal line, into the problem of stabbing a dynamically changing set of segments. The latter problem is solved using a (dynamic) segment tree.

2 Two constrained axis-parallel squares

We are given a set P of n points in the plane, and wish to find two axis-parallel squares, centered at points of P, whose union covers (contains) P, such that the area of the larger square is minimal. We first transform the corresponding decision problem into a constrained 2-piercing problem, which we solve in $O(n \log n)$ time. We then apply the algorithm of Frederickson and Johnson [8] for the optimization.

2.1 The decision algorithm

The decision problem is stated as follows: Given a set P of n points, are there two constrained axis-parallel squares, each of a given area $\mathcal{A}$, whose union covers P. We present an $O(n \log n)$ algorithm for solving the decision problem.

We adopt the notation of [20] (see also [12, 17]). Denote by $\mathcal{R}$ the set of axis-parallel squares of area $\mathcal{A}$ centered at the points of P. $\mathcal{R}$ is p-pierceable if there exists a set $\mathcal{X}$ of p points which intersects each of the squares in $\mathcal{R}$. The set $\mathcal{X}$ is called a piercing set for $\mathcal{R}$. Notice that $\mathcal{X}$ is a piercing set for $\mathcal{R}$ if and only if the union of the axis-parallel squares of area $\mathcal{A}$ centered at the points of $\mathcal{X}$ covers P. $\mathcal{R}$ is p-constrained pierceable if there exists a piercing set of p points which is contained in P. Thus, solving the decision problem is equivalent to determining whether $\mathcal{R}$ is 2-constrained pierceable.

We first compute the rectangle $R = \cap \mathcal{R}$. If R is not empty then $\mathcal{R}$ is 1-pierceable, and we check whether it is also 1-constrained pierceable by checking whether P has a point in R. If $\mathcal{R}$ is 1-constrained pierceable then we are done, so assume that it is not. If $\mathcal{R}$ was not found to be 1-pierceable, then we apply the linear time algorithm of [20] (see also [5]) to check whether $\mathcal{R}$ is 2-pierceable. If $\mathcal{R}$ is neither 1-pierceable nor 2-pierceable, then obviously $\mathcal{R}$ is not 2-constrained pierceable and we are done. Assume therefore that $\mathcal{R}$ is 2-pierceable (or 1-pierceable).

Assume $\mathcal{R}$ is 2-constrained pierceable, and let $p_1, p_2 \in P$ be a pair of piercing points for $\mathcal{R}$. We assume that p_1 lies to the left of and below p_2. (The case where p_1 lies to the left of and above p_2 is treated analogously.) We next show that $\mathcal{R}$ can be divided into two subsets $\mathcal{R}_1, \mathcal{R}_2$, such that (i) $p_1 \in \cap \mathcal{R}_1$, $p_2 \in \cap \mathcal{R}_2$, and (ii) $\mathcal{R}_1$ (alternatively $\mathcal{R}_2$) can be represented in a way that will assist us in the search for p_1 and p_2.

Denote by $X_{\mathcal{R}}$ the centers of the squares in $\mathcal{R}$ (the points in P) sorted by their x-coordinate (left to right), and by $Y_{\mathcal{R}}$ the centers of the squares in $\mathcal{R}$ sorted by their y-coordinate (low to high). We now claim:

Claim 1 *If p_1 and p_2 are as above, then $\mathcal{R}$ can be divided into two subsets $\mathcal{R}_1$ and $\mathcal{R}_2$, $p_1 \in \cap \mathcal{R}_1$, $p_2 \in \cap \mathcal{R}_2$, such that $\mathcal{R}_1$ can be represented as the union of two subsets $\mathcal{R}_1^x$ and $\mathcal{R}_1^y$ (not necessarily disjoint, and one of them might be empty), where the centers of squares of $\mathcal{R}_1^x$ form a consecutive subsequence of the list $X_{\mathcal{R}}$, starting from its beginning, and the centers of squares of $\mathcal{R}_1^y$ form a consecutive subsequence of $Y_{\mathcal{R}}$, starting from the list's beginning.*

Proof. We construct the sets $\mathcal{R}_1^x$ and $\mathcal{R}_1^y$, and put $\mathcal{R}_1 = \mathcal{R}_1^x \cup \mathcal{R}_1^y$ and $\mathcal{R}_2 = \mathcal{R} - \mathcal{R}_1$. We then show that indeed $p_1 \in \cap \mathcal{R}_1$ and $p_2 \in \cap \mathcal{R}_2$.

We consider the centers in $Y_{\mathcal{R}}$, one by one, in increasing order, until a center is encountered whose corresponding square A is not pierced by p_1. $\mathcal{R}_1^y$ consists of all squares in $Y_{\mathcal{R}}$ below A (i.e., preceding A in $Y_{\mathcal{R}}$). A might be the first square in $Y_{\mathcal{R}}$, in which case $\mathcal{R}_1^y$ is empty. We now find the location of the x-coordinate of the center of A in $X_{\mathcal{R}}$, and start moving from this point leftwards, i.e., in decreasing order. Thus moving, we either encounter a square, call it B, that is **higher** than A and is not pierced by p_2, or we do not.

If we do not encounter such a square B (which is clearly the case if the bottom edge of A lies above p_1), then put $\mathcal{R}_1^x = \emptyset$, otherwise $\mathcal{R}_1^x$ consists of all squares in $X_{\mathcal{R}}$ to the left of B including B.

It remains to show that $p_1 \in \cap \mathcal{R}_1$ and that $p_2 \in \cap \mathcal{R}_2$. We assume that the square B exists, which is the slightly more difficult case. We first show the former assertion, i.e., $p_1 \in \cap \mathcal{R}_1$. The fact that p_2 is not in B implies that p_2 lies to the right of the right edge of B, because B cannot lie below p_2 since it is higher than A which is already pierced by p_2. Therefore none of the squares in $\mathcal{R}_1^x$ is pierced by p_2 thus $p_1 \in \cap \mathcal{R}_1^x$. By our construction, $p_1 \in \cap \mathcal{R}_1^y$, so together we have $p_1 \in \cap \mathcal{R}_1$. Now consider a square $C \in \mathcal{R}_2$, $C \neq A$. C is higher than A, because it is not in $\mathcal{R}_1^y$. Therefore if C is not pierced by p_2, then C must lie to the left of A but then it is in $\mathcal{R}_1^x$ and thus not in $\mathcal{R}_2$.

The claim above reveals a monotonicity property that allows us to design an efficient algorithm for the decision problem. We employ a technique, due to Sharir [18], that resembles searching in monotone matrices; for a recent application and refinement of this technique, see [6]. Let M be an $n \times n$ matrix whose rows correspond to $X_{\mathcal{R}}$ and whose columns correspond to $Y_{\mathcal{R}}$. An entry M_{xy} in the matrix is defined as follows. Let D_x be the set of squares in $\mathcal{R}$ such that the x-coordinate of their centers is smaller or equal to x, and let D_y be the set of squares in $\mathcal{R}$ such that the y-coordinate of their centers is smaller or equal to y. Let $D^l_{xy} = D_x \cup D_y$ and $D^r_{xy} = (\mathcal{R} - D^l_{xy})$

$$
M_{xy} = \begin{cases}
\text{`YY'} & \text{if both } D^r_{xy} \text{ and } D^l_{xy} \text{ are 1-constrained pierceable} \\
\text{`YN'} & \text{if } D^r_{xy} \text{ is 1-constrained pierceable but } D^l_{xy} \text{ is not} \\
\text{`NY'} & \text{if } D^r_{xy} \text{ is not 1-constrained pierceable but } D^l_{xy} \text{ is} \\
\text{`NN'} & \text{if neither } D^r_{xy} \text{ nor } D^l_{xy} \text{ is 1-constrained pierceable.}
\end{cases}
$$

In order to apply Sharir's technique the lines and columns of M^1 must be non-decreasing (assuming 'Y' > 'N'), and the lines and columns of M^2 must be non-increasing, where M^i is the matrix obtained from M by picking from each entry only the i'th letter, $i = 1, 2$. In our case this property clearly holds, since, for example, if for some x_0 and y_0, $M^1_{x_0,y_0} =$'Y', then for any $x' \geq x_0$ and $y' \geq y_0$, $M^1_{x',y'} =$'Y'. Thus we can determine whether M contains an entry 'YY' by inspecting only $O(n)$ entries in M, advancing along a *connected* path within M [6]. For each entry along this path, we need to determine whether D^z_{xy} is 1-constrained pierceable, $z \in \{l, r\}$. This can be done easily in $O(\log n)$ time by maintaining dynamically the intersection $\cap D^z_{xy}$, and utilizing a standard orthogonal range searching data structure of size $O(n \log n)$ [4]. Thus in $O(n \log n)$ time we can determine whether M contains a 'YY' entry.

Theorem 2. *Given a set P of n input points and area $\mathcal{A}$, one can find two constrained axis-parallel squares of area $\mathcal{A}$ each that cover P in time $O(n \log n)$ using $O(n \log n)$ space.*

We have just found whether a set of equal-sized squares is 2-pierceable by two of their centers. For the optimization, we shrink these squares as much as possible, so that they remain 2-constrained pierceable.

2.2 Optimization

For solving the optimization problem we observe that each L_∞ distance (multiplied by 2 and squared) can be a potential area solution. We can represent all L_∞ distances as in [9] by sorted matrices. We sort all the points of P in x and y directions. Entry (i, j) in the matrix M_1 stores the value $4(x_j - x_i)^2$, where x_i, x_j are the x-coordinates of the points with indices i, j in the sorted x-order, and, similarly, entry (i, j) in the matrix M_2 stores the value $4(y_j - y_i)^2$, where y_i, y_j are the y-coordinates of the points with indices i, j in the sorted y-order. We then apply the Frederickson and Johnson algorithm [8] to M_1 and M_2 and obtain the smallest value in the matrices for which the decision algorithm answers "Yes" and thus obtain the optimal solution. We have shown:

Theorem 3. *Given a set P of n input points, one can find two constrained axis-parallel squares that cover all the input points such that the size of the larger square is minimized in $O(n \log^2 n)$ time using $O(n \log n)$ space.*

3 Two constrained mutually-parallel squares

In this section we deal with the following problem. Given a set P of n points in the plane, find a pair of mutually-parallel constrained squares whose union contains P, so as to minimize the area (equivalently, the side length) of the larger square. The problem where the squares are not constrained was recently solved by Jaromczyk and Kowaluk [10] in $O(n^2)$ time using $O(n^2)$ space.

We first solve the decision problem for squares with a given area $\mathcal{A}$ in time $O(n^2 \log^2 n)$ and $O(n^2)$ space. For the optimization, we present a parallel version of another algorithm (solving a different problem), to which we apply Megiddo's parametric search [14] to obtain an $O(n^2 \log^4 n)$ time and $O(n^2)$ space optimization algorithm.

3.1 The decision algorithm

For each of the input points, $p_i \in P$, draw an axis-aligned square s_i of area $\mathcal{A}$ centered at p_i. Denote by U_i the set of points in P that are not covered by s_i. If for some i there is a constrained axis-aligned square which covers U_i, then we are done. Otherwise, we rotate the squares simultaneously about their centers, stopping to check coverage at certain *rotation events* and halting when P is covered by two of the current squares. A *rotation event* occurs whenever a point of P enters or leaves a square s_i, $i = 1 \ldots n$. When a square s_i rotates by $\frac{\pi}{2}$ from its initial axis-aligned position, every point of P enters and leaves s_i at most once. Thus, the number of rotation events for s_i is $O(n)$. We can precompute all the rotation events in $O(n^2)$ time with $O(n^2)$ space. We sort the rotation events according to their corresponding angles.

We compute the initial convex hulls for each $U_i, i = 1, \ldots, n$ (i.e., at orientation $\theta = 0$). Assume that at the current rotation event a point p_j enters s_i. The set U_i and its convex hull are updated as p_j leaves U_i, and we check whether there exists a constrained cover of P involving s_i and one of the other constrained squares (that must cover U_i).

We explain how this is done. First we find the tangents to the convex hull of U_i that are parallel to the sides of s_i. They define a rectangle R_i that covers U_i. If R_i has a side of length greater than $\sqrt{\mathcal{A}}$, then none of the other $n - 1$ constrained squares covers U_i. Otherwise we define a *search region* S which is the locus of all points of L_∞ distance at most $\frac{\sqrt{\mathcal{A}}}{2}$ from all four sides of R_i, and search for a point of P in S. (Clearly S is a rectangle whose sides are parallel to the sides of s_i.)

We perform orthogonal range searching to determine whether there is a point of P in S. Assume we have computed all the rotation events and have $O(n^2)$ rectangular search regions associated with them. We use a range search tree as follows. Denote by L the list of all $O(n^2)$ lines passing through pairs of points in P. Let Q consist of all the slopes of lines in L that lie in the range $[0, \pi/2)$, and of all the slopes in the range $[0, \pi/2)$ of lines that are perpendicular to lines in L. We sort Q, obtaining the sorted sequence $\{\alpha_1, \alpha_2, \ldots\}$. We rotate the axes so that the x-axis has slope α_1, and compute a range search tree for P with respect to the rotated axes, storing just the labels of the points of P in the tree. For the search regions whose (appropriate) sides have slope between α_1 and α_2, we perform a usual range search with this tree. Before considering the next search regions, we rotate the axes some more until the x-axis has slope α_2, and update the range search tree accordingly: Assuming the leaves of the main structure in the range tree are sorted by x-coordinate, and the leaves in the secondary trees are sorted by y-coordinate. If when moving from α_1 to α_2 a swap occured in the x-order of one pair of points, then we swap the (labeling of the) points in the main structure and in the appropriate secondary structures; if the swap occured in the y-order, then we swap the labeling in the appropriate secondary structures. Now we can proceed with the search ranges whose (appropriate) sides have slope between α_2 and α_3. And so on.

When a point p_j leaves s_i, we perform a similar update and range search; we add p_j to U_i, update the convex hull of U_i and search for a point of P in the newly defined range S.

We analyze the time and space required for the decision algorithm. The total number of rotation events is $O(n^2)$. They can be precomputed and sorted in $O(n^2 \log n)$ time with $O(n^2)$ space. Similarly Q can be obtained within the same bounds. Merging the two sets of slopes (rotation events and Q) is done in time $O(n^2)$. Initially computing the convex hulls for all sets U_i takes $O(n^2 \log n)$ time with $O(n^2)$ space. Applying the data structure and algorithm of Overmars and van Leeuwen [16], each update of a convex hull takes $O(\log^2 n)$ time, so in total $O(n^2 \log^2 n)$ time and $O(n^2)$ space. Our range searching algorithm takes $O(\log^2 n)$ time per query and per update, after spending $O(n \log n)$ preprocessing time and using $O(n \log n)$ space (notice that this is the total space requirement for the range searching), and we perform $O(n^2)$ queries and updates. We have shown:

Theorem 4. *Given a set P of n points and an area $\mathcal{A}$, one can decide whether P can be covered by two constrained mutually-parallel squares, each of area $\mathcal{A}$, in $O(n^2 \log^2 n)$ time and $O(n^2)$ space.*

3.2 Optimization

Having provided a solution to the decision problem, we now return to the minimization problem. The number of candidate square sizes is $O(n^4)$ (see below and Figure 1). The candidate sizes are determined by either

- A point of P as a center of a square (see Figure 1(i)–(iv)) and either (i) another point of P on a corner of this square, or (ii) two points of P on parallel sides of the square, or (iii) two points of P on one side of the square, or (iv) two points of P on adjacent sides of the square, or

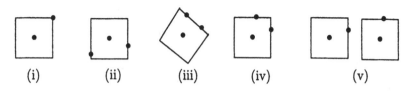

Fig. 1. Critical events that determine candidate square sizes. Cases (i) – (iv) involve a single square, and case (v) two squares.

– Two points of P as centers of two squares and another point of P on the boundary of each of the squares (Figure 1(v)).

In order to apply the Megiddo optimization scheme we have to parallelize our decision algorithm. However, the range searching of the decision algorithm is not parallelizable, so, as in [1], we come up with an auxiliary problem whose parallel version will generate the optimal solution to our problem.

The auxiliary problem: Given a set P of $n > 2$ points and a fixed size d, produce the set S of all strips of width d that contain a point of P on each of their boundaries. Let $\bar{S}$ be the set of strips S except that the assigned slope of a strip $s_i \in \bar{S}$ is the slope of the strip $s_i \in S$ plus $\pi/2$ modulo π. Let $\mathcal{S} = S \cup \bar{S}$. The problem is to sort the strips in $\mathcal{S}$ according to their (assigned) slope.

Clearly not all pairs of points in P define strips in S. The pairs of points in P whose distance is smaller than d will not generate the required width strip. For every pair of points in P whose distance from each other is larger than d, there are exactly two slopes for which the width of the strip, with a point of this pair on each of its boundaries, is d. We add the strips in these slopes (and their $\bar{S}$ corresponding strips) to S. Reporting the sorted order of $\mathcal{S}$ can be done in $O(n^2 \log n)$ time, and a parallel algorithm with $O(n^2)$ processors will sort the list in $O(\log n)$ time [3].

Assume we (generically) apply this parallel sort algorithm for the optimal square size d^*. For this we augment the algorithm, as in [1], by a preliminary stage that disposes of the cases in which the width of the strip is exactly the distance between two points of P, and those in which the width is the distance between two points multiplied by $\sqrt{2}/2$. We call these distances *special distances*. We can afford to list all these $O(n^2)$ strip widths, sort them, and perform a binary search for d^* over them, applying our decision algorithm of the previous subsection at each of the comparisons. This results in an initial closed interval of real numbers, I_0, that contains the optimal square size d^*, and none of the just computed special sizes is contained in its interior. Clearly d^* is bounded away from zero.

Consider now a single parallel stage. In such a stage we perform $O(n^2)$ comparisons, each comparison involving two pairs of points. There are two cases: (a) the two compared strips belong to S (identically both are in $\bar{S}$), and (b) one strip is of S and the other of $\bar{S}$. Let one such comparison involve the pairs (p_1, p_2) and (p_3, p_4). In order to resolve this comparison, we must compute for the point pair (p_1, p_2) the slopes of the two strips of width d^* that have p_1 on one boundary of the strip and p_2 on the other. Similarly, we compute the slopes of the two strips of width d^* through (p_3, p_4). Then we sort the four strips by their slopes. Of course, we do not know d^*, so we compute the (at most two) *critical values* d where the sorted order of the four strips changes, namely, for case (a) above, where the two strips are parallel, and for case (b), when the two strips are perpendicular to each other. We do this for all $O(n^2)$ critical value comparisons. Now we apply the decision algorithm of the subsection above to perform a binary search over the $O(n^2)$ critical values that were computed. Thus we find an interval $I \subseteq I_0$ where d^* resides, resolve all the comparisons of this parallel stage, and proceed to the next parallel stage.

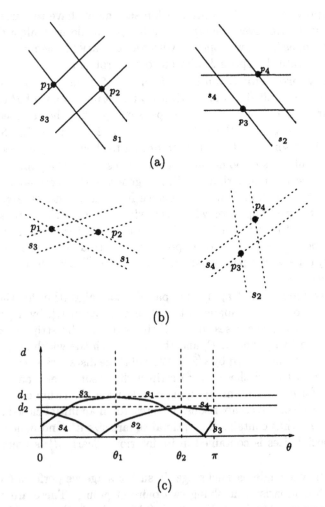

Fig. 2. Slope ordering for the comparison of (p_1, p_2) and (p_3, p_4): (a) strips s_1 and s_2 are parallel for some d, (b) the ordering of the slopes at d^*, (c) d as a function of θ

What does resolving mean here? See Figure 2 which depicts case (a). If the comparison was made for two pairs of points (p_1, p_2) and (p_3, p_4) then, if the distance between a pair of points, $d_1 = (p_1, p_2)$ or $d_2 = (p_3, p_4)$, is smaller than the smaller endpoint of the current interval I then this pair will not have a strip of width d^* and it is omitted from the rest of the sort. If the distance is larger than the smaller endpoint of I then the slope ordering of the four strips at d^* is uniquely determined as follows. In Figure 2(a) the strips s_1 and s_2 are parallel at some width d', and in Figure 2(b) we plot the strips of width d^* for the two pairs of points. In Figure 2(c) we graph d as a function of $\theta \in [0, \pi)$ for the two pairs of points. The graph of $d = d_1 \cos(\theta - \theta_1)$ achieves its maximum at (θ_1, d_1), and similarly the graph of $d = d_2 \cos(\theta - \theta_2)$ achieves its maximum at (θ_2, d_2), where θ_1 (θ_2) is the angle that the line perpendicular to the line through (p_1, p_2) $((p_3, p_4))$ makes with the positive x-axis. It is easy to see that for every d each pair of points has two strips and that the two functions intersect at two points. We split the domain of definition of each function to two parts, one in which the function strictly increases and one in which it strictly decreases. In Figure 2(a) and 2(b) the strip s_1 corresponds to the decreasing half of the function in Figure 2(c) and s_3 to the increasing half. Similarly with the strips of (p_3, p_4), s_2 corresponds to the increasing half and s_4 to the decreasing half. Thus the order of the strips at d^* is the order in which the line $d = d^*$ intersects their functions, and the width values at the intersection points of the two functions consist of the critical values for these two pairs of points.

For case (b) assume the pair (p_1, p_2) belongs to a strip of $\bar{S}$. We simply cyclically shift the function of (p_1, p_2) (of S) by $\pi/2$. The intersection points of the functions are now at two values of d' where the two strips are perpendicular to each other, and all the rest of the argument is analogous.

Note: We have to be a little more careful here about the notion of the domain of definition of the functions, and we might want to break the segments also at $\theta = 0$. This is a slight formality that we neglect since it does not change anything in the analysis.

The closed interval I is always guaranteed to contain d^* but we need to show that a comparison is made where $d = d^*$.

Claim 5 *If d^* is not one of the special distances then the slope order of the strips changes as d changes from values slightly smaller than d^* to values slightly larger than d^*.*

Note that at some stage the optimal solution will appear on the boundary of the interval I computed at that stage (it could even appear on the boundary of I_0). However, once it appears, it will remain one of the endpoints of all subsequently computed intervals. At the end, we run the decision algorithm for the left endpoint of the final interval. If the answer is positive, then this endpoint is d^*, otherwise d^* is the right endpoint of the final interval.

Theorem 6. *Let P be a set of n points, we can find a pair of mutually-parallel constrained squares whose union covers P and such that the area of the larger square is minimized in $O(n^2 \log^4 n)$ time and $O(n^2)$ space.*

4   Two constrained general squares

In this section the squares may rotate independently. We first state a subproblem whose solution is used as a subroutine in the full solution. Then we present an algorithm for solving the decision problem. This algorithm is used to perform a binary search over the sorted set of potential solutions, producing the solution to the optimization problem.

The subproblem: Given a set P of n points in the plane and a point q, find the minimum area square that is centered at q and that covers P. The square may rotate.

The algorithm for solving the subproblem is as follows. Assume q is the origin. Let θ be an angle in $[0, \frac{\pi}{2})$. Consider the projections, $x_i(\theta)$ and $y_i(\theta)$, of a point $p_i \in P$ on the x-axis and y-axis, after rotating the axes by θ. If the distance between p_i and q is r_i, and the angle between the vector p_i and the x-axis at its initial position is θ_i, then for an angle θ we have

$$x_i(\theta) = r_i \cos(\theta_i - \theta) \text{ and } y_i(\theta) = r_i \sin(\theta_i - \theta) .$$

A square centered at q rotated by angle θ that has p_i on its boundary is of side length $z_i(\theta) = 2 \times \max\{|x_i(\theta)|, |y_i(\theta)|\}$. Note that it is enough to rotate the axes by angle $\theta, 0 \leq \theta < \frac{\pi}{2}$ to get all possible sizes of squares centered at q that have p_i on their boundary.

Observe the plane (θ, z), on which we graph both $z_{2i-1}(\theta) = |x_i(\theta)|$ and $z_{2i}(\theta) = |y_i(\theta)|$, $i = 1, \ldots n$. We call the set of these $2n$ functions E_q. It is easy to see that every two functions z_j and z_k intersect at most twice. The upper envelope of the functions in E_q denotes, for each θ, the size $2z(\theta)$ of the smallest square (centered at q and rotated by θ) that covers P, and the point (or two points) of P corresponding to the function (or two functions) that attains (attain) the maximum at this θ is the point (are the two points) of P on the boundary of the square. The lowest point on this envelope gives the angle, the size, and the point(s) that determine the minimal square. The upper envelope, and the lowest point on it, can be computed in $O(n \log n)$ time [19], and this is the runtime of the solution of the subproblem above.

For the two squares decision problem we repeat some notations and ideas from the previous section. Let s_i be a square of the given area $\mathcal{A}$ centered at $p_i \in P$. We define rotation events for s_i as the angles at which points of P enter or leave s_i. Denote by U_i the set of points not covered by s_i at the current rotation angle. Using the subproblem described above, we find the smallest constrained square that covers U_i, by computing n sets E_j, where E_j is the set of $2|U_i|$ functions associated with the center point p_j.

We describe our algorithm for determining whether one of the constrained centers is some fixed point $p_i \in P$. Then we apply this algorithm for each of the points in P. Initially, at $\theta = 0$, we construct all the sets E_j, so that each set contains only the functions that correspond to the points in the initial U_i. The rotation events for this phase are those caused by a point of P entering or leaving s_i. At each rotation event we update U_i and all the sets E_j. We then check whether there is a point on the upper envelope of one of the E_j's which is below the line $z = \sqrt{A}$. If a point (θ_0, z_0), $z_0 \leq \sqrt{A}$, was found on the upper envelope of some E_j, then the square s_i at its current position, and the square s_j at angle θ_0 consist of a solution to the decision problem.

Updating the upper envelopes corresponding to the sets E_j turns out to be time consuming, therefore we transform the problem of determining whether one of the upper envelopes has a low enough point to a segment stabbing problem as follows. Observe one set E_j. If we draw a horizontal line at $z = \sqrt{A}$, then each function curve in E_j is cut into at most three continuous subcurves, at most two of them are below the line. We project all the subcurves of E_j that are below the line on the θ-axis, obtaining a set of segments. Assume the number of points in U_i is k, then if (and only if) there is a point θ_0 on the θ-axis that is *covered* by $2k$ segments then there is a square of the required size, of orientation θ_0, centered at p_j which covers the points of U_i.

We construct a segment tree T_j [15] with $O(n)$ leaves (for the segments obtained from all potential curves in E_j). Each node in the tree contains, besides the standard segment information, the maximum cover of the node (namely, the largest number of segments that can be stabbed in the range of the node, for details see [15]). The root of the tree contains the *maximum cover* of the whole range $0 \leq \theta < \frac{\pi}{2}$. The size of one tree is $O(n)$ and each update is performed in time $O(\log n)$. Initially, at $\theta = 0$, we insert into T_j the segments corresponding to the curves of the points in U_i, and check whether the maximum cover equals twice the cardinality of U_i. One update to U_i involves at most two segment updates in T_j.

We consider the time and space complexity of the algorithm. For one point p_i as a candidate center, the initial trees T_j are constructed in time $O(n^2 \log n)$, occupying $O(n^2)$ space. There are $O(n)$ rotation events for s_i, and an update to one T_j is performed in $O(\log n)$ time, totaling $O(n^2 \log n)$ time for all rotation events and all T_j's. The space requirement is $O(n^2)$. Applying the algorithm sequentially for all i in $\{1, \ldots, n\}$ gives $O(n^3 \log n)$ runtime, while the space remains $O(n^2)$.

For the optimization problem, we perform for each i as above the following. Consider $p_i \in P$ as one of the solution centers. The corresponding square is defined either by another point of P in its corner, or by two points of P. So we compute the $O(n^2)$ potential area sizes with p_i as a center. We sort the area sizes and apply binary search to find the smallest area squares that cover P with p_i as one of the centers in the solution. At each of the $O(\log n)$ search steps, we apply the decision algorithm above (just with p_i as one of the centers). We perform this search for all $i \in \{1, \ldots, n\}$. We have shown:

Theorem 7. *Given a set P of n input points we can find a pair of general constrained squares whose union covers P and such that the area of the larger square is minimized in $O(n^3 \log^2 n)$ time and $O(n^2)$ space.*

Acknowledgments
We express our thanks to Paul Chew for helpful discussions.

References

1. P. Agarwal and M. Sharir, "Planar geometric location problems", *Algorithmica* 11 (1994), 185–195.
2. P. Agarwal, M. Sharir, E. Welzl, "The discrete 2-center problem", *Proc. 13th ACM Symp. on Computational Geometry*, 147–155, 1997.
3. R. Cole, "Parallel merge sort", *SIAM J. Computing* 17(4) (1988), 770–785.
4. M. de Berg, M. van Kreveld, M. Overmars and O. Schwartzkopf, *Computational Geometry, Algorithms and Applications*, Springer-Verlag, 1997.
5. L. Danzer and B. Grünbaum, "Intersection properties of boxes in R^{d}", *Combinatorica* 2(3) (1982), 237–246.
6. O. Devillers and M.J. Katz, "Optimal line bipartitions of point sets", *Int. J. Comput. Geom. and Appls*, to appear.
7. D. Eppstein, "Faster construction of planar two-centers", *Proc. 8th ACM-SIAM Symp. on Discrete Algorithms*, 131–138, 1997.
8. G.N. Frederickson and D.B. Johnson, "Generalized selection and ranking: sorted matrices", *SIAM J. Computing* 13 (1984), 14–30.
9. A. Glozman, K. Kedem and G. Shpitalnik, "Efficient solution of the two-line center problem and other geometric problems via sorted matrices", *Proc. 4th Workshop Algorithms Data Struct.*, Lecture Notes in Computer Science (955), 26–37, 1995.
10. J.W. Jaromczyk and M. Kowaluk, "Orientation independent covering of point sets in R^2 with pairs of rectangles or optimal squares", *Proc. European Workshop on Computational Geometry*, Lecture Notes in Computer Science (871), 71–78, 1996.
11. J.W. Jaromczyk and M. Kowaluk, "The two-line center problem from a polar view: A new algorithm and data structure", *Proc. 4th Workshop Algorithms Data Struct.*, Lecture Notes in Computer Science (955), 13–25, 1995.
12. M.J. Katz and F. Nielsen, "On piercing sets of objects", *Proc. 12th ACM Symp. on Computational Geometry*, 113–121, 1996.
13. M.J. Katz and M. Sharir, "An expander-based approach to geometric optimization", *SIAM J. Computing* 26 (1997), 1384–1408.
14. N. Megiddo, "Applying parallel computation algorithms in the design of serial algorithms", *J. ACM* 30 (1983), 852–865.
15. K. Mehlhorn, *Multi-dimensional Searching and Computational Geometry*, in "Data Structures and Algorithms", vol. 3, Springer-Verlag, 1984.
16. M.H. Overmars and J. van Leeuwen, "Maintenance of configurations in the plane", *J. Comput. Syst. Sci.* 23 (1981), 166–204.
17. M. Segal, "On the piercing of axis-parallel rectangles and rings", *Proc. 5th European Symp. on Algorithms*, Lecture Notes in Computer Science (1284), 430–442, 1997.
18. M. Sharir, "A near-linear algorithm for the planar 2-center problem", *Proc. 12th ACM Symp. on Computational Geometry*, 106–112, 1996.
19. M. Sharir and P. Agarwal, *Davenport-Shintzel sequences and their applications*, Cambridge University Press, New-York, 1995.
20. M. Sharir and E. Welzl, "Rectilinear and polygonal p-piercing and p-center problems", *Proc. 12th ACM Symp. on Computational Geometry*, 122–132, 1996.

Worst-Case Efficient External-Memory Priority Queues*

Gerth Stølting Brodal[1,**] and Jyrki Katajainen[2,***]

[1] Max-Planck-Institut für Informatik, Im Stadtwald, D-66123 Saarbrücken, Germany
[2] Datalogisk Institut, Københavns Universitet, Universitetsparken 1,
DK-2100 København Ø, Denmark

Abstract. A priority queue Q is a data structure that maintains a collection of elements, each element having an associated priority drawn from a totally ordered universe, under the operations INSERT, which inserts an element into Q, and DELETEMIN, which deletes an element with the minimum priority from Q. In this paper a priority-queue implementation is given which is efficient with respect to the number of block transfers or I/Os performed between the internal and external memories of a computer. Let B and M denote the respective capacity of a block and the internal memory measured in elements. The developed data structure handles any intermixed sequence of INSERT and DELETEMIN operations such that in every disjoint interval of B consecutive priority-queue operations at most $c \log_{M/B} \frac{N}{M}$ I/Os are performed, for some positive constant c. These I/Os are divided evenly among the operations: if $B \geq c \log_{M/B} \frac{N}{M}$, one I/O is necessary for every $B/(c \log_{M/B} \frac{N}{M})$th operation and if $B < c \log_{M/B} \frac{N}{M}$, $\frac{c}{B} \log_{M/B} \frac{N}{M}$ I/Os are performed per every operation. Moreover, every operation requires $O(\log_2 N)$ comparisons in the worst case. The best earlier solutions can only handle a sequence of S operations with $O(\sum_{i=1}^{S} \frac{1}{B} \log_{M/B} \frac{N_i}{M})$ I/Os, where N_i denotes the number of elements stored in the data structure prior to the ith operation, without giving any guarantee for the performance of the individual operations.

1 Introduction

A *priority queue* is a data structure that stores a set of elements, each element consisting of some *information* and a *priority* drawn from some totally ordered universe. A priority queue supports the operations:

---

* The full version has appeared as Technical Report 97/25, Department of Computer Science, University of Copenhagen, Copenhagen, 1997.

** Supported by the Carlsberg foundation under grant No. 96-0302/20. Partially supported by the ESPRIT Long Term Research Program of the EU under contract No. 20244 (project ALCOM-IT). Email: brodal@mpi-sb.mpg.de.

*** Supported partially by the Danish Natural Science Research Council under contract No. 9400952 (project Computational Algorithmics). Email: jyrki@diku.dk.

INSERT(x): Insert a new element x with an arbitrary priority into the data structure.

DELETEMIN(): Delete and return an element with the minimum priority from the data structure. In the case of ties, these are broken arbitrarily. The precondition is that the priority queue is not empty.

Priority queues have numerous applications, a few listed by Sedgewick [28] are: sorting algorithms, network optimization algorithms, discrete event simulations and job scheduling in computer systems. For the sake of simplicity, we will not hereafter make any distinction between the elements and their priority.

In this paper we study the problem of maintaining a priority queue on a computer with a two-level memory: a fast *internal memory* and a slow *external memory*. We assume that the computer has a processing unit, the *processor*, and a collection of hardware, the *I/O subsystem*, which is responsible for transferring data between internal and external memory. The processor together with the internal memory can be seen as a traditional random access machine (RAM) (see, e.g., [3]). In particular, note that the processor can only access data stored in internal memory. The capacity of the internal memory is assumed to be bounded so it might be necessary to store part of the data in external memory. The I/O subsystem transfers the data between the two memory levels in blocks of a fixed size.

The behavior of algorithms on such a computer system can be characterized by two quantities: *processor performance* and *I/O performance*. By the processor performance we mean the number of primitive operations performed by the processor. Our measure of processor performance is the number of element comparisons carried out. It is straightforward to verify that the total number of other (logical, arithmetical, etc.) operations required by our algorithms is proportional to that of comparisons. Assuming that the elements occupy only a constant number of computer words, the total number of primitive operations is asymptotically the same as that of comparisons. Our measure of I/O performance is the number of block transfers or *I/Os* performed, i.e., the number of blocks read from the external memory plus the number of blocks written to the external memory. Our main goal is to analyze the total work carried out by the processor and the I/O subsystem during the execution of the algorithms.

The *system performance*, i.e., the total elapsed execution time when the algorithms are run on a real computer, depends heavily on the realization of the computer. A real computer may have multiple processors (see, e.g., [18]) and/or the I/O subsystem can transfer data between several disks at the same time (cf. [2, 25, 30]), the processor operations (see, e.g., [27]) and/or the I/Os (cf. [19]) might be pipelined, but the effect of these factors is not considered here. It has been observed that in many large-scale computations the increasing bottleneck of the computation is the performance of the I/O subsystem (see, e.g., [15, 26]), increasing the importance of I/O efficient algorithms.

When expressing the performance of the priority-queue operations, we use the following parameters:

B: the number of elements per block,

M: the number of elements fitting in internal memory, and

N: the number of elements currently stored in the priority queue; more specifically, the number of elements stored in the structure just prior to the execution of INSERT or DELETEMIN.

We assume that each block and the internal memory also fit some pointers in addition to the elements, and $B \geq 1$ and $M \geq 23B$. Furthermore, we use $\log_a n$ as a shorthand notation for $\max(1, \ln n / \ln a)$, where ln denotes the natural logarithm.

Several priority-queue schemes, such as implicit heaps [33], leftist heaps [12, 20], and binomial queues [9, 31] have been shown to permit both INSERT and DELETEMIN with worst-case $O(\log_2 N)$ comparisons. Some schemes, such as implicit binomial queues [10] guarantee worst-case $O(1)$ comparisons for INSERT and $O(\log_2 N)$ comparisons for DELETEMIN. Also any kind of balanced search trees, such as AVL trees [1] or red-black trees [16] could be used as priority queues. However, due to the usage of explicit or implicit pointers the performance of these structures deteriorates on a two-level memory system. It has been observed by several researchers that a d-ary heap performs better than the normal binary heap on multi-level memory systems (see, e.g., [22, 24]). For instance, a B-ary heap reduces the number of I/Os from $O(\log_2 \frac{N}{B})$ (cf. [4]) to $O(\log_B \frac{N}{B})$ per operation [24]. Of course, a B-tree [8, 11] could also be used as a priority queue, with which a similar I/O performance is achieved. However, in a virtual-memory environment a B-ary heap seems to be better than a B-tree [24].

When a priority queue is maintained in a two-level memory, it is advantageous to keep the small elements in internal memory and the large elements in external memory. Hence, due to insertions large elements are to be moved from internal memory to external memory and due to deletions small elements are to be moved from external memory to internal memory. Assuming that we maintain two buffers of B elements in internal memory, one for INSERTs and one for DELETEMINs, at most every Bth INSERT and DELETEMIN will cause a buffer overflow or underflow. Several data structures take advantage of this kind of buffering. Fishspear, developed by Fischer and Paterson [14], can be implemented by a constant number of push-down stacks, implying that any sequence of S INSERT and DELETEMIN operations requires at most $O(\sum_{i=1}^{S} \frac{1}{B} \log_2 N_i)$ I/Os, where N_i denotes the size of the data structure prior to the ith operation. Wegner and Teuhola [32] realized that a binary heap, in which each node stores B elements, guarantees $O(\log_2 \frac{N}{B})$ I/Os for every Bth INSERT and every Bth DELETEMIN operation in the worst case.

The above structures assume that the internal memory can only fit $O(B)$ elements, i.e., a constant number of blocks. Even faster solutions are possible if the whole capacity of the internal memory is utilized. Arge [5, 6] introduced an (a,b)-tree structure that can be used to carry out any sequence of S INSERT and DELETEMIN operations with $O(\frac{S}{B} \log_{M/B} \frac{S}{M})$ I/Os. Fadel et al. [13] gave a heap structure with a similar I/O performance but their bound depends on the size profile, not on S. Their heap structure can handle any sequence of S operations with $O(\sum_{i=1}^{S} \frac{1}{B} \log_{M/B} \frac{N_i}{M})$ I/Os, where N_i denotes the size of the data structure prior to the ith operation. The number of comparisons required when handling the sequence is $O(\sum_{i=1}^{S} \log_2 N_i)$. When this data structure is used for sorting N elements, both the processor and I/O performance match the well-known lower bounds $\Omega(\frac{N}{B} \log_{M/B} \frac{N}{M})$ I/Os [2] and $\Omega(N \log_2 N)$ comparisons (see, e.g., [20]), which are valid for all comparison-based algorithms.

To achieve the above bounds—as well as our bounds—the following facilities must be provided:

1. we should know the capacity of a block and the internal memory beforehand,
2. we must be able to align elements into blocks, and
3. we must have a full control over the replacement of the blocks in internal memory.

There are operating systems that provide support for these facilities (see, e.g., [17, 21, 23]).

The tree structure of Arge and the heap structure of Fadel et al. do not give any guarantees for the performance of individual operations. In fact, one INSERT or DELETEMIN can be extremely expensive, the cost of handling the whole sequence being an upper bound. Therefore, it is risky to use these structures in on-line applications. For large-scale real-time discrete event simulations and job scheduling in computer systems it is often important to have a guaranteed worst-case performance.

We describe a new data structure that gives worst-case guarantees for the cost of individual operations. Basically, our data structure is a collection of sorted lists that are incrementally merged. This idea is borrowed from a RAM priority-queue structure of Thorup [29]. Thorup used two-way merging in his internal data structure but we use multi-way merging since it behaves better in an external-memory environment. As to the processor and I/O performance, our data structure handles any intermixed sequence of operations as efficiently as the heap structure by Fadel et al. [13]. In every disjoint interval of B consecutive priority-queue operations our data structure requires at most $c \log_{M/B} \frac{N}{M}$ I/Os, for some positive constant c. These I/Os are divided evenly among the operations. If $B \geq c \log_{M/B} \frac{N}{M}$, one I/O is necessary for every $B/(c \log_{M/B} \frac{N}{M})$th priority-queue operation, and if $B < c \log_{M/B} \frac{N}{M}$, $\frac{c}{B} \log_{M/B} \frac{N}{M}$ I/Os are performed per every priority-queue operation. Moreover, every operation requires $O(\log_2 N)$ comparisons in the worst case.

2 Basic Data Structure

The basic components of our priority-queue data structure are sorted lists. When new elements arrive, these are added to a list which is kept in internal memory and sorted incrementally. If the capacity of internal memory is exceeded due to insertions, a fraction of the list containing the recently inserted elements is transferred to external memory. To bound the number of lists in external memory we merge the existing lists. This merging is related to the merging done by the external mergesort algorithm [2]. One particular list that is kept in internal memory contains the smallest elements. If this list is exhausted due to deletions, new smallest elements are extracted from the lists in external memory. Because we are interested in worst-case bounds the merging is accomplished incrementally. A similar idea has been applied by Thorup [29] to construct RAM priority queues but instead of two-way merging we rely on multi-way merging.

Before giving the details of the data structure, let us recall the basic idea of external mergesort which sorts N elements with $O(\frac{N}{B} \log_{M/B} \frac{N}{M})$ I/Os [2]. First, the given N elements are partitioned into $\Theta(N/M)$ lists each of length $\Theta(M)$. Second, each of the lists are read into internal memory and sorted, requiring $O(N/B)$ I/Os in total. Third, $\Theta(M/B)$ sorted lists of shortest length are repeatedly merged until only one sorted list remains containing all the elements. Since each element takes part in $O(\log_{M/B} \frac{N}{M})$ merges, the total number of I/Os is $O(\frac{N}{B} \log_{M/B} \frac{N}{M})$.

Our data structure consists of two parts: an *internal part* and an *external part*. The data structure takes two parameters K and m, where K is a multiple of B, $9K + 5B \leq M$, and $m = K/B$. The internal part of the data structure stores $O(K)$ elements and is kept all the time in internal memory. The external part is a priority queue which permits the operations:

BATCHINSERT$_K(X)$: Insert a sorted list X of K elements into the external-memory data structure.

BATCHDELETEMIN$_K()$: Delete the K smallest elements from the data structure in external-memory.

Both of these operations require at most $O(\frac{K}{B} \log_m \frac{N}{K})$ I/Os and $O(K \log_2 \frac{N}{K})$ comparisons in the worst case. For every Kth operation on the internal part we do at most one batch operation involving K elements on the external part of the data structure.

The internal part of the data structure consists of two sorted lists MIN and NEW of length at most $3K$ and $2K$, respectively. We represent both MIN and NEW as a balanced search tree that permits insertions and deletions of elements with $O(\log_2 K)$ comparisons. The rôle of MIN is to store the current at most $3K$ smallest elements in the priority queue whereas the intuitive rôle of NEW is to store the at most $2K$ recently inserted elements. All elements in MIN are smaller than the elements in NEW and the elements in the external part of the data structure, i.e., the overall minimum element is the minimum of MIN.

The external part of the data structure consists of sorted lists of elements. Each of these lists has a *rank*, which is a positive integer, and we let R denote the *maximum rank*. In Sect. 3.4 we show how to guarantee that $R \leq \log_m \frac{N}{K} + 2$. The lists with rank i, $i \in \{1, \ldots, R\}$, are $L_i^1, L_i^2, \ldots, L_i^{n_i}, \overline{L}_i^1, \overline{L}_i^2, \ldots, \overline{L}_i^{\overline{n}_i}$, and $\overline{L}_i$.

For each rank i, we will incrementally merge the lists $\overline{L}_i^1, \ldots, \overline{L}_i^{\overline{n}_i}$ and append the result of the merging to the list $\overline{L}_i$. The list $\overline{L}_i$ contains the already merged part of $\overline{L}_i^1, \ldots, \overline{L}_i^{\overline{n}_i}$, and all elements in $\overline{L}_i$ are therefore smaller than those in $\overline{L}_i^1, \ldots, \overline{L}_i^{\overline{n}_i}$. When the incremental merge of the lists $\overline{L}_i^j$ finishes, the list $\overline{L}_i$ will be promoted to a list with rank $i + 1$, provided that $\overline{L}_i$ is sufficiently long, and a new incremental merge of lists with rank i is initiated by making $L_i^1, \ldots, L_i^{n_i}$ the new $\overline{L}_i^j$ lists. The details of the incremental merge are given in Sect. 3.2.

We guarantee that the length of each of the external lists is a multiple of B. An external list L containing $|L|$ elements is represented by a single linked list of $|L|/B$ blocks, each block storing B elements plus a pointer to the next block, except for the last block which stores a null pointer. There is one exception to this representation. The last block of $\overline{L}_i$ does not store a null pointer, but a pointer to the first block of $\overline{L}_i^1$ (if $\overline{n}_i = 0$, the last block of $\overline{L}_i$ stores a null pointer). This allows us to avoid updating the last block of $\overline{L}_i$ when merging the lists $\overline{L}_i^1, \ldots, \overline{L}_i^{\overline{n}_i}$.

In the following, we assume that pointers to all the external lists are kept in internal memory together with their sizes and ranks. If this is not possible, it is sufficient to store this information in a single linked list in external memory. This increases the number of I/Os required by our algorithms only by a small constant.

In Sect. 3 we describe how BATCHINSERT$_K$ and BATCHDELETEMIN$_K$ operations are accomplished on the external part of the data structure, and in Sect. 4 we describe how the external part can be combined with the internal part of the data structure to achieve a worst-case efficient implementation of INSERT and DELETEMIN operations.

3 Maintenance of the External Part

3.1 The MERGE$_K$ Procedure

The heart of our construction is the procedure MERGE$_K(X_1, X_2, \ldots, X_\ell)$, which incrementally merges and removes the K smallest elements from the sorted lists $X_1, \ldots, X_\ell$. All list lengths are assumed to be multiples of B. After the merging of the K smallest elements we rearrange the remaining elements in the X_i lists such that the lists still have lengths which are multiples of B. We allow MERGE$_K$ to make the X_i lists shorter or longer. We just require that the resulting X_i lists remain sorted. For the time being, we assume that the result of MERGE$_K$ is stored in internal memory.

The procedure MERGE$_K$ is implemented as follows. For each list X_i we keep the block containing the current minimum of X_i in internal memory. In internal memory we maintain a heap [33] over the current minima of all the lists. We use the heap to find the next element to be output in the merging process. Whenever an element is output, it is the current minimum of some list X_i. We remove the element from the heap and the list X_i, and insert the new minimum of X_i into the heap, provided that X_i has not become empty. If necessary, we read the next block of X_i into internal memory.

After the merging phase, we have from each list X_i a partially filled block B_i in internal memory. Let $|B_i|$ denote the number of elements left in block B_i. Because we have merged K elements from the blocks read and K is a multiple of B, $\sum_{i=1}^{\ell} |B_i|$ is also a multiple of B. Now we merge the remaining elements in the

B_i blocks in internal memory. This merge requires $O(\ell B \log_2 \ell)$ comparisons. Let $\hat{X}$ denote the resulting list and let B_j be the block that contained the maximum element of $\hat{X}$. Finally, we write $\hat{X}$ to external memory such that X_j becomes the list consisting of $\hat{X}$ concatenated with the part of X_j that already was stored in external memory. Note that X_j remains sorted.

In total, MERGE$_K$ performs at most $K/B + \ell$ I/Os for reading the prefixes of $X_1, \ldots, X_\ell$ (for each list X_i, we read at most one block of elements that do not take part in the merging) and at most ℓ I/Os for writing $\hat{X}$ to external memory. The number of comparisons required for MERGE$_K$ for each of the $K + \ell B$ elements read into internal memory is $O(\log_2 \ell)$. Hence, we have proved

Lemma 1. MERGE$_K(X_1, \ldots, X_\ell)$ *performs at most* $2\ell + K/B$ *I/Os and* $O((K + \ell B) \log_2 \ell)$ *comparisons. The number of elements to be kept in internal memory by* MERGE$_K$ *is at most* $K + \ell B$. *If the resulting list is written to external memory incrementally, only* $(\ell + 1)B$ *elements have to be kept in internal memory simultaneously.*

3.2 Batch Insertions

To insert a sorted list of K elements into the external part of the data structure we increment n_1 by one and let $L_1^{n_1}$ contain the K new elements, and apply the procedure MERGESTEP(i), for each $i \in \{1, \ldots, R\}$.

The procedure MERGESTEP(i) does the following. If $\bar{n}_i = 0$, the incremental merge of lists with rank i is finished, and we make $\bar{L}_i$ the list $L_{i+1}^{n_{i+1}+1}$, provided that $|\bar{L}_i| \geq Km^i$. Otherwise, we let $\bar{L}_i$ be the list $L_i^{n_i+1}$ because the list is too short to be promoted. Finally, we initiate a new incremental merge by making the lists $L_i^1, \ldots, L_i^{n_i}$ the new $\bar{L}_i^j$ lists. If $\bar{n}_i > 0$, we concatenate $\bar{L}_i$ with the result of MERGE$_K(\bar{L}_i^1, \ldots, \bar{L}_i^{\bar{n}_i})$, i.e., we perform K steps of the incremental merge of $\bar{L}_i^1, \ldots, \bar{L}_i^{\bar{n}_i}$. Note that, by writing the first block of the merged list on the place occcupied earlier by the first block of $\bar{L}_i^1$, we do not need to update the pointer in the previous last block of $\bar{L}_i$.

The total number of I/Os performed in a batched insertion of K elements is K/B for writing the K new elements to external memory and by Lemma 1 at most $2(\bar{n}_i + K/B)$ for incrementally merging the lists with rank i. The number of comparisons for rank i is $O((\bar{n}_i B + K) \log_2 \bar{n}_i)$. The maximum number of elements to be stored in internal memory for batched insertions is $(\bar{n}_{\max} + 1)B$, where $\bar{n}_{\max} = \max\{\bar{n}_1, \ldots, \bar{n}_R\}$. To summarize, we have

Lemma 2. *A sorted list of K elements can be inserted into the external part of the data structure by performing* $(1 + 2R)K/B + 2\sum_{i=1}^{R} \bar{n}_i$ *I/Os and performing* $O(\sum_{i=1}^{R}(\bar{n}_i B + K) \log_2 \bar{n}_i)$ *comparisons. At most* $(\bar{n}_{\max} + 1)B$ *elements need to be stored in internal memory.*

3.3 Batch Deletions

The removal of the K smallest elements from the external part of the data structure is carried out in two steps. In the first step the K smallest elements are located. In the second step the actual deletion is accomplished.

Let $\mathcal{L}$ be one of the lists L_i^1 or $\overline{L}_i$, for some i, or an empty list. We will guarantee that the list $\mathcal{L}$ contains the K smallest elements of the lists considered so far. Initially $\mathcal{L}$ is empty. By performing $L_i^1 \leftarrow \text{MERGE}_K(L_i^1, \ldots, L_i^{n_i}) \cdot L_i^1$, L_i^1 now contains the K smallest elements of $L_i^1, \ldots, L_i^{n_i}$. The procedure SPLITMERGE_K takes two sorted lists as its arguments and returns (the name of) one of the lists. If the first argument is an empty list, then the second list is returned. Otherwise, we require that the length of both lists to be at least K and we rearrange the K smallest elements of both lists as follows. The two prefixes of length K are merged and split among the two lists such that the lists remain sorted and the length of the lists remain unchanged. One of the lists will now have a prefix containing K elements which are smaller than all the elements in the other list. The list with this prefix is returned. For each rank $i \in \{1, \ldots R\}$, we now carry out the assignments $L_i^1 \leftarrow \text{MERGE}_K(L_i^1, \ldots, L_i^{n_i}) \cdot L_i^1$, $\mathcal{L} \leftarrow \text{SPLITMERGE}_K(\mathcal{L}, L_i^1)$, and $\mathcal{L} \leftarrow \text{SPLITMERGE}_K(\mathcal{L}, \overline{L}_i)$.

It is straightforward to verify that after performing the above, the prefix of the list $\mathcal{L}$ contains the K smallest elements in the external part of the data structure. We now delete the K smallest elements from list $\mathcal{L}$, and if $\mathcal{L}$ is $\overline{L}_i$ we perform $\text{MERGESTEP}(i)$ once.

By always keeping the prefix of $\mathcal{L}$ in internal memory the total number of I/Os for the deletion of the K smallest elements (without the call to MERGESTEP) is $(4R-1)(K/B) + 2\sum_{i=1}^{R} n_i$ because, for each rank i, $n_i + 2(K/B)$ blocks are to be read into internal memory and all blocks except the K/B blocks holding the smallest elements should be written back to external memory. The number of comparisons for rank i is $O((K + Bn_i) \log_2 n_i)$. The additional call to MERGESTEP requires at most $K/B + \overline{n}_i$ additional block reads and block writes, and $O((K + B\overline{n}_i) \log_2 \overline{n}_i)$ comparisons. Let $n_{\max} = \max\{n_1, \ldots, n_R\}$ and $\overline{n}_{\max} = \max\{\overline{n}_1, \ldots, \overline{n}_R\}$. The maximum number of elements to be stored in internal memory for the batched minimum deletions is $2K + B\max\{n_{\max}, \overline{n}_{\max}\}$.

Lemma 3. *The K smallest elements can be deleted from the external part of the data structure by performing at most $4R(K/B) + 2\sum_{i=1}^{R} n_i + \overline{n}_{\max}$ I/Os and $O(\sum_{i=1}^{R}(K + n_iB) \log_2 n_i + (K + B\overline{n}_{\max}) \log_2 \overline{n}_{\max})$ comparisons. At most $2K + B\max\{n_{\max}, \overline{n}_{\max}\}$ elements need to be stored in internal memory.*

3.4 Bounding the Maximum Rank

We now describe a simple approach to guarantee that the maximum rank R of the external data structure is bounded by $\log_m N/K + 2$. Whenever insertions cause the maximum rank to increase, this is because of $\text{MERGESTEP}(R-1)$ has finished an incremental merge resulting a list of length Km^{R-1}, which implies that $R \leq \log_m \frac{N}{K} + 1$. The problem we have to consider is how to decrement R when deletions are performed.

Our solution is the following. Whenever MERGESTEP(R) finishes the incremental merge of lists with rank R, we check if the resulting list $\overline{L}_R$ is very small. If $\overline{L}_R$ is very small, i.e., $|\overline{L}_R| < Km^{R-1}$, and there are no other list of rank R, we make $\overline{L}_R$ a list with rank $R - 1$ and decrease R.

To guarantee that the same is done also in the connection with batched minimum deletions, we always call after each BATCHDELETEMIN$_K$ operation, described in Sect. 3.3, MERGESTEP(R) k times (for $m \geq 4$ it turns out that $k = 1$ is sufficient, and for $m = 3$ or $m = 2$ it is sufficient to let $k = 2$ or $k = 3$). It can be proved that this guarantees $R \leq \log_m \frac{N}{K} + 2$.

3.5 Resource Bounds for the External Part

In the previous discussion we assumed that n_i and $\overline{n}_i$ where sufficiently small, such that we could apply MERGESTEP to the L_i^j and $\overline{L}_i^j$ lists. Let m' denote a maximum bound on the merging degree. It can be proved that $m' \leq 5 + 2m$, for $m \geq 2$. Because $m = K/B$ it follows that the maximum rank is at most $\log_{K/B} \frac{N}{K} + 2$ and that the maximum merging degree is $5 + 2K/B$. From Lemmas 2 and 3 it follows that the number of I/Os required for inserting K elements or deleting the K smallest elements is at most $O(\frac{K}{B} \log_{K/B} \frac{N}{K})$ and the number of comparisons required is $O(K \log_2 \frac{N}{K})$. The maximal number of elements to be stored in internal memory is $4K + 5B$.

4 Internal Buffers and Incremental Batch Operations

We now describe how to combine the buffers *NEW* and *MIN* represented by binary search trees with the external part of the priority-queue data structure. We maintain the invariant that $|MIN| \geq 1$, provided that the priority queue is nonempty. Recall that we also required that $|MIN| \leq 3K$ and $|NEW| \leq 2K$.

We first consider INSERT(x). If x is less than or equal to the maximum of *MIN* or all elements of the priority queue are stored in *MIN*, we insert x into *MIN* with $O(\log_2 K)$ comparisons. If *MIN* exceeds its maximum allowed size, $|MIN| = 3K + 1$, we move the maximum of *MIN* to *NEW*. Otherwise, x is larger than the maximum of *MIN* and we insert x into *NEW* with $O(\log_2 K)$ comparisons. The implementation of DELETEMIN deletes and returns the minimum of *MIN*. Both operations require at most $O(\log_2 K)$ comparisons.

There are two problems with the implementation of INSERT and DELETEMIN. Insertions can cause *NEW* to become too big and deletions can make *MIN* empty. Therefore, for every Kth priority-queue operation we perform one batch insertion or deletion. If $|NEW| \geq K$, we remove K elements from *NEW* one by one and perform BATCHINSERT$_K$ on the removed elements. If $|NEW| < K$ and $|MIN| \leq 2K$, we instead increase the size of *MIN* by moving K small elements to *MIN* as follows. First, we perform a BATCHDELETEMIN$_K$ operation to extract the K least elements from the external part of the data structure. The

K extracted elements are inserted into NEW one by one, using $O(K \log_2 K)$ comparisons. Second, we move the K smallest elements of NEW to MIN one by one. If $|NEW| < K$ and $|MIN| > 2K$, we do nothing but delay the batch operation until $|MIN| = 2K$ or $|NEW| = K$. Each batch operation requires at most $O(\frac{K}{B} \log_{K/B} \frac{N}{K})$ I/Os and at most $O(K(\log_2 \frac{N}{K} + \log_2 K)) = O(K \log_2 N)$ comparisons.

By doing one of the above described batch operations for every Kth priority-queue operation it is straightforward to verify that $|NEW| + (3K - |MIN|) \leq 2K$, provided that the priority queue contains at least K elements, implying $|NEW| \leq 2K$ and $|MIN| \geq K$, because each batch operation decreases the left-hand side of the equation by K.

The idea is now to perform a batch operation incrementally over the next K priority-queue operations. Let N denote the number of elements in the priority queue, when the corresponding batch operation is initiated. Notice that N can at most be halved while performing a batch operation, because $N \geq 2K$ prior to the batch operation. Because $|MIN| \geq K$ when a batch operation is initiated, it is guaranteed that MIN is nonempty while incrementally performing the batch operation over the next K priority-queue operations.

Because a batch operation requires at most $O(\frac{K}{B} \log_{K/B} \frac{N}{K})$ I/Os and at most $O(K \log_2 N)$ comparisons, it is sufficient to perform at most $O(\log_2 N)$ comparisons of the incremental batch operation per priority-queue operation and operations if $B \geq c \log_{M/B} \frac{N}{M}$, one I/O for every $B/(c \log_{M/B} \frac{N}{M})$th priority-queue operation and if $B < c \log_{M/B} \frac{N}{M}$, $\frac{c}{B} \log_{M/B} \frac{N}{M}$ I/Os for every priority-queue operation, for some positive constant c, to guarantee that the incremental batch operation is finished after K priority-queue operations.

Because $|MIN| \leq 3K$, $|NEW| \leq 2K$, and a batched operation at most requires $4K + 5B$ elements to be stored in internal memory, we have the constraint that $9K + 5B \leq M$. Let now $K = \lfloor (M - 5B)/9 \rfloor$. Recall that we assumed that $M \geq 23B$, and therefore, $K \geq 2B$. Since $M > K$, $O(\log_{M/B} \frac{N}{M}) = O(\log_{K/B} \frac{N}{M})$. Hence, we have proved the main result of this paper.

Main theorem *There exists an external-memory priority-queue implementation that supports* INSERT *and* DELETEMIN *operations with worst-case* $O(\log_2 N)$ *comparisons per operation. If* $B \geq c \log_{M/B} \frac{N}{M}$*, one I/O is necessary for every* $B/(c \log_{M/B} \frac{N}{M})$th *operation and if* $B < c \log_{M/B} \frac{N}{M}$*,* $\frac{c}{B} \log_{M/B} \frac{N}{M}$ *I/Os are performed per every operation, for some positive constant* c*.*

5   Concluding Remarks

We have presented an efficient priority-queue implementation which guarantees a worst-case bound on the number of comparisons and I/Os required for the individual priority-queue operations. Our bounds are comparison based.

If the performance bounds are allowed to be amortized, the data structure can be simplified considerably, because no list merging and batch operation is required to be incrementally performed. Then no $\overline{L}_i$ and $\overline{L}_i^j$ lists are required, and we can satisfy $1 \leq |MIN| \leq K$, $|NEW| \leq K$, and $n_i < m$ by always (completely) merging exactly m lists of equal rank, the rank of a list L being $\lfloor \log_m \frac{|L|}{K} \rfloor$.

What if the size of the elements or priorities is not assumed to be constant? That is, express the bounds as a function of N and the length of the priorities. How about the priorities having variable lengths? Initial research in this direction has been carried out by Arge et al. [7], who consider the problem of sorting strings in external memory.

References

1. G. M. Adel'son-Vel'skiĭ and E. M. Landis. An algorithm for the organization of information. *Soviet Mathematics*, volume 3, pages 1259–1263, 1962.
2. A. Aggarwal and J. S. Vitter. The input/output complexity of sorting and related problems. *Communications of the ACM*, volume 31, pages 1116–1127, 1988.
3. A. V. Aho, J. E. Hopcroft, and J. D. Ullman. *The Design and Analysis of Computer Algorithms*. Addison-Wesley Publishing Company, Reading, 1974.
4. T. O. Alanko, H. H. A. Erkiö, and I. J. Haikala. Virtual memory behavior of some sorting algorithms. *IEEE Transactions on Software Engineering*, volume SE-10, pages 422–431, 1984.
5. L. Arge. The buffer tree: A new technique for optimal I/O-algorithms. In *Proceedings of the 4th Workshop on Algorithms and Data Structures*, Lecture Notes in Computer Science 955, Springer, Berlin/Heidelberg, pages 334–345, 1995.
6. L. Arge. Efficient external-memory data structures and applications. BRICS Dissertation DS-96-3, Department of Computer Science, University of Aarhus, Århus, 1996.
7. L. Arge, P. Ferragina, R. Grossi, and J. S. Vitter. On sorting strings in external memory. In *Proceedings of the 29th Annual ACM Symposium on Theory of Computing*, ACM Press, New York, pages 540–548, 1997.
8. R. Bayer and E. M. McCreight. Organization and maintenance of large ordered indexes. *Acta Informatica*, volume 1, pages 173–189, 1972.
9. M. R. Brown. Implementation and analysis of binomial queue algorithms. *SIAM Journal on Computing*, volume 7, pages 298–319, 1978.
10. S. Carlsson, J. I. Munro, and P. V. Poblete. An implicit binomial queue with constant insertion time. In *Proceedings of the 1st Scandinavian Workshop on Algorithm Theory*, Lecture Notes in Computer Science 318, Springer-Verlag, Berlin/Heidelberg, pages 1–13, 1988.
11. D. Comer. The ubiquitous B-tree. *ACM Computing Surveys*, volume 11, pages 121–137, 1979.
12. C. A. Crane. Linear lists and priority queues as balanced trees. Technical Report STAN-CS-72-259, Computer Science Department, Stanford University, Stanford, 1972.
13. R. Fadel, K. V. Jakobsen, J. Katajainen, and J. Teuhola. Heaps and heapsort on secondary storage. To appear in *Theoretical Computer Science*.
14. M. J. Fischer and M. S. Paterson. Fishspear: A priority queue algorithm. *Journal of the ACM*, volume 41, pages 3–30, 1994.
15. G. A. Gibson, J. S. Vitter, J. Wilkes *et al.* Strategic directions in storage I/O issues in large-scale computing. *ACM Computing Surveys*, volume 28, pages 779–793, 1996.

118

16. L. J. Guibas and R. Sedgewick. A dichromatic framework for balanced trees. In *Proceedings of the 19th Annual Symposium on Foundations of Computer Science*, IEEE, New York, pages 8–21, 1978.

17. K. Harty and D. R. Cheriton. Application-controlled physical memory using external page-cache management. In *Proceedings of the 5th International Conference on Architectural Support for Programming Languages and Operating Systems, ACM SIGPLAN Notices*, volume 27, number 9, pages 187–197, 1992.

18. J. JáJá. *An Introduction to Parallel Algorithms*. Addison-Wesley Publishing Company, Reading, 1992.

19. B. H. H. Juurlink and H. A. G. Wijshoff. The parallel hierarchical memory model. In *Proceedings of the 4th Scandinavian Workshop on Algorithm Theory*, Lecture Notes in Computer Science 824, Springer-Verlag, Berlin/Heidelberg, pages 240–251, 1994.

20. D. E. Knuth. *The Art of Computer Programming*, volume 3/ *Sorting and Searching*. Addison-Wesley Publishing Company, Reading, 1973.

21. K. Krueger, D. Loftesness, A. Vahdat, and T. Anderson. Tools for the development of application-specific virtual memory management. In *Proceedings of the 8th Annual Conference on Object-Oriented Programming Systems, Languages, and Applications, ACM SIGPLAN Notices*, volume 28, number 10, pages 48–64, 1993.

22. A. LaMarca and R. E. Ladner. The influence of caches on the performance of heaps. *The ACM Journal of Experimental Algorithmics*, volume 1, article 4, 1996.

23. D. McNamee and K. Amstrong. Extending the Mach external pager interface to accommodate user-level block replacement policies. Technical Report 90-09-05, Department of Computer Science and Engineering, University of Washington, Seattle, 1990.

24. D. Naor, C. U. Martel, and N. S. Matloff. Performance of priority queue structures in a virtual memory environment. *The Computer Journal*, volume 34, pages 428–437, 1991.

25. M. H. Nodine and J. S. Vitter. Large-scale sorting in parallel memories. In *Proceedings of the 3rd ACM Symposium on Parallel Algorithms and Architectures*, ACM Press, New York, pages 29–39, 1991.

26. Y. N. Patt. Guest editor's introduction: The I/O subsystem — A candidate for improvement. *IEEE Computer*, volume 27, number 3, pages 15–16, 1994.

27. D. A. Patterson and J. L. Hennessy. *Computer Organization & Design: The Hardware/Software Interface*. Morgan Kaufmann Publishers, San Francisco, 1994.

28. R. Sedgewick. *Algorithms*. Addison-Wesley Publishing Company, Reading, 1983.

29. M. Thorup. On RAM priority queues. In *Proceedings of the 7th Annual ACM-SIAM Symposium on Discrete Algorithms*, ACM, New York and SIAM, Philadelphia, pages 59–67, 1996.

30. J. S. Vitter and E. A. M. Shriver. Algorithms for parallel memory I: Two-level memories. *Algorithmica*, volume 12, pages 110–147, 1994.

31. J. Vuillemin. A data structure for manipulating priority queues. *Communications of the ACM*, volume 21, pages 309–315, 1978.

32. L. M. Wegner and J. I. Teuhola. The external heapsort. *IEEE Transactions on Software Engineering*, volume 15, pages 917–925, 1989.

33. J. W. J. Williams. Algorithm 232, Heapsort. *Communications of the ACM*, volume 7, pages 347–348, 1964.

Simple Confluently Persistent Catenable Lists

(Extended Abstract)

Haim Kaplan[1], Chris Okasaki[2]*, and Robert E. Tarjan[3]**

[1] AT&T labs, 180 Park Ave, Florham Park, NJ. hkl@research.att.com
[2] School of Computer Science, Carnegie Mellon University, Pittsburgh, PA 15213. cokasaki@cs.cmu.edu
[3] Department of Computer Science, Princeton University, Princeton, NJ 08544 and InterTrust Technologies Corporation, Sunnyvale, CA 94086. ret@cs.princeton.edu.

Abstract. We consider the problem of maintaining persistent lists subject to concatenation and to insertions and deletions at both ends. Updates to a persistent data structure are nondestructive—each operation produces a new list incorporating the change while keeping intact the list or lists to which it applies. Although general techniques exist for making data structures persistent, these techniques fail for structures that are subject to operations, such as catenation, that combine two or more versions. In this paper we develop a simple implementation of persistent double-ended queues with catenation that supports all deque operations in constant amortized time.

1 Introduction

Over the last fifteen years, there has been considerable development of *persistent* data structures, those in which not only the current version, but also older ones, are available for access (*partial persistence*) or updating (*full persistence*). In particular, Driscoll, Sarnak, Sleator, and Tarjan [5] developed efficient general methods to make pointer-based data structures partially or fully persistent, and Dietz [3] developed an efficient general method to make array-based structures fully persistent.

These general methods support updates that apply to a single version of a structure at a time, but they do not accommodate operations that combine two different versions of a structure, such as set union or list catenation. Driscoll, Sleator, and Tarjan [4] coined the term *confluently persistent* for fully persistent structures that support such combining operations. An alternative way to obtain persistence is to use strictly functional programming (By strictly functional we mean that lazy evaluation, memoization, and other such techniques are not

---

* Supported by the Advanced Research Projects Agency CSTO under the title "The Fox Project: Advanced Languages for Systems Software", ARPA Order No. C533, issued by ESC/ENS under Contract No. F19628-95-C-0050.
** Research at Princeton University partially supported by NSF Grant No. CCR-9626862.

allowed). For list-based data structure design, strictly functional programming amounts to using only the LISP functions CAR, CONS, CDR. Strictly functional data structures are automatically persistent, and indeed confluently persistent.

A simple but important problem in data structure design that makes the issue of confluent persistence concrete is that of implementing persistent double-ended queues (deques) with catenation. A series of papers [4, 2] culminated in the work of Kaplan and Tarjan [8], who developed a confluently persistent implementation of deques with catenation that has a worst-case constant time and space bound for any deque operation, including catenation. The Kaplan-Tarjan data structure and its precursors obtain confluent persistence by being strictly functional.

If all one cares about is persistence, strictly functional programming is unnecessarily restrictive. In particular, Okasaki [12, 11, 13] observed that the use of lazy evaluation in combination with memoization can lead to efficient functional (but not strictly functional) data structures that are confluently persistent. In order to analyze such structures, Okasaki developed a novel kind of debit-based amortization. Using these techniques and weakening the time bound from worst-case to amortized, he was able to considerably simplify the Kaplan-Tarjan data structure, in particular to eliminate its complicated skeleton that encodes a tree extension of a redundant digital numbering system.

In this paper we explore the problem of further simplifying the Kaplan-Tarjan result. We obtain a confluently persistent implementation of deques with catenation that has a constant amortized time bound per operation. Our structure is substantially simpler than the original Kaplan-Tarjan structure, and even simpler than Okasaki's structure: whereas Okasaki requires efficient persistent deques without catenation as building blocks, our structure is entirely self-contained. Furthermore our analysis uses a standard credit-based approach. As compared to Okasaki's method, our method requires an extension of the concept of memoization: we allow any expression to be replaced by an equivalent expression.

The remainder of this extended abstract consists of five sections. In Section 2, we introduce terminology and concepts. In Section 3, we illustrate our approach by developing a persistent implementation of deques without catenation. In Section 4, we develop our solution for deques with catenation. We conclude in Section 5 with some remarks and open problems.

2 Preliminaries

The objects of our study are lists. As in [8] we allow the following operations on lists:

MAKELIST(x): return a new list containing the single element x.

PUSH(x, L): return a new list formed by adding element x to the front of list L.

POP(L):   return a pair whose first component is the first element on list L and whose second component is a list containing the second through last elements of L.

INJECT(L, x):   return a new list formed by adding element x to the back of list L.

EJECT(L):   return a pair whose first component is a list containing all but the last element of L and whose second component is the last element of L.

CATENATE(L, R):   return a new list formed by catenating L and R, with L first.

We seek implementations of these operations (or specific subsets of them) on persistent lists: any operation is allowed on any previously constructed list or lists at any time. For discussions of various forms of persistence see [5]. A *stack* is a list on which only PUSH and POP are allowed. A *queue* is a list on which only INJECT and POP are allowed. A *steque* (*stack-ended queue*) is a list on which only PUSH, POP, and INJECT are allowed. Finally, a *deque* (*double-ended queue*) is a list on which all four operations PUSH, POP, INJECT, and EJECT are allowed. For any of these four structures, we may or may not allow catenation. If catenation is allowed, PUSH and INJECT become redundant, since they are special cases of catenation, but it is sometimes convenient to treat them as separate operations because they are easier to implement than general catenation.

We say a data structure is *strictly functional* if it can be built and manipulated using the LISP functions CAR, CONS, CDR. That is, the structure consists of a set of immutable nodes, each either an atom or a node containing two pointers to other nodes, with no cycles of pointers. The nodes we use to build our structures actually contain a fixed number of fields; reducing our structures to two fields per node by adding additional nodes is straightforward. Various nodes in our structure represent lists. To obtain our results, we extend strict functionality by allowing, in addition to CAR, CONS, CDR, the operation of replacing a node in a structure by another node representing the same list. Such a replacement can be performed in an imperative setting by replacing all the fields in the node, for instance in LISP by using REPLACA and REPLACD. Replacement can be viewed as a generalization of memoization. In our structures, any node is replaced at most twice, which means that all our structures can be implemented in a write-once memory. (It is easy to convert an algorithm that overwrites any field only a fixed constant number of times into a write-once algorithm, with only a constant factor loss of efficiency.)

To perform amortized analysis, we use a standard potential-based framework. We assign to each configuration of the data structure (the totality of nodes currently existing) a *potential*. We define the amortized cost of an operation to be its actual cost plus the net increase in potential caused by performing the operation. In our applications, the potential of an empty structure is zero and the potential is always non-negative. It follows that, for any sequence of operations starting with an empty structure, the total actual cost of the operations is bounded above by the sum of their amortized costs. See the survey paper [14] for a more complete discussion of amortized analysis.

3 Noncatenable Deques

In this section we describe an implementation of persistent noncatenable deques with a constant amortized time bound per operation. The structure is based on the analogous Kaplan-Tarjan structure [8] but is much simpler. The result presented here illustrates our technique for doing amortized analysis of a persistent data structure. At the end of the section we comment on the relation between the structure proposed here and previously existing solutions.

3.1 Representation

Here and in subsequent sections we say a data structure is *over* a set A if it stores elements from A. Our representation is recursive. It is built from bounded-size deques called *buffers*, each containing at most three elements. Buffers are of two kinds: *prefixes* and *suffixes*. A nonempty deque d over A is represented by an ordered triple consisting of a prefix over A, denoted by $pr(d)$; a (possibly empty) *child deque* of ordered *pairs* over A, denoted by $c(d)$; and a suffix over A, denoted by $sf(d)$. Each pair consists of two elements from A. The child deque $c(d)$, if nonempty, is represented in the same way. We define the set of *descendants* $\{c^i(d)\}$ of a deque d in the standard way—namely, $c^0(d) = d$ and $c^{i+1}(d) = c(c^i(d))$, provided $c^i(d)$ and $c(c^i(d))$ exist.

The order of elements in a deque is defined recursively to be the one consistent with the order of each triple, each buffer, each pair, and each child deque. Thus, the order of elements in a deque d is first the elements of $pr(d)$, then the elements of each pair in $c(d)$, and finally the elements of $sf(d)$.

In general the representation of a deque is not unique—the same sequence of elements may be represented by triples that differ in the sizes of their prefixes and suffixes, as well as in the contents and representations of their descendant deques. Whenever we refer to a deque d we actually mean a particular representation of d, one that will be clear from the context.

The pointer structure for this representation is straightforward: a node representing a deque d contains pointers to $pr(d)$, $c(d)$, and $sf(d)$. Note that, since the node representing $c^i(d)$ contains a pointer to $c^{i+1}(d)$, the pointer structure of d is essentially a linked list of its descendants. By *overwriting* $pr(d)$, $c(d)$, or $sf(d)$ with a new prefix, child deque, or suffix respectively, we mean assigning a new value to the corresponding pointer field in d. As discussed in Section 2, we will always overwrite fields in such a way that the sequence of elements stored in d remains the same and the change is only in the representation of d. By *assembling* a deque from a prefix p, a child deque y, and a suffix s, we mean creating a new node with pointers to p, y, and s.

3.2 Operations

We describe in detail only the implementation of POP; the detailed implementations of the other operations are similar. Each operation on a buffer is implemented by creating an appropriately modified new copy.

POP(d): If $pr(d)$ is empty and $c(d)$ is nonempty, then let $((x, y), c') = \text{POP}(c(d))$ and $p' = \text{INJECT}(y, \text{INJECT}(x, pr(d)))$. Overwrite $pr(d)$ with p' and $c(d)$ with c'. Then if $pr(d)$ is nonempty, perform $(x, p) = \text{POP}(pr(d))$, return x as the item component of the result, and assemble the deque component of the result from p, $c(d)$, and $sf(d)$. Otherwise, the only part of d that is nonempty is its suffix. Perform $(x, s) = \text{POP}(sf(d))$ and return x together with a deque assembled from an empty prefix, an empty child deque, and s.

Note that the implementation of POP is recursive: POP can call itself once. The implementation of EJECT is symmetric to the implementation of POP. The implementation of PUSH is as follows. Check whether the prefix contains three elements; if so, recursively push a pair onto the child deque. Once the prefix contains at most two elements, add the new element to the front of the prefix. INJECT is symmetric to PUSH.

3.3 Analysis

We call a buffer *red* if it contains zero or three elements, and *green* if it contains one or two elements. A node representing a deque can be in one of three possible states: rr, if both of its buffers are red; gr, if one buffer is green and the other red; and gg, if both buffers are green. We define $\#rr$, $\#gr$, and $\#gg$ to be the numbers of nodes in states rr, gr, and gg, respectively. Note that deques can share descendants. For instance, d and $d' = \text{POP}(d)$ can both contain pointers to the same child deque. We count each shared node only once, however. We define the potential Φ of a collection of deques to be $3 * (\#rr) + \#gr$.

To analyze the amortized cost of POP, we assume that the actual cost of a call to POP, excluding the recursive call, is one. Thus if a top level POP invokes POP recursively $k - 1$ times, the total actual cost is k.

Assume that a top level POP invokes $k - 1$ recursive POPs. The ith invocation of POP, for $1 \leq i \leq k - 1$, overwrites $c^{i-1}(d)$, changing its state from rr to gr or from gr to gg. Then it assembles its result, which creates a new node whose state (gr or gg) is identical to the state of $c^{i-1}(d)$ after the overwriting. In summary, the ith recursive call to POP, $1 \leq i \leq k - 1$, replaces an rr node with two gr nodes or a gr node with two gg nodes, and in either case decreases the potential by one. The last call, POP($c^{k-1}(d)$), creates a new node that can be in any state, and so increases the potential by at most three. Altogether, the k invocations of POP increase the potential by at most $3 - (k - 1)$. Since the actual cost is k, the amortized cost is constant.

A similar analysis shows that the amortized cost of PUSH, INJECT, and EJECT is also constant. Thus we obtain the following theorem.

Theorem 1. *Each of the operations* PUSH, POP, INJECT, *and* EJECT *on the data structure defined in this section takes* $O(1)$ *amortized time.*

3.4 Related Work

The structure just described is based on the Kaplan-Tarjan structure of [8, Section 3], but simplifies it in three ways. First, the skeleton of our structure (the sequence of descendants) is a stack; in the Kaplan-Tarjan structure, this skeleton must be partitioned into a stack of stacks in order to support worst-case constant-time operations (via a redundant binary counting mechanism). Second, the recursive changes to the structure to make its nodes green are one-sided, instead of two-sided: in the original structure, the stack-of-stacks mechanism requires coordination to keep both sides of the structure in related states. Third, the maximum buffer size is reduced, from five to three. In the special case of a steque, the maximum size of the suffix can be further reduced, to two. In the special case of a queue, both the prefix and the suffix can be reduced to maximum size two.

There is an alternative, much older approach that uses incremental recopying to obtain persistent deques with worst-case constant-time operations. See [8] for a discussion of this approach. The incremental recopying approach yields an arguably simpler structure than the one presented here, but our structure generalizes to allow catenation, which no one knows how to implement efficiently using incremental recopying. Also, our structure can be extended to support access, insertion, and deletion d positions away from the end of a list in $O(\log d)$ amortized time, by applying the ideas in [9].

4 Catenable Deques

In this section we show how to extend our ideas to support catenation. Specifically, we describe a data structure for catenable deques that achieves an $O(1)$ amortized time bound for PUSH, POP, INJECT, EJECT, and CATENATE. Our structure is based upon an analogous structure of Okasaki [13], but simplified to use constant-size buffers.

4.1 Representation

We use three kinds of buffers: *prefixes*, *middles*, and *suffixes*. A nonempty deque d over A is represented either by a suffix $sf(d)$ or by a 5-tuple that consists of a prefix $pr(d)$, a left deque of triples $ld(d)$, a middle $md(d)$, a right deque of triples $rd(d)$, and a suffix $sf(d)$. A *triple* consists of a *first middle buffer*, a deque of triples, and a *last middle buffer*. One of the two middle buffers in a triple must be nonempty, and in a triple that contains a nonempty deque both middles must be nonempty. All buffers and triples are over A. A prefix or suffix in a 5-tuple contains three to six elements, a suffix in a suffix-only representation contains one to eight elements, a middle in a 5-tuple contains exactly two elements, and a nonempty middle buffer in a triple contains two or three elements.

The order of elements in a deque is the one consistent with the order of each 5-tuple, each buffer, each triple, and each recursive deque. The pointer structure is again straightforward, with the nodes representing 5-tuples or triples containing one pointer for each field.

4.2 Operations

We describe only the functions PUSH, POP, and CATENATE, since INJECT is symmetric to PUSH and EJECT is symmetric to POP. We begin with PUSH.

PUSH(x, d):
Case 1: Deque d is represented by a 5-tuple.
1) If $|pr(d)| = 6$ then create two new prefixes p' and p'' where p' contains the first four elements of $pr(d)$ and p'' contains the last two elements of $pr(d)$. Overwrite $pr(d)$ with p' and $ld(d)$ with the result of PUSH($(p'', \emptyset, \emptyset), ld(d)$).
2) Let $p = $ PUSH($x, pr(d)$) and assemble the result from p, $ld(d)$, $md(d)$, $rd(d)$, and $sf(d)$.
Case 2: Deque d is represented by a suffix only.
If $|sf(d)| = 8$, then create a prefix p containing the first three elements of $sf(d)$, a middle m containing the fourth and fifth elements of $sf(d)$, and a new suffix s containing the last three elements of $sf(d)$. Overwrite $pr(d)$, $md(d)$, and $sf(d)$ with p, m, and s, respectively. Let $p' = $ PUSH(x, p) and assemble the result from p', $\emptyset$, m, $\emptyset$, and s. If $|sf(d)| < 8$, let $s' = $ PUSH($x, sf(d)$) and represent the result by s' only.

Note that PUSH (and INJECT) creates a valid deque even when given a deque in which the prefix (or suffix, respectively) contains only two elements. Such deques may exist transiently during a POP (or EJECT), but are immediately passed to PUSH (or INJECT) and then discarded.

CATENATE(d_1, d_2):
Case 1: Both d_1 and d_2 are represented by 5-tuples.
Let y be the first element in $pr(d_2)$, and let x be the last element in $sf(d_1)$. Create a new middle m containing x followed by y. Partition the elements in $sf(d_1) - \{x\}$ into at most two buffers s_1' and s_1'' each containing two or three elements in order, with s_1'' possibly empty. Let $ld_1' = $ INJECT($(md(d_1), rd(d_1), s_1'), ld(d_1)$). If $s_1'' \neq \emptyset$ then Let $ld_1'' = $ INJECT($(s_1'', \emptyset, \emptyset), ld_1'$); otherwise, let $ld_1'' = ld_1'$. Similarly partition the elements in $pr(d_1) - \{y\}$ into at most two prefixes p_2' and p_2'' each containing two or three elements in order, with p_2' possibly empty. Let $rd_2' = $ PUSH($(p_2'', ld(d_2), md(d_2)), rd(d_2)$). If $p_2' \neq \emptyset$ let $rd_2'' = $ PUSH($(p_2', \emptyset, \emptyset), rd_2'$); otherwise, let $rd_2'' = rd_2'$. Assemble the result from $pr(d_1)$, ld_1'', m, rd_2'', and $sf(d_2)$.
Case 2: d_1 or d_2 is represented by a suffix only.
Push or inject the elements of the suffix-only deque one by one into the other deque.

In order to define the POP operation, we define a NÄIVE-POP procedure that simply pops its argument without making sure that the result is a valid deque.

NÄIVE-POP(d): If d is represented by a 5-tuple, let $(x, p) = \text{POP}(pr(d))$ and return x together with a deque assembled from p, $ld(d)$, $md(d)$, $rd(d)$, and $sf(d)$. If d consists of a suffix only, let $(x, s) = \text{POP}(sf(d))$ and return x together with a deque represented by s only.

POP(d):
If deque d is represented by a suffix only, or if $|pr(d)| > 3$, then perform $(x, d') = \text{NÄIVE-POP}(d)$ and return (x, d'). Otherwise, carry out the appropriate one of the following three cases to increase the size of $pr(d)$; then perform $(x, d') = \text{NÄIVE-POP}(d)$ and return (x, d').
Case 1: $|pr(d)| = 3$ and $ld(d) \neq \emptyset$.
Inspect the first triple t in $ld(d)$. If either the first nonempty middle buffer in t contains 3 elements or t contains a nonempty deque, then perform $(t, l) = \text{NÄIVE-POP}(ld(d))$; otherwise, perform $(t, l) = \text{POP}(ld(d))$. Let $t = (x, d', y)$ and w.l.o.g. assume that x is nonempty if t consists of only one nonempty middle buffer. Apply the appropriate one of the following two subcases.
Case 1.1: $|x| = 3$.
Pop the first element of x and inject it into $pr(d)$. Let x' be the buffer obtained from x after the pop and let p' be the buffer obtained from $pr(d)$ after the inject. Overwrite $pr(d)$ with p' and overwrite $ld(d)$ with the result of $\text{PUSH}((x', d', y), l)$.
Case 1.2: $|x| = 2$.
Inject all the elements from x into $pr(d)$ to obtain p'. Then, if d' and y are null, overwrite $pr(d)$ with p' and overwrite $ld(d)$ with l. If on the other hand, d' and y are not null, let $l' = \text{CATENATE}(d', \text{PUSH}((y, \emptyset, \emptyset), l))$, and overwrite $pr(d)$ with p' and $ld(d)$ with l'.
Case 2: $|pr(d)| = 3$, $ld(d) = \emptyset$, and $rd(d) \neq \emptyset$.
Inspect the first triple t in $rd(d)$. If either the first nonempty middle buffer in t contains 3 elements or t contains a nonempty deque, then perform $(t, r) = \text{NÄIVE-POP}(rd(d))$; otherwise, perform $(t, r) = \text{POP}(rd(d))$. Let $t = (x, d', y)$ and w.l.o.g. assume that x is nonempty if t consists of only one nonempty middle buffer. Apply the appropriate one of the following two subcases.
Case 2.1: $|x| = 3$.
Pop an element from $md(d)$ and inject it into $pr(d)$, Let m be the buffer obtained from $md(d)$ after the pop and p the buffer obtained from $pr(d)$ after the inject. Pop an element from x and inject it into m to obtain m'. Let x' be the buffer obtained from x after the pop, and let $r' = \text{PUSH}((x', d', y), r)$. Overwrite $pr(d)$, $ld(d)$, $md(d)$, and $rd(d)$ with p, $\emptyset$, m', and r' respectively.
Case 2.2: $|x| = 2$
Inject the two elements in $md(d)$ into $pr(d)$ to obtain p. Overwrite $pr(d)$, $md(d)$, and $rd(d)$ with p, x, and r', where $r' = r$ if d' and y are empty and $r' = \text{CATENATE}(d', \text{PUSH}((y, \emptyset, \emptyset), r))$ otherwise.
Case 3: $|pr(d)| = 3$, $ld(d) = \emptyset$, and $rd(d) = \emptyset$.
If $|sf(d)| = 3$, then combine $pr(d)$, $md(d)$, and $sf(d)$ into a single buffer s and overwrite the representation of d with a suffix-only representation using s. Other-

wise, overwrite $pr(d)$, $md(d)$, and $sf(d)$ with the results of shifting one element from the middle to the prefix, and one element from the suffix to the middle.

4.3 Analysis

We call a prefix or suffix in a 5-tuple *red* if it contains either three or six elements and *green* otherwise. We call a suffix in a suffix-only representation *red* if it contains eight elements and *green* otherwise. The prefix of a suffix-only deque is considered to have the same color as the suffix. A node representing a deque can be in one of three states: rr, if both the prefix and suffix are red, gr, if one buffer is green and the other red, or gg, if both buffers are green. We define the potential Φ of a collection of deques exactly as in the previous section: $\Phi = 3 * (\#rr) + \#gr$ where $\#rr$ and $\#gr$ are the numbers of nodes that are in states rr and gr, respectively.

The amortized costs of PUSH and INJECT are $O(1)$ by an argument identical to that given in the analysis of POP in the previous section. CATENATE calls PUSH and INJECT a constant number of times and assembles a single new node, so its amortized cost is also $O(1)$.

Finally, we analyze POP. Assume that a call to POP recurs to depth k. By an argument analogous to the one given in the analysis of POP in the previous section, each of the first $k - 1$ calls to POP pays for itself by decreasing the potential by one. The last call to POP may invoke PUSH or CATENATE, and excluding this invocation has a constant amortized cost. Since the amortized cost of PUSH and CATENATE is constant, we conclude that the the amortized cost of POP is constant.

In summary we have proved the following theorem:

Theorem 2. *Our deque representation supports* PUSH, POP, INJECT, EJECT, *and* CATENATE *in* $O(1)$ *amortized time.*

4.4 Related Work

The structure presented in this section is analogous to the structures of [13, Section 8] and [7, Section 9] but simplifies them as follows. First, the buffers are of constant size, whereas in [13] and [7] they are noncatenable deques. Second, the skeleton of the present structure is a binary tree, instead of a tree extension of a redundant digital numbering system as in [7]. The amortized analysis uses the standard potential function method of [14] rather than the more complicated debit mechanism used in [13].

For catenable steques (EJECT is not allowed) we have a simpler structure that has a stack as its skeleton rather than a binary tree. It is based on the same recursive decomposition of lists as in [8, Section 4]. Our new structure simplifies the structure of [8] because we use constant size buffers rather than noncatenable stacks, and our pointer structure defines a stack rather than a stack of stacks. We will describe this structure in the full version of the paper.

5 Further Results and Open Questions

If the universe A of elements over which deques are constructed has a total order, we can extend the structures described here to support an additional heap order based on the order on A. Specifically, we can support the additional operation of finding the minimum element in a deque (but not deleting it) while preserving a constant amortized time bound for every operation, including finding the minimum. We merely have to store with each buffer, each deque, and each pair or triple the minimum element in it. For related work see [1, 2, 6, 10].

We can also support a *flip* operation on deques. A flip operation reverses the linear order of the elements in the deque: the ith from the front becomes the ith from the back, and vice-versa. For the noncatenable deques of Section 3, we implement flip by maintaining a *reversal bit* that is flipped by a flip operation. If the reversal bit is set, a push becomes an inject, a pop becomes an eject, an inject becomes a push, and an eject becomes a pop. To support catenation as well as flip we use reversal bits at all levels. We must also symmetrize the definition in Section 4 to allow a deque to be represented by a prefix only, and extend the various operations to handle this possibility. The interpretation of reversal bits is cumulative. That is, if d is a deque and x is a deque inside of d, x is regarded as being reversed if an odd number of reversal bits are set to 1 along the path of actual pointers in the structure from the node for d to the node for x. Before performing catenation, if the reversal bit of either or both of the two deques is 1, we push such bits down by flipping such a bit of a deque x to 0, flipping the bits of all the deques to which x points, and swapping the appropriate buffers and deques (the prefix and suffix exchange roles, as do the left deque and right deque). We do such push-downs of reversal bits by assembling new deques, not by overwriting the old ones.

We have devised an alternative implementation of catenable deques in which the sizes of the prefixes and suffixes are between 3 and 5 instead of 3 and 6. To achieve this we have to use two additional pointers in each node. For a node that represents a deque d, one additional pointer, if not null, points to the result of POP(d); and the other, if not null, points to the result of EJECT(d). The implementation of push and catenate is essentially as in Section 4. The changes in pop (and eject) are as follows. While popping a deque d with a prefix of size 3, if the pointer to POP(d) is not null we read the result from there. Otherwise, we carry out a sequence of operations as in Section 4 but instead of overwriting the buffers of d before creating the result we create the result and record it in the additional pointer field of the node representing d. Using a more complicated potential function than the one used in Section 4 we can show that this implementation runs in $O(1)$ amortized time per operation.

One direction for future research is to find a way to simplify our structures further. Specifically, consider the following alternative representation of catenable deques, which uses a single recursive subdeque rather than two such subdeques. A nonempty deque d over A is represented by a triple that consists of a prefix $pr(d)$, a (possibly empty) child deque of triples $c(d)$, and a suffix $sf(d)$. A *triple* consists of a nonempty *prefix*, a deque of triples, and a nonempty *suffix*, or just of a nonempty prefix or suffix. All buffers and triples are over A. The operations PUSH, POP, INJECT, and EJECT have implementations similar to their implementations in Section 4. The major difference is in the implementation of CATENATE, which for this structure requires a call to POP. Specifically, let d_1 and d_2 be two deques to be catenated. CATENATE pops $c(d_1)$ to obtain a triple (p, d', s) and a new deque c, injects $(s, c, sf(d_1))$ into d' to obtain d'' and then pushes $(p, d'', pr(d_2))$ onto $c(d_2)$ to obtain c'. The final result is assembled from $pr(d_1)$, c', and $sf(d_2)$. It is an open question whether this algorithm runs in constant amortized time per operation for any constant upper and lower bounds on the buffer sizes.

Another research direction is to design a confluently persistent representation of sorted lists such that accesses or updates d positions from an end take $O(\log d)$ time, and catenation takes $O(1)$ time. The best structure so far developed for this problem has a doubly logarithmic catenation time [9]; it is strictly functional, and the time bounds are worst-case.

References

1. A. L. Buchsbaum, R. Sundar, and R. E. Tarjan. Data structural bootstrapping, linear path compression, and catenable heap ordered double ended queues. *SIAM J. Computing*, 24(6):1190–1206, 1995.

2. A. L. Buchsbaum and R. E. Tarjan. Confluently persistant deques via data structural bootstrapping. *J. of Algorithms*, 18:513–547, 1995.

3. P. F. Dietz. Fully persistent arrays. In *Proceedings of the 1989 Workshop on Algorithms and Data Structures (WADS'89)*, pages 67–74. Springer, 1995. LNCS 382.

4. J. Driscoll, D. Sleator, and R. Tarjan. Fully persistent lists with catenation. *Journal of the ACM*, 41(5):943–959, 1994.

5. J. R. Driscoll, N. Sarnak, D. Sleator, and R. Tarjan. Making data structures persistent. *J. of Computer and System Science*, 38:86–124, 1989.

6. Hania Gajewska and Robert E. Tarjan. Deques with heap order. *Information Processing Letters*, 12(4):197–200, 1986.

7. H. Kaplan. *Purely functional lists*. PhD thesis, Department of Computer Science, Princeton University, Princeton, NJ 08544, 1997.

8. H. Kaplan and R. E. Tarjan. Persistent lists with catenation via recursive slow-down. In *Proceedings of the 27th Annual ACM Symposium on Theory of Computing (Preliminary Version)*, pages 93–102. ACM Press, 1995. Complete version submitted to Journal of the ACM.

9. H. Kaplan and R. E. Tarjan. Purely functional representations of catenable sorted lists. In *Proceedings of the 28th Annual ACM Symposium on Theory of Computing*, pages 202–211. ACM Press, 1996.

10. S. R. Kosaraju. An optimal RAM implementation of catenable min double-ended queues. In *Proc. 5th ACM-SIAM Symposium on Discrete Algorithms*, pages 195–203, 1994.

11. C. Okasaki. Amortization, lazy evaluation, and persistence: Lists with catenation via lazy linking. In *Proc. 36th Symposium on Foundations of Computer Science*, pages 646–654. IEEE, 1995.

12. C. Okasaki. Simple and efficient purely functional queues and deques. *J. Functional Progamming*, 5(4):583–592, 1995.
13. C. Okasaki. *Purely functional data structures.* PhD thesis, School of Computer Science, Carnegie Mellon University, Pittsburgh, PA 15213, 1996.
14. R. E. Tarjan. Amortized computational complexity. *SIAM J. Algebraic Discrete Methods*, 6(2):306–318, 1985.

Improved Upper Bounds for Time-Space Tradeoffs for Selection with Limited Storage

Venkatesh Raman[1] and Sarnath Ramnath[2]

[1] The Institute of Mathematical Sciences,
Chennai 600 113, India
vraman@imsc.ernet.in
[2] Department of Computer Science,
St Cloud State University,
720, 4th ave S., St Cloud MN 56303.
sarnath@eeyore.stcloud.msus.edu

Abstract. We consider the problem of finding an element of a given rank in a totally ordered set given in a read-only array, using limited extra storage cells. We give new algorithms for various ranges of extra space. Our upper bounds improve the previously known bounds in the range of space s such that s is $o(\lg^2 n)$ and $s \geq c \lg \lg n / \lg \lg \lg n$ for some constant c. We also give faster algorithms to find small ranks.

1 Introduction

Finding the median or an element of any given rank in a totally ordered set is a fundamental problem in data processing and in the computational complexity of comparison based problems. It is also a well studied theoretical problem. We consider the problem when the given data resides in a read-only array (and hence cannot be moved within the array). This is of practical importance in today's CD-ROM databases in which each record may have several keys and the database may be sorted in one of them, while we may want to select an item of a particular rank in some other key. The restriction on the data movements also gives rise to interesting time-space tradeoffs for selection.

Munro and Paterson [3] were the first to consider the time-space tradeoff for selection. They designed algorithms assuming sequential access to the input, and their measure of time was the number of passes over the input. Frederickson [2] later looked at the problem in a random access model, and reworked the cost in terms of the number of comparisons made. He also improved some of the upper bounds. The current best upper bound when s, the extra space available, is $\Omega(\lg^2 n)^1$ is $O(n \lg n / \lg s + n \lg^* s)$. Munro and Raman [4] later considered the problem when very few ($O(\lg n)$) extra cells were available. The problem seems to become harder with this further limitation on extra space. Their algorithm made $O(2^s s! n^{1+1/s})$ comparisons using $1 \leq s \leq O(\sqrt{\frac{\lg n}{\lg \lg n}})$

---

[1] lg denotes logarithm to the base 2 throughout the paper.

additional storage locations. In the intermediate range, the best known algorithm has complexity $O(n^2/s)$ [3]. In contrast the only lower bound known for the problem is $\Omega(n \lg n / \lg s)$ [2] due to an adversary strategy given by Munro and Paterson [3].

In this paper we give new algorithms when the space range is $O(\lg^2 n)$. When s is $\Omega(\lg n)$ but $o(\lg^2 n)$, our algorithm improves drastically the best known algorithm in this range which has the time complexity $O(n^2/s)$. When s is $O(\lg n)$, our algorithm (which takes $O(sn^{1+1/s} \lg n)$ time) improves that of Munro and Raman [4] as long as $2^s s! > s \lg n$ (this happens for a broad range of s, for example, when $s \geq c \lg \lg n / \lg \lg \lg n$ for some positive constant c).

Throughout the paper, we assume that all input values are distinct. When we describe recursive algorithms, we omit the end condition, and the base value of the recurrence relations for the time or space complexity analysis. The end conditions are usually obvious, and assuming a small constant for the base value will suffice for solving the recurrence relations. Our measure of time is the number of comparisons made by the algorithm, and of space is the number of extra locations besides the input array used by the algorithm. These locations, occasionally refered to as cells or registers are capable of holding an input element, or an index to a location of the input array or any value upto n.

In the next section, we give our new algorithm for selection that uses $\Theta(\lg n)$ space and $O(n \lg^2 n)$ time. This algorithm uses a technique to select an approximate median pair - a pair of elements that partitions the given list into three parts each of size at most half the original size. This algorithm may be of independent interest for selection problems.

In Section 3, the algorithm that uses $\Theta(\lg n)$ space is generalized to situations where more (or less) extra space is available, and also to the scenario when multiple ranks have to be selected in the list residing in read-only memory. Section 4 gives improved algorithms to find small ranks. These algorithm are quite efficient to find an item of small rank and their running time match that of the best known algorithm when the item to be selected is the median element. Section 5 gives a summary of our results and lists an open problem.

2   Selection Using $\Theta(\lg n)$ Space

Let A be the given array of n keys in read-only memory in which we want to find the kth smallest element for some k. Let lo and hi be the smallest and the largest element in A respectively. We define *an approximate median pair* for A as a pair of elements in A, l and u, $l < u$ such that the number of key values in A in the ranges $[lo...l)$, $(l..u)$ and $(u..hi]$ are all strictly less than $\lceil n/2 \rceil$. The partitions created by l and u will be called *buckets*.

Our first observation is that from the approximate median pairs of the two halves of the given list, we can find an approximate median pair of the entire list in linear time using constant space.

Lemma 1. *Given two arrays A_1 and A_2 in a read-only memory with approximate median pairs (l_1, u_1) and (l_2, u_2) respectively, an approximate median pair for $A_1 \cup A_2$ can be computed in $O(|A_1| + |A_2|)$ time using $O(1)$ space.*

Proof. First compare all elements of A_1 and A_2 with l_1, l_2, u_1 and u_2, and compute the number of elements of $A_1 \cup A_2$ in each of the five buckets in the partition induced by l_1, l_2, u_1 and u_2. Since each of these buckets can contain items from at most one bucket of A_1 and at most one bucket of A_2, the number of items in each of these buckets is strictly less than $\lceil |A_1|/2 \rceil + \lceil |A_2|/2 \rceil$.

First we claim that this number (of items in each bucket) is actually strictly less than $\lceil (|A_1| + |A_2|)/2 \rceil$. This claim is obvious if at most one of $|A_1|$ and $|A_2|$ is odd. But, if both $|A_1|$ and $|A_2|$ are odd, then the number of elements of A_1 in the corresponding bucket is at most $(|A_1| - 1)/2$ and the number of elements from A_2 that can be in that bucket is at most $(|A_2| - 1)/2$; so together there can be at most $(|A_1| + |A_2|)/2 - 1$ elements in that bucket, and the claim follows.

Now scan these four elements in the increasing order of their values, and let l be the last of these elements such that the number of elements of $A_1 \cup A_2$ less than l is strictly less than $\lceil (|A_1| + |A_2|)/2 \rceil$. Let u be the element (out of l_1, l_2, u_1, u_2) that immediately succeeds l in the increasing order (if l is the last of these four, choose $u = l$ and l to be the element preceding it in the sorted order of l_1, l_2, u_1, u_2). We claim that l and u form an approximate median pair for the set $A_1 \cup A_2$. Since the number of elements in each of the five buckets is less than $\lceil (|A_1| + |A_2|)/2 \rceil$, it suffices to show that the number of elements greater than u is at most $\lceil (|A_1| + |A_2|)/2 \rceil$. If, in our initial choice of l, l was the last out of the four in order, then the claim follows easily. Otherwise, from the way u was chosen, the number of elements less than u is at least $\lceil (|A_1| + |A_2|)/2 \rceil$. It follows that the number of elements greater than u is strictly less than $\lceil (|A_1| + |A_2|)/2 \rceil$. $\blacksquare$

If A_1 contains the first $\lfloor n/2 \rfloor$ and A_2, the remaining $\lceil n/2 \rceil$ elements of the given list, then by recursively computing the approximate median pairs of A_1 and A_2 and computing from them, an approximate median pair for A using Lemma 1, the following Lemma follows.

Lemma 2. *Given a list of n elements in a read-only memory, an approximate median pair for the list can be found in $O(n \lg n)$ time using $O(\lg n)$ extra space.*

Once we find an approximate median pair, by comparing all elements with that pair and keeping a count of the bucket sizes, we can figure out in which bucket the element we are seeking (the element of rank k for an appropriate k) lies. As the size of each bucket is at most $\lceil n/2 \rceil$, we can recurse on those elements of interest. However, since we are working in read-only memory, we cannot bring together all these elements of interest. The trick to identify them is to use a pair of elements called *filters* (using the terminology of Munro and Paterson [3]). Let us call an element *active* if it is a possible candidate for the element we seek (i.e. it has not been ruled out by our algorithm). We call a pair of elements x and y, $x < y$, *filters*, if every active element of the input array is greater than x and smaller than y. Initially all the input elements are active, and the minimum and maximum element (assuming the element we seek is not either of these) form the initial filters. At each iteration of the algorithms we propose, several active elements of the list are eliminated (which is recorded by changing the filter values). Note however, that the entire list has to be scanned to identify the active elements.

After finding an approximate median pair, one or both the elements of the pair replace one or both (respectively) elements of the filters (if one of the elements of this pair is not what we seek). It can be seen that Lemma 2 can be generalized to the following.

Lemma 3. *Given a list of n elements in a read-only memory in which only m of them are active, an approximate median pair (l, u), $l < u$, that partitions the active elements into three buckets, each of which has size less than $\lceil m/2 \rceil$ can be found in $O(n \lg m)$ time using $O(\lg m)$ extra space.*

The next result follows immediately.

Theorem 1. *Given a list of n elements in a read-only memory, the kth smallest element of the list, for any k, can be found in $O(n \lg^2 n)$ time using $O(\lg n)$ extra space.*

Note that this improves the earlier known algorithm [4] which has a runtime of $n \lg^{\omega(1)} n$ when $O(\lg n)$ indices are available. Though that algorithm uses only $O(\sqrt{\lg n / \lg \lg n})$ indices, it was not clear how to beat the runtime using $\Theta(\lg n)$ indices.

3 Some Generalizations of the Main Algorithm

We generalize our selection algorithm in three directions: using less space, using more space and for selecting multiple ranks.

3.1 Algorithm Using $O(\lg n)$ Space

Let p, $p \geq 2$ be an integer parameter We explain how we can find an approximate median pair using $O(\lg n / \lg p)$ space albeit taking $O(np \lg n / \lg p)$ time.

Divide the array into p blocks $A_1, A_2, ... A_p$ of size $\lceil n/p \rceil$ each (except for the last which may have fewer elements). Find an approximate median pair of the first block recursively in the first step. We describe the algorithm inductively. In the ith step, $1 \leq i \leq p - 1$ we have an approximate median pair for the block defined by $A_1 \cup A_2 \cup ... \cup A_i$. In the $(i + 1)$st step, find an approximate median pair of the block A_{i+1} recursively and then find an approximate median pair for the block defined by $A_1 \cup A_2 \cup ... \cup A_{i+1}$ using the approximate median pair for A_{i+1} and that of $A_1 \cup A_2 \cup ... \cup A_i$ using Lemma 1. Thus in p steps, we will have an approximate median pair of the entire array. The total time required $T(n)$ satisfies the recurrence: $T(n) = pT(n/p) + O(n/p)(2 + 3 + ... + p - 1)$ or $T(n) = pT(n/p) + O(np)$ which solves to $T(n) = O(np \lg n / \lg p)$. The space requirement, $S(n)$ satisfies the recurrence: $S(n) = S(n/p) + O(1)$ which solves to $S(n)$ is $O(\lg n / \lg p)$. Note that the constant factors in the 'O' terms are independent of p. We now have the following lemma.

Lemma 4. *Given a list of n elements in a read-only memory, an approximate median pair of the list can be found in $O(np \lg n / \lg p)$ time using $O(\lg n / \lg p)$ extra space where $p \geq 2$ is any integer parameter.*

It is easy to see the following generalization of the above lemma and the theorem that follows.

Lemma 5. *Given a list of n elements in a read-only memory out of which only m are active, an approximate median pair of the active elements can be found in $O(np \lg m / \lg p)$ time using $O(\lg m / \lg p)$ extra space where $p \geq 2$ is any integer parameter.*

Theorem 2. *Given a list of n elements in a read-only memory, the kth smallest element for any k can be found in $O(np \lg^2 n / \lg p)$ time using $O(\lg n / \lg p)$ extra space where $p \geq 2$ is any integer parameter.*

By letting $p = \lfloor n^{1/s} \rfloor$, we get the following corollary.

Corollary 1. *Given a list of n elements in a read-only memory, the kth smallest element of the list can be found in $O(sn^{1+1/s} \lg n)$ time using $O(s)$ extra space where $s \geq 1$ is any parameter.*

Note that this algorithm improves drastically the algorithm of Munro and Raman [4], when the space available s is such that $2^s s! > s \lg n$ (this happens, for example, when $s \geq c \lg \lg n / \lg \lg \lg n$ for some positive constant c).

3.2 Algorithm Using $\Omega(\lg n)$ Space

If we have more space available, we can compute, instead of an approximate median pair, a set of $(2p - 2$, say) splitters which partitions the given array into $2p - 1$ buckets ($p \geq 2$), each containing less than $\lceil n/p \rceil$ items. We then combine this idea with that of merging several blocks at a time, in order to get an improvement in the running time. The idea is to first generalize Lemma 1.

Lemma 6. *Let $R_1, R_2, ... R_p$ be p arrays in a read-only memory, with each R_i, $1 \leq i \leq p$, having $2p - 2$ splitters $r_1^i, r_2^i, ..., r_{2p-2}^i$ such that every bucket in the partition induced by these splitters contains fewer than $\lceil |R_i|/p \rceil$ items. We can compute a set of $2p - 2$ splitters $s_1, s_2, ..., s_{2p-2}$ for $S = R_1 \cup R_2 \cup ... \cup R_p$ such that each bucket of S contains fewer than $\lceil |S|/p \rceil$ items, in $O(|S| \lg p)$ time using $O(p^2)$ space.*

Proof. We describe an algorithm that computes the splitters. Sort the splitters of the R_i's and then do a binary search on these splitters for each item in S, so that for every pair of adjacent splitters (x, y) (in sorted order) we know the number of items of S that lie between x and y. This takes $O(|S| \lg p)$ time. Let lo and hi be the smallest and the largest element of the list S. Scan the sorted array of splitters and choose $s_1, s_2, ..., s_{2p-2}$ as follows:

- Choose s_1 to be the largest splitter such that the number of items of S in the range $[lo..s_1)$ is less than $\lceil |S|/p \rceil$. Choose s_2 be the splitter immediately following s_1 in the sorted order of the splitters.
- For $i = 1$ to $(p-2)$, choose s_{2i+1} to be the largest splitter such that the number of items in the bucket $(s_{2i}..s_{2i+1})$ is less than $\lceil |S|/p \rceil$; choose s_{2i+2} to be the splitter immediately following s_{2i+1} in the sorted order of the splitters.

All buckets of the form $(s_{2i}..s_{2i+1})$ contain fewer than $\lceil |S|/p \rceil$ items, by construction. Consider a bucket of the form $(s_{2i}..s_{2i+1})$. It is formed by choosing adjacent pairs of splitters. Since each of the R_i's can contribute at most only one of their buckets to this bucket of S, the bucket size must be less than $\sum_{i=1}^{p} \lceil |R_i|/p \rceil$. We show, in fact, that the bucket size is less than $\lceil |S|/p \rceil$. Let a_i be the size of the bucket in R_i that contributes to the bucket in question, then we know that $a_i < \lceil |R_i|/p \rceil$ and so $a_i \le \lceil |R_i|/p \rceil - 1$. Therefore,

$$\sum_{i=1}^{p} a_i \le \sum_{i=1}^{p} (\lceil |R_i|/p \rceil) - p$$
$$\le \sum_{i=1}^{p} (|R_i|/p + (p-1)/p) - p$$
$$= \sum_{i=1}^{p} (|R_i|/p) - 1 = (|S|/p - 1)$$
$$< \lceil |S|/p \rceil$$

and the claim follows. Now all that remains to show is that the bucket $(s_{2p-2}..hi]$ contains fewer than $\lceil |S|/p \rceil$ items. By our selection process, every range of the form $(s_{2i}..s_{2i+2})$ contains at least $\lceil |S|/p \rceil$ items. Therefore the range $(lo..s_{2p-2})$ contains at least $(p-1) \times \lceil |S|/p \rceil$ items, and therefore we are left with no more than $\lceil |S|/p \rceil$ items to be placed in the last bucket. It follows then that the bucket $(s_{2p-2}..hi]$ contains fewer than $\lceil |S|/p \rceil$ items. ∎

Now, by dividing the given list into p blocks of size $\lceil n/p \rceil$ each, we can find the $2p-2$ splitters of each block recursively. Then we can combine the splitters of the p blocks to find $2p-2$ splitters for the given list using the previous lemma. The time $T(n)$ taken for this satisfies $T(n) = pT(n/p) + O(n \lg p)$ which solves to $O(n \lg n)$. The space requirement $S(n)$ satisfies $S(n) = S(n/p) + O(p^2)$ which solves to $O(p^2 \lg n / \lg p)$. Thus we have the following lemma.

Lemma 7. *Given a list of n elements in a read-only memory, a sequence of $2p-2$ elements of the list with the property that the number of elements in each bucket is at most $\lceil n/p \rceil$ can be found in $O(n \lg n)$ time using $O(p^2 \lg n / \lg p)$ extra space where $p \ge 2$ is an integer parameter.*

The above lemma generalizes to the following:

Lemma 8. *Given a list of n elements in a read-only memory out of which m are active, a sequence of $2p - 2$ elements of the list with the property that the number of elements in each bucket is at most $\lceil m/p \rceil$ can be found in $O(m \lg m + n \lg m / \lg p)$ time using $O(p^2 \lg m / \lg p)$ extra space where $p \geq 2$ is an integer parameter.*

Proof. Note that in the proof of Lemma 6, only the active elements need to be compared with the splitters of the blocks. However, to identify the active elements $2n$ comparisons (with the filters) are made in the worst case. Thus the recurrence, as in the previous lemma, for time and space in this scenario, satisfy $T(m) = pT(m/p) + O(n + m \lg p)$ which solves to $O(n \lg m / \lg p + m \lg m)$. The space requirement $S(m)$ satisfies $S(m) = S(m/p) + O(p^2)$ which solves to $O(p^2 \lg m / \lg p)$. ∎

Now, after the first step where we find the $2p - 2$ splitters for the list, we compare all active elements with the splitters and identify the bucket to recurse and appropriately modify the filters. Since there will be at most $\lceil n/p \rceil$ active elements, the selection can be completed in $O(\lg n / \lg p)$ iterations. The following theorem follows from the above lemma.

Theorem 3. *Given a list of n elements in a read-only memory, the kth smallest element of the list can be found using $O(n \lg n + n \lg^2 n / \lg^2 p)$ time using $O(p^2 \lg n / \lg p)$ extra space where $p \geq 2$ is an integer parameter.*

When $p = \lceil \sqrt{\lg n \lg \lg n} \rceil$, this gives an $O(n \lg^2 n / (\lg \lg n)^2)$ algorithm using $O(\lg^2 n)$ indices, which is a $\lg n / \lg \lg n$ factor worse than the algorithm of Frederickson [2] which takes $O(n \lg n / \lg \lg n)$ time using the same space. However when the number of available indices is $o(\log^2 n)$ and $\Omega(\lg n)$, ours is the best known algorithm.

3.3 Multiselect and Sorting with Minimum Data Movement

In this section, we consider the problem of selecting from the given list in a read-only memory, elements whose ranks are $r_1, r_2, ... r_m$ where $1 \leq r_1 < r_2 < ... r_m \leq n$. First we outline a method to solve the following problem.

Given a list A of n elements in a read-only memory and a set of k intervals $[x_1, y_1], [x_2, y_2], ... [x_k, y_k]$, $x_1 \leq y_1 \leq x_2 \leq y_2 ... \leq x_k \leq y_k$, find the approximate median pairs of the interval i, for all $1 \leq i \leq k$; interval i consists of all elements of the given list that lie between x_i and y_i.

The idea is to partition the given list A into two parts A_1 and A_2 and to recursively compute approximate median pairs of the intervals in A_1 as well as in A_2. Then we combine (as in Lemma 1) the answers to obtain an approximate median pairs of all intervals in A. We get the following result (see [5] for details).

Lemma 9. *Given a list A of n elements in a read-only memory and a set of k intervals $[x_1, y_1], [x_2, y_2], ... [x_k, y_k]$, approximate median pairs of the elements of A in each interval can be found in $O(n \lg n \lg k)$ time using $O(k \lg n)$ extra space.*

By repeated applications of the above lemma, we can do a multiselect in a read-only memory, giving us the following result(see [5] for details).

Theorem 4. *Given a list A of size n in a read-only memory, and m integers, $1 \le r_1 < r_2 < .... < r_m \le n$, the items $a_1, ...a_m$ with respective ranks in A as $r_1, r_2, ...r_m$ can be found in $O(n \lg^2 n \lg m)$ time using $O(m \lg n)$ extra space.*

Though 'multiselect' is a problem of independent interest, we show that we can use our multiselect algorithm to sort a given list with the minimum number of data moves. Though $\Omega(n)$ moves are required to sort any given list of n elements in the worst case, the lower bound may differ based on the given input. In [4] an exact lower bound for the number of data moves required to sort any *given* list was given, and some cycle chasing algorithms were developed for sorting with this exact number of moves. We can use our multiselect algorithm to chase several cycles simultaneously, yielding the following result (see [5] for details).

Theorem 5. *An array of size n can be sorted with the minimum number of data moves using $O((n^2 \lg^2 n \lg m)/m)$ comparisons and $O(m \lg n)$ space.*

This drastically improves the algorithm of Munro and Raman (Theorem 8 of [4]) for sorting with minimum data movement when m is $\omega(1)$. For example, when $O(\sqrt{n})$ indices are available, our algorithm uses $O(n^{1+1/2} \lg^4 n)$ comparisons whereas the algorithm in [4] takes $\Theta(n^2)$ comparisons.

4 Faster Algorithms for Selecting Small Ranks

In this section we develop selection algorithms whose time or space complexity is a function of k, the rank of the item to be selected. These are particularly useful for selecting items of small rank. When the rank reaches $n/2$, the complexity of these algorithms match those in the earlier sections.

We begin with the following easy theorem.

Theorem 6. *The kth smallest item from a list of size n in a read-only memory can be found in $O(n)$ time using $O(k)$ space.*

Proof. The algorithm is as follows:

- Partition the given list into $\lceil n/k \rceil$ blocks of size k each. Copy the elements of the first block, and designate the k items as the *first sample*.
- In the ith step, $2 \le i \le \lceil n/k \rceil$, copy the ith block and find the kth smallest among those in the ith block and the $(i-1)$th sample using any linear selection algorithm. Choose those elements smaller than and including the kth smallest found as the ith sample.

It is easily seen that *ith sample* contains the k smallest items from the first i blocks; it follows therefore that the kth (or the largest) item in the $\lceil n/k \rceil$th sample is the answer. Each step takes $O(k)$ time using $O(k)$ storage cells establishing the time and space bounds. ∎

The space requirement of this algorithm can be reduced if we do not store all the k smallest items seen so far, but only some evenly spaced sample (of size $2p-2$ for some $p \geq 2$, say) from this set. Once again, we shall merge samples as we go along, discarding elements in a careful manner. The details are as follows:

1. Partition the given list into $\lceil n/k \rceil$ blocks of size k each. ¿From within the first block choose a sample of size $2p-2$ such that each of the $2p-1$ buckets contains less than $\lceil k/p \rceil$ items using Lemma 7. Sort these $2p-2$ items; designate this sequence as the *first sample*. Along with each of the sample item, associate a weight which indicates *approximately* the number of elements in the bucket preceding it. (For the *first* sample these weights are the *exact* number of elements in the preceding bucket.)

2. For $i := 2 \ldots \lceil n/k \rceil$, compute the ith sample as follows: pick a sample of $2p-2$ items from the ith block (as described in step 1) along with the weights and merge those elements with the $(i-1)$st sample. Note that each item in the merged sequence has an associated weight (though the weights do *not* necessarily denote *exactly* the number of items in the bucket preceding it among the blocks considered so far, except in the first sample). Now pick the ith sample as follows: walk along the merged sequence computing the total weight until we reach an item a_i such that total weight up to a_{i-1} is less than $\lceil k/p \rceil$ and weight up to a_i is at least $\lceil k/p \rceil$. Include both a_{i-1} and a_i in the sample, reset the weight of a_{i-1} to the cumulative weight up to that element, and start with a_{i+1}, setting total weight back to zero. Continue this process until the sum of the weights of all the sample items chosen is at least k.

3. $max :=$ largest item in the $\lceil n/k \rceil$th sample. Designate all items less than max as *active* items by appropriately modifying the filters. Recursively select the kth item from this sample.

Lemma 10. *The number of items of the given list less than the largest item in the $\lceil n/k \rceil$th sample is at most $k + O(n/p)$ and at least k.*

Proof. The lower bound follows trivially as the cumulative weight at an item gives a lower bound for the number of items less than the element. The upper bound follows from the following two observations:

- In the ith sample, each bucket contains less than $i * \lceil k/p \rceil$ items among the first i blocks.
- The maximum element of the lth sample is less than at most $k + l\lceil k/p \rceil$ items of the first l blocks.

See [5] for details. ∎

The algorithm from the previous section (Lemma 7) enables us to find the required sample in $O(k \lg k)$ time using $O(p^2 \lg k / \lg p)$ space for each block. We can therefore find max in $O(n \lg k)$ time using $O(p^2 \lg k / \lg p)$ space. Thus we have

Lemma 11. *Given a list of n elements in a read-only memory, an element whose rank is at least k and at most $k + O(n/p)$ can be found in $O(n \lg k)$ time using $O(p^2 \lg k / \lg p)$ space where $(k+1)/2 \geq p \geq 2$ is an integer parameter.*

Furthermore, if there are only m active elements in the list, then the number of blocks in the first step would be $\lceil m/k \rceil$. Applying Lemma 8 for each block, we have

Lemma 12. *Given a list of n elements in a read-only memory out of which only m elements are active, an element whose rank in the set of active elements is at least k and at most $k + O(m/p)$ can be found using $O(m \lg k + n \lg k / \lg p)$ time using $O(p^2 \lg k / \lg p)$ extra space where $(k+1)/2 \geq p \geq 2$ is an integer parameter.*

Thus applying the above lemma $O(\lg n / \lg p)$ times, the number of active elements will reduce to $O(k \lg n / \lg p)$. Applying the above lemma another $O(\lg k / \lg p)$ times, the number of active elements will reduce further to $O(k \lg k / \lg p)$. At this point, applying a version of Theorem 3 (when the number of active elements is $O(k \lg k / \lg p)$), we get the following theorem.

Theorem 7. *The kth smallest item from a list of n elements in read-only memory can be selected in $O(n \lg k + n \lg n \lg k / \lg^2 p)$ time using $O(p^2 \lg k / \lg p)$ extra space where $(k+1)/2 \geq p \geq 2$ is an integer parameter.*

Setting $p = 2$ (we find approximate median pairs in each block), we have

Corollary 2. *The kth smallest item from a list of n elements in read-only memory can be selected in $O(n \lg n \lg k)$ time using $O(\lg k)$ extra space.*

If the amount of extra space s, is $o(\lg k)$, then we can find an approximate median pair of each block of size k using the algorithm of Lemma 4 (set p to $\lfloor n^{1/s} \rfloor$ there) rather than Lemma 2. In that case, the time taken to find an approximate median pair for each block would be $O(sk^{1+1/s})$ and to find max is $O(nsk^{1/s})$. This yields the following theorem.

Theorem 8. *The kth smallest item from a list of n elements in a read-only memory can be found in $O(nsk^{1/s} \lg n)$ time using $O(s)$ space where $1 \leq s \leq \lg k$ is a parameter.*

Note that the bounds in the above two theorems are better for small k, and match the bounds in the last section if k reaches $n/2$.

4.1 An $O(n \lg^* n)$ Algorithm with $\Theta(n^\epsilon)$ Space

Suppose we want to find the kth smallest element in the given list using $\Theta(\sqrt{n \lg n})$ space. If k is $O(\sqrt{n \lg n})$, then Theorem 6 would give an $O(n)$ algorithm. Otherwise, Theorem 7 in the last section would yield an $O(n \lg n)$ algorithm. Here we derive a new sampling technique which is applied to obtain an $O(n \lg^* n)$ time algorithm for this special case, matching the best known algorithm [2] for this range of space.

Our algorithm selects the kth smallest element in $O(cn)$ time using $O(\sqrt{n \lg n})$ space, as long as k is no more than $n / \lg^{(c)} n$, for some parameter c. (Here, $\lg^{(c)} n$ denotes the cth iterate of the lg function.) In particular, if c is a constant, we have a linear time algorithm. The value c could at most be $\lg^* n$, in which case we get an $O(n \lg^* n)$ algorithm.

In order to facilitate the discussion, we ignore various necessary integer rounding up and down. This does not affect the asymptotic analysis.

The main idea behind the algorithm is captured in the following lemma.

Lemma 13. *Given a list containing n items in a read-only memory and an integer k, $1 \leq k \leq n$, we can find an item of rank at least k and at most $2k + \sqrt{n \lg n}$ in $O(n)$ time using $O(\sqrt{n \lg n})$ space.*

Proof. The algorithm is as follows:

Step 1: Divide the list into roughly $\sqrt{n / \lg n}$ blocks of size approximately $\sqrt{n \lg n}$ each. For each block, choose a $\lg n$ sized sample $x_0, x_1, x_2, ... x_{\lg n}$, where x_i is the item in the block that has a rank 2^i. Assign $weight(x_i) = 2^{i-1}$, $i > 0$, and $weight(x_0) = 1$.

Step 2: This gives us $\sqrt{n / \lg n}$ samples of size $\lg n$ each. Collect all these samples and sort them. Let the sorted sequence be $y_1, ... y_m$, where $m = \sqrt{n \lg n}$. Find the least i such that $\Sigma_{j=1}^{i}(weight(y_j)) \geq k$. Choose y_i as the required element.

The space bound is obvious. The time bound follows from the fact that each sample in Step 1 can be computed in $O(\sqrt{n \lg n})$ time and $O(\sqrt{n \lg n})$ space by first copying the indices of those elements into the extra storage, and then doing a standard multiselect.

We need to establish that y_i satisfies the required property. Observe that the *weight* of each item y in the sample represents the size of a set of elements from the given list that are of value at most y, and all these sets are pairwise disjoint. It immediately follows that $rank(y_i) \geq k$.

Assume without loss of generality, that y_i belongs to the sample selected from the first block. We shall derive an upper bound on the number of items in all of the other blocks that could have value at most y_i. Consider some block j, $2 \leq j \leq \sqrt{n / \lg n}$, with the sample elements $x_0^j, ... x_p^j$ where $p = \lg n$. Let $x_{l_j}^j$ be such that $x_{l_j}^j \leq y_i \leq x_{l_j+1}^j$. Clearly, $\Sigma_{j=2}^{p}\Sigma_{m=1}^{l_j}(weight(x_m^j)) \leq k$ from the way y_i was chosen. ¿From the manner in which the samples are chosen, it follows that $\Sigma_{j=2}^{p}\Sigma_{m=1}^{l_j+1}(weight(x_m^j)) \leq 2k$. Now, since $x_{l_j+1}^j \geq y_i$, no item in block j that is of value greater than $x_{l_j+1}^j$ can have value at most y_i. It follows that at most $2k$ items from the blocks $2, 3, ... p$ can have value at most y_i. Since the first block contains at most $\sqrt{n \lg n}$ items, the result follows. ∎

It follows therefore that we can reduce the number of active items to $2k + \sqrt{n \lg n}$ within the time and space bounds given in the above lemma. At this point, we can switch over to the algorithm presented in [2] and select the desired item using no more than c iterations, where c is a parameter such that $k \leq n/\lg^{(c)} n$. We therefore get the following result:

Lemma 14. *The kth smallest item can be selected from a list of size n in a read-only memory in $O(cn)$ time using $O(\sqrt{n \lg n})$ space, if $k \leq n/\lg^{(c)} n$.*

More generally we have,

Theorem 9. *The kth smallest item can be selected from a list of size n in a read-only memory in $O(n(\lg^* n - \lg^*(n/k)))$ time and $O(\sqrt{n \lg n})$ space.*

5 Conclusions

We have considered the selection problem when the given list of elements reside in a read-only memory and hence cannot be moved within the list. We gave two families of algorithms using $O(s)$ indices, one for the range $1 \leq s \leq \lg n$ and the other for the range $\lg n \leq s \leq \lg^2 n$. Though our algorithms are slightly worse than that of Munro and Raman [4] when s is a constant, they are the best known for large values of s (roughly in the range $o(\lg^2 n) > s \geq c(\lg \lg n)/\lg \lg \lg n$ for some positive constant c). The immediate open question is whether our bounds are tight. Though near optimal bounds ([3], [2]) are known for selection in read-only memory when the space available is $\Omega(\lg^2 n)$, the correct bounds for the lower range of space are not clear.

We also developed improved algorithms to select small ranks. These algorithms are quite efficient to find an element of small rank, and they match the performance of the best known selection algorithm in this model when the item to be selected is the median element. Our algorithm in the last section can be used to develop a fast parallel algorithm in the EREW PRAM to select items of small rank. This algorithm takes $O(\lg n(\lg^* n - \lg^*(n/k)))$ time using $O(n)$ work and $O(n/\lg n)$ processors on an EREW PRAM to select the kth smallest item from a list of n items. This bound matches Cole's $O(\lg n \lg^* n)$ time bound[1] when we are seeking the median element. See [5] for details.

References

1. R. J. Cole, An optimally efficient selection algorithm, *Information Processing Letters* **26** (6) (1988) 295 – 299.
2. G. N. Frederickson, Upper bounds for time-space trade-offs in sorting and selection, *Journal of Computer and System Sciences* **34** (1987) 19 – 26.
3. J. I. Munro and M. S. Paterson, Selection and sorting with limited storage, *Theoretical Computer Science* **12** (1980) 315 – 325.
4. J. I. Munro and V. Raman, Selection from read-only memory and sorting with minimum data movement, *Theoretical Computer Science* **165** (1996) 311 – 323.
5. V. Raman and S. Ramnath, "Improved Upper Bounds for Time-Space Tradeoffs for Selection with Limited Storage", TR98/04/17, Institute of Mathematical Sciences, Chennai, India.

Probabilistic Data Structures for Priority Queues (Extended Abstract)

R.Sridhar, K.Rajasekar and C.Pandu Rangan *

Department of Computer Science and Engineering, IIT Madras 600036, India.

Abstract. We present several simple probabilistic data structures for implementing priority queues. We present a data structure called simple bottom-up sampled heap (SBSH), supporting insert in $O(1)$ expected time and delete, delete minimum, decrease key and meld in $O(\log n)$ time with high probability. An extension of SBSH called BSH1, supporting insert and meld in $O(1)$ worst case time is presented. This data structure uses a novel "buffering technique" to improve the expected bounds to worst-case bounds. Another extension of SBSH called BSH2, performing insert, decrease key and meld in $O(1)$ amortized expected time and delete and delete minimum in $O(\log n)$ time with high probability is also presented. The amortized performance of this data structure is comparable to that of Fibonacci heaps (in probabilistic terms). Moreover, unlike Fibonacci heaps, each operation takes $O(\log n)$ time with high probability, making the data structure suitable for real-time applications.

Keywords : priority queue, probabilistic data structures, decrease key, meld, skip list, bottom-up sampling, amortization, buffering technique

1   Introduction

The implementation of priority queues is a classical problem in data structures [2, 3, 4, 5, 6, 8]. Priority queues are extensively used in many algorithms for various applications like network optimization and task scheduling [8].

Deterministic data structures achieving best performance in the amortized and in the worst-case sense are reported in [8] and [3] respectively. These data structures support delete and delete minimum in $O(\log n)$ time and insert, decrease key and meld in $O(1)$ time. However the latter data structure is extremely complicated and may not be of much practical importance, as quoted in [3].

Probabilistic alternatives to deterministic data structures are better in terms of simplicity and constant factors involved in actual run-time. Skip lists [10] and randomized treaps [1] are examples of data structures supporting dictionary maintenance with performance comparable to more complicated deterministic data structures like AVL trees [7] and red-black trees [13]. Although skip lists are proposed to implement the dictionary operations, it is not hard to see that delete minimum can be executed in $O(1)$ expected time as the minimum is stored

---

* Corresponding Author. E-mail : rangan@iitm.ernet.in

as the first item in the skip list. However, no probabilistic data structure implementing all the operations supported by (say) Fibonacci heaps, in comparable (probabilistic) time has been reported.

We first propose a data structure called simple bottom-up sampled heap (SBSH) that uses bottom up sampling, as in skip lists. We then propose two extensions of SBSH called BSH1 and BSH2. The performance of these data structures are compared with existing deterministic counterparts in table 1.

Table 1. Comparison of various data structures for priority queues

	delete minimum	delete	insert	decrease key	meld
Binomial heaps [6] (**worst case**)	$O(\log n)$	$O(\log n)$	$O(\log n)$	$O(\log n)$	$O(\log n)$
Fibonacci heaps [8] (**amortized**)	$O(\log n)$	$O(\log n)$	$O(1)$	$O(1)$	$O(1)$
Fast meldable heaps [2] (**worst case**)	$O(\log n)$	$O(\log n)$	$O(1)$	$O(\log n)$	$O(1)$
Brodal [3] (**worst case**)	$O(\log n)$	$O(\log n)$	$O(1)$	$O(1)$	$O(1)$
SBSH (this paper)	$O(\log n)$ whp	$O(\log n)$ whp	$O(1)$ exp	$O(\log n)$ whp	$O(\log n)$ whp
BSH1 (this paper)	$O(\log n)$ whp	$O(\log n)$ whp	$O(1)$	$O(\log n)$ whp	$O(1)$
BSH2 (this paper)	$O(\log n)$ whp	$O(\log n)$ whp	$O(1)$ amor	$O(1)$ amor	$O(1)$ amor

whp :with high probability exp : expected amor : Amortized expected

From table 1, we can conclude that if high probability bounds are considered to be as good as worst case bounds, SBSH and BSH1 have performances comparable to Binomial heaps [6] and Fast Meldable heaps [2] respectively. The amortized performance of BSH2 is comparable to that of Fibonacci heaps [8]. Moreover, BSH2 performs all operations in $O(\log n)$ time with high probability while Fibonacci heap may take $\Theta(n)$ time for delete minimum operation. This makes BSH2 useful in real-time applications.

2 Simple Bottom-up sampled heaps

In this section, we describe a simple data structure called simple bottom-up sampled heap (SBSH) that performs priority queue operations in time comparable to Binomial heaps [6] (see rows 1 and 5 of Table 1).

2.1 Properties

Let $S = \{x_1, x_2, \ldots, x_n\}$ be a set of *distinct* [2] real numbers, where $x_1 = min(S)$.

Consider a sequence of sets $S = S_1, S_2, \ldots$ obtained as follows. The minimum element x_1 is included in all these sets. An element x_i, $i > 1$ in set S_i is included in S_{i+1} if the toss of a p-biased coin is head. Let h be the smallest index such that $S_h = \{x_1\}$. Then the sequence $S_1, S_2, \ldots, S_h$ is said to be a bottom-up sample of S.

An equivalent way to obtain a bottom-up sample of S is to include each x_i, for $i \geq 2$, in sets $S_1, \ldots, S_{X_i}$ where X_i is a geometric random variable with $Pr[X_i = k] = p^{k-1}(1 - p)$ for all $k \geq 1$. The element x_1 is included in sets $S_1, \ldots, S_h$, where $h = 1 + max\{X_i \mid 2 \leq i \leq n\}$.

We construct a simple bottom-up sampled heap (SBSH) H from a bottom-up sample as follows. For an element $x_i \in S_j$ create a node labeled x_i^j. These nodes are arranged as a tree. All nodes of the form x_i^1 are called **leaf nodes**. The leaf node x_i^1 contains x_i. All other nodes (non-leaf nodes) do not contain any value. We call all nodes of the form x_1^i as **base nodes**.

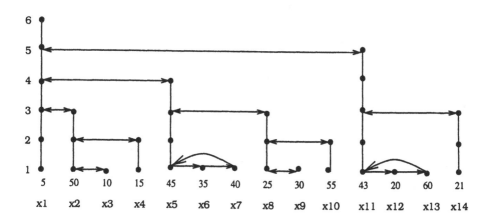

Fig. 1. A simple bottom-up sampled heap

A pointer to the node x_1^1 called bottom pointer is stored in $H.bottom$. This is used to access the heap H. Let $ht(H) = h$ denote the height of the data structure. The pointers connecting nodes in the tree are described below :

down : All non-leaf nodes x_i^j have $x_i^j.down = x_i^{j-1}$, and all leaf nodes have *down* pointer as NIL. For example, in Fig.1, $x_4^2.down = x_4^1$ and $x_5^1.down = $ NIL.

up : All nodes x_i^j such that $x_i \in S_{j+1}$ have $x_i^j.up = x_i^{j+1}$, and all other nodes have *up* pointer as NIL. In Fig.1, $x_{14}^2.up = x_{14}^3$ and $x_{10}^2.up = $ NIL.

---

[2] This assumption is made to simplify the presentation of our results. In general, our data structures can be extended to maintain a multi-set S

sibling : Let $S_j = \{x_{i_1}, x_{i_2}, \ldots, x_{i_k}\}$. The *sibling* pointer of a node x_{i_m} points to

1. x_{i_m}, if $j = h$.
2. $x_{i_{m+1}}$, if $m < k$ and $x_{i_{m+1}}$ is not present in S_{j+1}.
3. x_{i_l} otherwise, where l is the maximum index $\leq i$ such that x_{i_l} is present in S_{j+1}.

The sibling set of a node x, $sib(x)$ is defined as the set of nodes other than x that can be reached by traversing *sibling* pointers only. For example, $sib(x_5^3) = \{x_8^3\}$ and $sib(x_{11}^1) = \{x_{12}^1, x_{13}^1\}$.

Bdescendant : We define the set of descendants of a non-leaf node x, $DescSet(x)$, as the set of leaf nodes that can be reached from $x.down$ by a sequence of *down* and *sibling* pointers.

The pointer *Bdescendant* of a non-leaf node, x points to the leaf node with minimum value among all the nodes in $DescSet(x)$. The *Bdescendant* pointer of any leaf node is NIL. In Fig.1, $x_5^4.Bdescendant = x_8^1$ and $x_{11}^5.Bdescendant = x_{12}^1$.

Using this best descendant pointer we can define the weight of any node x_i^j as

$$w(x_i^j) = \begin{cases} w(Bdescendant(x_i^j)) & \text{if } j > 1 \\ x_i & \text{if } j = 1 \end{cases}$$

In Fig.1, $w(x_8^3) = 25$ and $w(x_2^2) = 10$.

Definition 1. The parent function of a node x in a SBSH is defined as

$$parent(x) = \begin{cases} \text{NIL} & \text{if } x.up = \text{NIL and } x.sibling = x \\ x.up & \text{if } x.up \neq \text{NIL} \\ parent(x.sibling) & \text{otherwise} \end{cases}$$

Notice that $parent(x_1^h) = \text{NIL}$. Since all nodes have exactly one sibling with non-NIL *up* pointer, the *parent* function can be evaluated on all nodes. From the definition of sibling set, we observe that $(y \in sib(x)) \equiv (parent(x) = parent(y))$.

Let us define $parent^j$ as the *parent* function composed with itself j times, i.e.

$$parent^j(x) = \begin{cases} parent(parent^{j-1}(x)) & \text{if } j > 0 \\ x & \text{if } j = 0 \end{cases}$$

Observation 2. *A leaf node x belongs to $DescSet(x_i^j)$ iff $parent^{j-1}(x)$ is x_i^j.*

From the above observation, we can give a recursive definition for *DescSet* as

$$DescSet(x) = \begin{cases} \bigcup_{y \in sib(x.down) \cup \{x.down\}} DescSet(y) & \text{if } x.down \neq \text{NIL} \\ x & \text{if } x.down = \text{NIL} \end{cases} \quad (1)$$

Thus, the best descendant of a node x can be computed as the minimum of the best descendants of nodes in the sibling linked list of $x.down$.

2.2 Operations

We shall now consider the various priority queue operations on simple bottom-up sampled heaps. The decrease key operation is performed as a delete and an insert operation.

INSERT. When an element z is to be inserted in the SBSH, we repeatedly toss a coin till a failure is encountered. If we obtain k consecutive successes, then we create a leaf node z^1 and non-leaf nodes $z^2, \ldots, z^{k+1}$ and connect z^j with z^{j+1} by both *up* and *down* pointers for all $1 \leq j \leq k$. Then, we insert this column of nodes next to the column of *base nodes*. For all $1 \leq j \leq k$, the *base node* x_1^j, is detached from its siblings and z^j takes the place of x_1^j in the list. The node z^{k+1} is inserted next to x_1^{k+1} in the sibling list of x_1^{k+1}. The *Bdescendant* pointers of the nodes z^j for all $2 \leq j \leq k+1$ are evaluated using (1).

While inserting an element, care must be taken in the following special cases :

1. If $z < w(H.bottom)$, then swap z and $w(H.bottom)$.
2. If $k + 1 > ht(H) - 1$, then create new levels.

DELETE. The delete operation removes an element x_i given the pointer to the leaf node x_i^1. Notice that all nodes of the form x_i^j, where $j \geq 1$ have to be removed. Deletion consists of the following steps :

1. If $x_i^1 = H.bottom$, Then Call DELETEMIN(H) and exit
2. **FindLeft** : This operation finds out the node x_{i-1}^1. This is done by traversing the *up* pointer till NIL is encountered. Let the last node encountered be x_i^k. The right-most descendant of the left sibling [3] of x_i^k is x_{i-1}^1.
3. **Remove** : This operation removes all nodes of the form x_i^j, where $j \geq 1$. The sibling linked list of x_i^j is merged with the sibling linked list of $parent^{j-1}(x_{i-1}^1)$ for all $1 \leq j \leq k$. The *Bdescendant* pointers of all nodes of the form $parent^{j-1}(x_{i-1}^1)$ are updated during the traversal.
4. **Cascade** : Notice that nodes of the form $parent^j(x_{i-1}^1)$, where $j > k$ may have incorrect *Bdescendant* pointers. The cascade operation updates the *Bdescendant* pointer of these nodes, till a node with no change in its *Bdescendant* pointer is encountered.

DELETEMIN. The delete minimum operation involves the following operations

1. Finding the second minimum : The second minimum node can be found by finding the node with minimum weight of *Bdescendant*, among all siblings of *base nodes*. The correctness of this operation follows from (1).
2. Removing the minimum element : Let leaf node x contain the second minimum element. We set $w(H.bottom)$ to $w(x)$ and call $Delete(H, x)$.

---

[3] The node y such that $y.sibling = x_i^k$

MELD. The meld operation combines two heaps destructively to form another heap containing elements of these heaps put together.

The minimum of one heap (say H_1) is inserted into the other heap H_2. The base nodes of H_1 are removed from bottom to top and at each level its siblings are attached to the rightmost node in the corresponding level in H_2, creating new levels in H_2 if necessary. Changes in *Bdescendant* pointers may be cascaded upwards. $H_2.bottom$ is returned as the new bottom of the heap.

2.3 Analysis

As the data structure presented is probabilistic, we analyze the random variables using notions of expected value, *high probability* and *very high probability*. An event is said to occur with high probability if it occurs with probability at least $1-n^{-k}$, for all $0 < k = O(1)$. An event is said to occur with very high probability if it occurs with probability at least $1 - k^{-n}$, for all $1 < k = O(1)$.

Every element in an SBSH has its height as a geometric random variable. Moreover every sibling linked list contains a geometric random number of nodes. Using Chernoff bounds [11] to analyze the sum of independent geometric random variables, we can prove the following theorem.

Theorem 3. *SBSH is a data structure on pointer machines that*

1. *has height $O(\log n)$ with high probability*
2. *performs insert operation in $O(1)$ expected time*
3. *performs all operations in $O(\log n)$ time with high probability*
4. *has $O(n)$ space complexity with very high probability.*

3 BSH1

In this section, we present an extension of SBSH called BSH1. First, we present a lazy meld routine. Then, we improve $O(1)$ expected bounds on runtime to constant worst case bounds.

3.1 Performing meld operation in a lazy manner

We now present an extension of SBSH that performs meld operation in $O(1)$ expected time. We define an extra pointer, *top* for the data structure. This points to the node x_1^h. We maintain many simple bottom-up sampled heaps, called sub-heaps. The *bottom* and *top* pointers of one of these sub-heaps called primary sub-heap are used to access the data structure. Other sub-heaps (secondary sub-heaps) are accessible through **representative nodes**. The following property holds for every sub-heap in the data structure.

Property 4. *The leaf node storing the minimum in a sub-heap cannot be a representative node.*

Any leaf node other than those corresponding to the minimum element of a sub-heap in the data structure can be a representative node.

When a meld operation is performed on heaps H_1 and H_2, without melding the two heaps, the heap with larger minimum (say H_2 w.l.g) is represented by a node referred to as "representative" node, and this node is inserted into H_1. The *down* pointer of this node is set to $H_2.top$. Since representative nodes are always leaf nodes, the condition $x.down$ = NIL $\lor$ $x.representative = True$ specifies that node x is a leaf node. Similarly, $H_2.top.up$ is set to the corresponding representative node. It can be seen that the condition for a node x to be a topmost *base node* in any sub-heap is $x.up$ = NIL $\lor$ $x.up.representative = True$.

Since we perform the meld operation in a lazy manner, the other operations have to be modified appropriately. Insert operation is performed as in SBSH. Thus we perform insert and meld operations in $O(1)$ expected time.

The following steps are performed by DELETE(H, x).

1. If $x = H.bottom$, Then Call DELETEMIN(H) and exit
2. Delete x as in SBSH. The cascade operation is performed till we encounter a node y whose *Bdescendant* pointer is not changed or is a topmost *base node* in a sub-heap.
3. If $parent(y).representative = True$, then it means that the delete operation is performed on the minimum element in a secondary sub-heap. In this case, we remove the representative node $y.up$ and meld the sub-heap belonging to y with the primary sub-heap as in SBSH.

During the operation DELETEMIN(H), the following steps are performed and the initial value of $w(H.bottom)$ is reported.

1. Find the leaf node (say x) corresponding to the second minimum and replace $w(H.bottom)$ by $w(x)$.
2. If $x.down$ = NIL, then the node x is a non-representative node. In this case DELETE(H, x) is performed. Otherwise, DELETE($H, x.down.Bdescendant$) is executed. The correctness of this operation follows from property 4.

Lemma 5. *After any sequence of operations, property 4 is satisfied by the data structure.*

Lemma 6. *The number of representative nodes in the data structure is less than the number of non-representative nodes*

From the above lemma, we can infer that the asymptotic space complexity of the data structure is same as that of SBSH.

3.2 Buffered update

In this section we will improve the $O(1)$ expected bounds on insert and meld operations to $O(1)$ worst-case bounds, using "buffered update".

Maintaining *minsibling* **pointer.** BSH1 is based on the heap described in Sect. 3.1. An additional pointer, *minsibling* is defined for every *base node*. The pointer $x.minsibling$ will point to the minimum node in $\bigcup_{y \in sib(x)} DescSet(y)$. This will cause slight modifications in DELETE and DELETEMIN operations. Every time a node is removed from the sibling linked list of a *base node*, the *minsibling* pointer of the *base node* will be updated using the *Bdescendant* pointers of the siblings.

We also maintain a pointer *lastsibling* in each *base node* x which points to the node y having $y.sibling = x$. During insert operation, either a *base node* loses all its siblings, or the inserted node is added to the base node's sibling linked list. In the first case, the *minsibling* pointer is used to compute the *Bdescendant* of the inserted node and the *minsibling* pointer is set to NIL. In the second case, the *minsibling* pointer is set to the node with lesser weight among itself and *Bdescendant* of the inserted node. So at each level, the INSERT operation will take constant time (say one unit of work). Therefore the complexity of INSERT will be $O(X)$, where X is a geometric random variable.

Performing INSERT and MELD in constant time . Each sub-heap is associated with an auxiliary double ended queue accessible using $x.queue$, where x is the topmost *base node* of the sub-heap. It can be seen that $H.top.queue$ is the queue associated with the primary sub-heap.

These queues or buffers contain unfinished insertion tasks. Every entry e in the queue (buffer) will correspond to the insertion of an element in some number of levels (say l) in the heap. The entry e requires l units of work for its completion. We now present the constant time INSERT operation.

Function INSERT(z)

1   Create a leaf node z^1 and insert it next to $H.bottom$.
2   Add an entry to the queue corresponding to the insertion of non-leaf nodes.
3   Perform t units of work by removing tasks from the queue.
4   If the last task is not fully completed,
5   Place the remaining task in front of the queue.
6   Return a pointer to z^1

MELD operation involves the insertion of a node in the heap. Again, this can be done as given above. Therefore MELD can be done in constant time.

We now state a tail estimate on the work held in a buffer after any sequence of operations.

Lemma 7. *The work held in the buffer associated with any sub-heap is $O(1)$ expected and $O(\log n)$ with high probability provided $pt > 1$.*

DELETE **and** DELETEMIN. When deletion of an element (due to DELETE or DELETEMIN) from a sub-heap is done, all operations in the buffer associated with that sub-heap are performed. Once the buffer is emptied, the element is deleted as in Sect. 3.1. The extra overhead will be proportional to the work held in the buffer which is $O(\log n)$ with high probability (from lemma 7).

The space complexity of the buffers will be proportional to the number of nodes yet to be created in the data structure. Therefore the asymptotic bound on space complexity remains the same.

Theorem 8. *BSH1 is a data structure on pointer machines that*

1. *performs insert and meld operations in $O(1)$ worst-case time*
2. *performs delete, delete minimum and decrease key operations in $O(\log n)$ time with high probability*
3. *has $O(n)$ space complexity with very high probability.*

4 BSH2

In this section we present an extension of SBSH, called BSH2 that performs INSERT and DECREASEKEY operations in $O(1)$ expected time and DELETE, DELETEMIN and MELD in $O(\log n)$ time with high probability. In section 4.3, we prove that the amortized cost of INSERT, DECREASEKEY and MELD is $O(1)$ expected.

In addition to the *up, down, sibling* and *Bdescendant* pointers, we define a *Hancestor* pointer for all leaf nodes. This pointer points to the highest ancestor of the node having its *Bdescendant* pointer pointing to this leaf node. All non-leaf nodes have *Hancestor* pointer as NIL. Formally, a leaf node x has its highest ancestor as y if

1. $y.Bdescendant = x$
2. $parent(y) = $ NIL or $parent(y).Bdescendant \neq x$

The number of elements in a heap H is stored in $H.N$. Every node x stores its level number in $x.level$. An extendible array $H.Display$ stores the pointer to the *base node* at level j in $H.Display[j]$.

4.1 Randomizing the order of siblings

The following property is always satisfied by the data structure :

Property 9. *The probability that a node x in a sibling linked list is to the left of another node y in the same linked list [4] is $1/2$ and is independent of the weights of the nodes and the order in which nodes appear in other sibling linked lists.*

We define a SHUFFLE operation that restores property 9, after the insertion of a node x in the first position of a sibling linked list, i.e., $x.up$ is non-NIL. This places the first node of a sibling linked list in a randomly chosen position. The

---

[4] A node with non-NIL *up* pointer is encountered while traversing the sibling linked list from y to x

algorithm for SHUFFLE is given below.

Procedure SHUFFLE(x)

1   $k \leftarrow$ number of siblings of node x
2   Choose uniformly at random, an integer $i \in [0, k]$
3   If $i \neq 0$, Then
4           $parent(x).down \leftarrow x.sibling$
5   Insert x after i^{th} sibling in the circular linked list.

4.2 Operations

Operations DELETE and DELETEMIN are performed as in SBSH.

INSERT. Let z be the element to be inserted. The insertion operation is performed as in SBSH. Let nodes $z^1, z^2, \ldots, z^{k+1}$ correspond to the inserted element. Then the SHUFFLE operation is performed on each inserted node, i.e. $z^1, z^2, \ldots, z^{k+1}$. Therefore property 9 is satisfied.

DECREASEKEY. We now describe the DECREASEKEY(H, x, b) operation reducing the weight of a node x to b, in a heap H. Let the node y be the highest ancestor of x, i.e., $y = x.Hancestor$, and let z be the parent of y. Notice that $w(y)$ is equal to $w(x)$. Let the weight of z be c, and let y' be the sibling of y with weight c.

We now consider all cases that can arise while performing a DECREASEKEY operation and present the algorithm for each of these cases.

Case I : $b > c$
    We reduce $w(x)$ to b. Notice that no change is required in any of the *Bdescendant* and *Hancestor* pointers.
Case II : $b < c$ and y' is to the left of y in their sibling linked list
    We reduce $w(x)$ to c and call DECREASEKEY($H, z.Bdescendant, b$)
Case III : $b < c$ and y' is to the right of y in their sibling linked list
    We delete all nodes above y, changing *sibling*, *Bdescendant* and *Hancestor* pointers appropriately. We remove the sub-tree rooted at y and insert it next to $H.Display[y.level]$, as in the insert operation. In other words, non-leaf nodes are placed in levels $y.level + 1$ to $y.level + X$ next to the corresponding *base nodes*, where $Pr[X = k] = p^k(1 - p)$, for all $k \geq 0$.

Care should be taken to handle the following special cases.

1. If $x = H.bottom$, then we reduce $w(x)$ to b and exit.
2. If $b < w(H.bottom)$, then we swap b and $w(H.bottom)$ and continue the decrease key operation.

MELD. The meld operation in BSH2 is an extension of the meld operation in SBSH. Care is taken to maintain property 9.

Consider the operation $\text{MELD}(H_1, H_2)$ on two heaps H_1 and H_2. W.l.g we assume that H_1 is the heap with lesser number of elements, i.e., $H_1.N < H_2.N$. The following steps are performed :

1. An element with value $w(H_2.bottom)$ is inserted in H_1.
2. For all *base nodes* in H_2 from level 1 to $ht(H_1) - 1$, we merge the rightmost sibling linked list at level l in H_1 (say L_1) with the sibling linked list of the *base node* of H_2 at level l (say L_2). The order in which L_1 and L_2 are concatenated is random, i.e., the resulting list is $L_1 L_2$ or $L_2 L_1$ with equal probability.
3. If $ht(H_1) < ht(H_2)$, then we replace $H_1.bottom.Hancestor$ by H_2's *base node* at level $ht(H_1)$. Then, the first $ht(H_1)$ elements are updated in array $H_2.Display$ and array $H_1.Display$ is destroyed. We return H_2 as the resulting heap after setting $H_2.bottom$ to $H_1.bottom$.
4. If $ht(H_1) \geq ht(H_2)$, then the array $H_2.Display$ is destroyed and H_1 is returned as the resulting heap.

Lemma 10. *Property 9 is satisfied after any sequence of* INSERT, DELETE, DELETEMIN, DECREASEKEY *and* MELD *operations.*

From the above lemma, we can show that all operations defined for BSH2 are correct.

4.3 Analysis .

The runtime analysis of BSH2 is similar to that of SBSH. The overhead involved in "shuffling" the first node in a sibling linked list does not change the asymptotic bound on insert operation as

Lemma 11. *The shuffle operation takes $O(1)$ expected time.*

The following result on the complexity of DECREASEKEY operation can be inferred from property 9.

Lemma 12. *During the* DECREASEKEY *operation, the number of times case II arises is bounded by a geometric random variable.*

From the above lemma, we can show that decrease key operation takes $O(1)$ expected time.

The complexity of meld operation in BSH2 is $O(ht(H_1))$ expected. It can be noted that $ht(H_1) < ht(H_2)$ with probability atleast $1/2$. Using amortized analysis, we can exploit this property to obtain a constant expected amortized bound on insert, decrease key and meld operations.

Theorem 13. *BSH2 is a data structure implementable on RAM that*

1. *performs insert and decrease key operations in $O(1)$ expected time*

2. *performs all operations in* $O(\log n)$ *time with high probability*
3. *performs insert, decrease key and meld operations in amortized* $O(1)$ *expected time*
4. *has* $O(n)$ *space complexity with very high probability.*

5   Conclusion and Open Problems

In this paper, we have presented three simple and efficient data structures to implement single ended priority queues. In practice, these data structures may be better than their deterministic counterparts due to smaller constants in the asymptotic bounds.

Improving the amortized constant expected bounds in BSH2 to expected bounds or worst-case bounds, may result in a data structure theoretically comparable but practically better than Brodal's data structure [3].

References

1. C.R.Aragon and R.G. Seidel. *Randomized search trees.* Proc. 30th Ann. IEEE Symposium on Foundations of Computing, 540-545 (1989)
2. Gerth Stólting Brodal. *Fast meldable priority queues.* Proc. 4th International Workshop, WADS, 282-290 (1995)
3. Gerth Stólting Brodal. *Worst-case efficient priority queues.* Proc. 7th Ann. ACM Symposium on Discrete Algorithms, 52-58 (1996)
4. Giorgio Gambosi, Enrico Nardelli, Maurizio Talamo. *A Pointer-Free data structure for merging heaps and min-max heaps.* Theoritical Computer Science 84(1), 107-126 (1991)
5. James R. Driscoll, Harold N. Gabow, Ruth Shrairman and Robert E. Tarjan. *Relaxed Heaps : An alternative approach to Fibonacci Heaps with applications to parallel computing.* Comm. ACM 31(11), 1343-1354 (1988)
6. Jean Vuillemin. *A data structure for manipulating priority queues.* Comm. ACM 21(4), 309-315 (1978)
7. Knuth, D. *The Art of Computer Programming, Volume 3, Sorting and Searching.* Addison-Wesley, Reading, Mass., 1973
8. Michael L. Fredman and Robert E. Tarjan *Fibonacci heaps and their uses in improved network optimization algorithms.* Proc. 25th Annual Symposium on Foundations of Computer Science, 338-346 (1984)
9. Michiel Smid. *Lecture Notes : Selected Topics in Data Structures.* Max-Plank Institute for Informatics, Germany.
10. W. Pugh. *Skip lists : A probabilistic alternative to balanced trees.* Comm. ACM 33, 668-676 (1990)
11. P. Raghavan, *Lecture notes in randomized algorithms,* Technical Report RC15340, IBM J.J.Watson Research Center (1989).
12. Rolf Fagerberg, *A Note on Worst Case Efficient Meldable Priority Queues,* Technical Report, Odense University Computer Science Department Preprint 1996-12.
13. Thomas H. Cormen, Charles E. Leiserson, Ronald L. Rivest. *Introduction to Algorithms.* The MIT Press, Cambridge, Massachusetts (1989)

Extractors for Weak Random Sources and Their Applications

David Zuckerman[*]

Computer Science Dept., University of Texas at Austin, Austin TX 78712, USA
diz@cs.utexas.edu,
WWW home page: http://www.cs.utexas.edu/users/diz

Abstract. An extractor is an algorithm which, on input a long string from a defective random source and a short truly random string, outputs a long almost-random string. In this talk, we survey extractors for weak random sources and their applications.

Extended Abstract for Talk

Randomization is very useful in computer science (see e.g., [MR95]). Since a computer may not have access to truly random bits, it is natural to ask if access to weakly-random bits can be useful for randomized algorithms.

Extending models in [CG88,SV86], the most general model of a weak random source was studied in [Zuc90], where a lower bound on the min-entropy was imposed.

Definition 1. *The* min-entropy *of a distribution D on a probability space S is* $\min_{s \in S}\{-\log_2 D(s)\}$.

We would like an algorithm using a weak source to work for *all* sources which have min-entropy at least k (think of $k = n/2$, say, where n is the number of bits output by the source). The natural idea to use such a source is to convert these bad random bits into good random bits. It is easy to see that it is impossible to do this deterministically if $k \leq n-1$ (without knowing D) (see e.g., [Zuc96]). An extractor, first defined in [NZ96], is an algorithm which adds a small number of truly random bits and extracts many almost-random bits. We define an extractor after defining almost-random.

Definition 2. *The* statistical distance *(or variation distance) between two distributions D_1 and D_2 on the same space S is*

$$\max_{T \subseteq S} |D_1(T) - D_2(T)| = \frac{1}{2} \sum_{s \in S} |D_1(s) - D_2(s)|.$$

---

[*] Supported in part by NSF NYI Grant No. CCR-9457799, a David and Lucile Packard Fellowship for Science and Engineering, and an Alfred P. Sloan Research Fellowship.

Definition 3. $E : \{0,1\}^n \times \{0,1\}^t \to \{0,1\}^m$ *is a* (k, ϵ)-*extractor if, for x chosen according to any distribution on $\{0,1\}^n$ with min-entropy at least k, and y chosen uniformly at random from $\{0,1\}^t$, $E(x, y)$ is within statistical distance ϵ from the uniform distribution on $\{0,1\}^m$.*

The first explicit extractor construction was given in [NZ96], which used ideas from [Zuc90,Zuc96]. Improvements were given in [WZ95,SZ94,Zuc97,TS96] (related improvements were also given in [SSZ98,ACRT97]). The state of the art is three incomparable extractors. For min-entropy $k = \Omega(n)$ the extractor of [Zuc97] requires $t = O(\log(n/\epsilon))$ truly random bits to output $m = .99k$ almost-random bits. For $k = n^\gamma$, $\gamma < 1$, the extractor of [TS96] requires only slightly more than logarithmic bits to output $n^{\gamma'}$ almost-random bits. For any k, a different extractor in [TS96] requires $t = \text{polylog}(n/\epsilon)$ bits to output $m = k$ almost-random bits.

These extractors are very strong: for example, they give better expansion properties than can be proved using eigenvalues (see [WZ95]). For this reason, they have had many applications unrelated to weak random sources: expanders, superconcentrators, and non-blocking networks [WZ95], sorting and selecting in rounds [WZ95], pseudo-random generators for space-bounded computation [NZ96], unapproximability of clique [Zuc96], time versus space complexities [Sip88], leader election [Zuc97,RZ98], another proof that BPP $\subseteq$ PH [GZ97], and random sampling using few random bits [Zuc97]. We elaborate on this last application.

Definition 4. *A* (γ, ϵ)-*oblivious sampler is a deterministic algorithm which, on input a uniformly random n-bit string, outputs a sequence of d sample points $z_1, \ldots, z_d \in \{0,1\}^m$ such that for any function $f : \{0,1\}^m \to [0,1]$, we have $|\frac{1}{d}\sum_{i=1}^d f(z_i) - Ef| \leq \epsilon$ with probability $\geq 1 - \gamma$ (where $Ef = 2^{-m}\sum_{z \in \{0,1\}^m} f(z)$).*

Oblivious samplers and extractors turn out to be nearly equivalent views of the same combinatorial object. Not only do the extractor constructions yield randomness-optimal oblivious samplers, but the extractor view helps in proving other theorems about oblivious samplers. For example, it is relatively easy to see that although the ϵ error of the extractor is defined relative to $\{0,1\}$ functions, it also achieves this error with respect to more general $[0,1]$ functions. This is not as obvious for oblivious samplers, but follows as a corollary.

A more unintuitive theorem about oblivious samplers which can be proved using extractors is the following. An oblivious sampler which uses a constant factor times the optimal number of random bits can be converted to one which uses only 1.01, say, times the optimal number of random bits. The only disadvantage of the new sampler is that it uses a polynomially larger number of samples than the old sampler. For details see Sections 2.1, 2.2, and 2.4 of [Zuc97].

For those wishing to read more about extractors and their applications, we recommend starting with Nisan's excellent survey [Nis96].

References

[ACRT97] A. E. Andreev, A. E. F. Clementi, J. P. D. Rolim, and L. Trevisan. Weak random sources, hitting sets, and BPP simulations. In *Proceedings of the 38th Annual IEEE Symposium on Foundations of Computer Science*, pages 264–272, 1997.

[CG88] B. Chor and O. Goldreich. Unbiased bits from sources of weak randomness and probabilistic communication complexity. *SIAM Journal on Computing*, 17(2):230–261, 1988.

[GZ97] O. Goldreich and D. Zuckerman. Another proof that BPP ⊆ PH (and more). Technical Report TR97-045, Electronic Colloquium on Computational Complexity, 1997.

[MR95] R. Motwani and Prabhakar Raghavan. *Randomized Algorithms*. MIT Press, 1995.

[Nis96] N. Nisan. Extracting randomness: How and why – a survey. In *Proceedings of the 11th Annual IEEE Conference on Computational Complexity*, pages 44–58, 1996.

[NZ96] N. Nisan and D. Zuckerman. Randomness is linear in space. *Journal of Computer and System Sciences*, 52(1):43–52, 1996.

[RZ98] A. Russell and D. Zuckerman. Hyper-fast leader election protocols in the perfect information model. Unpublished manuscript, 1998.

[Sip88] M. Sipser. Expanders, randomness, or time vs. space. *Journal of Computer and System Sciences*, 36:379–383, 1988.

[SSZ98] M. Saks, A. Srinivasan, and S. Zhou. Explicit OR-dispersers with polylog degree. *Journal of the ACM*, 45:123–154, 1998.

[SV86] M. Santha and U. V. Vazirani. Generating quasi-random sequences from semi-random sources. *Journal of Computer and System Sciences*, 33:75–87, 1986.

[SZ94] A. Srinivasan and D. Zuckerman. Computing with very weak random sources. In *Proceedings of the 35th Annual IEEE Symposium on Foundations of Computer Science*, pages 264–275, 1994. To appear in SIAM Journal on Computing.

[TS96] A. Ta-Shma. On extracting randomness from weak random sources. In *Proceedings of the 28th Annual ACM Symposium on Theory of Computing*, pages 276–285, 1996.

[WZ95] A. Wigderson and D. Zuckerman. Expanders that beat the eigenvalue bound: Explicit construction and applications. *Combinatorica*. To appear. Revised version appears as Technical Report TR-95-21, Department of Computer Sciences, The University of Texas at Austin, June 1995. Preliminary version in *Proceedings of the 25th Annual ACM Symposium on Theory of Computing*, pages 245–251, 1993.

[Zuc90] D. Zuckerman. General weak random sources. In *Proceedings of the 31st Annual IEEE Symposium on Foundations of Computer Science*, pages 534–543, 1990.

[Zuc96] D. Zuckerman. Simulating BPP using a general weak random source. *Algorithmica*, 16:367–391, 1996.

[Zuc97] D. Zuckerman. Randomness-optimal oblivious sampling. *Random Structures and Algorithms*, 11:345–367, 1997.

Comparator Networks for Binary Heap Construction*

Gerth Stølting Brodal[1,**] and M. Cristina Pinotti[2,***]

[1] Max-Planck-Institut für Informatik, Im Stadtwald, D-66123 Saarbrücken, Germany
[2] Istituto di Elaborazione della Informazione, CNR, 56126 Pisa, Italy

Abstract. Comparator networks for constructing binary heaps of size n are presented which have size $\mathcal{O}(n \log \log n)$ and depth $\mathcal{O}(\log n)$. A lower bound of $n \log \log n - \mathcal{O}(n)$ for the size of any heap construction network is also proven, implying that the networks presented are within a constant factor of optimal. We give a tight relation between the leading constants in the size of selection networks and in the size of heap construction networks.

Introduction

The heap data structure, introduced in 1964 by Williams [17], has been extensively investigated in the literature due to its many applications and intriguing partial order. Algorithms for heap management — insertion, minimum deletion, and construction — have been discussed in several models of computation. For the heap construction algorithm, Floyd has given a sequential algorithm building the tree in a bottom-up fashion in linear time, which is clearly optimal. On the weak shared memory machine model, EREW-PRAM, Olariu and Wen can build a heap of size n in time $\mathcal{O}(\log n)$ and optimal work [14]. On the powerful CRCW-PRAM model, the best-known heap construction algorithm was given by Raman and Dietz and takes $\mathcal{O}(\log \log n)$ time [6]. The same time performance holds for the parallel comparison tree model [5]. Finally Dietz showed that $\mathcal{O}(\alpha(n))$, where $\alpha(n)$ is the inverse of Ackerman's function, is the expected time required to build a heap in the randomized parallel comparison tree model [5]. All the above parallel algorithms achieve optimal work $\mathcal{O}(n)$, and the time optimality of the deterministic algorithms can be argued by reduction from the selection of the minimum element in a set.

In this paper we address the heap construction problem for the simplest parallel model of computation, namely *comparator networks*. Comparator networks perform only comparison operations, which may occur simultaneously.

---

* This research was done while the first author was visiting the Istituto di Elaborazione della Informazione, CNR, Pisa.

** Supported by the Carlsberg foundation (Grant No. 96-0302/20). Partially supported by the ESPRIT Long Term Research Program of the EU under contract No. 20244 (ALCOM-IT). Email: brodal@mpi-sb.mpg.de.

*** Email: pinotti@iei.pi.cnr.it.

The most studied comparator networks are sorting and merging networks. In the early 1960's, Batcher proposed the odd-even merge algorithm to merge two sequences of n and m elements, $n \geq m$, which can be implemented by a merging network of size $\mathcal{O}((m + n) \log m)$. In the early 1970's Floyd [12] and Yao [18] proved the asymptotic optimality of Batcher's networks. The lower bound has recently been improved by Miltersen, Paterson and Tarui [13], closing the long-standing factor-of-two gap between upper and lower bounds. It is noteworthy to recall, that merge can be solved in the comparison tree model with a tree of depth $m + n - 1$.

Batcher also showed how his merge algorithm could be used to implement sorting networks with size $\mathcal{O}(n \log^2 n)$ and depth $\mathcal{O}(\log^2 n)$ to sort n inputs [12]. For a long time, the question remained open as to whether sorting networks with size $\mathcal{O}(n \log n)$ and depth $\mathcal{O}(\log n)$ existed. In 1983, Ajtai, Komlós and Szemerédi [1] presented sorting networks with size $\mathcal{O}(n \log n)$ and depth $\mathcal{O}(\log n)$ to sort n elements. This result, although partially unsatisfying due to big constants hidden by the $\mathcal{O}$-notation, reveals that the sorting problem requires the same amount of work in both comparison tree and comparator network models.

Selection, sorting and merging are strictly related problems. Several sequential algorithms with linear work have been discussed for selection. The first is due to Blum et al. [4] and requires $5.43n$ comparisons. This result was later improved by Schönhage et al. to $3n$ [16] and by Dor and Zwick to $2.95n$ [7,8]. Bent and John proved a lower bound of $2n$ for this problem [3]. Dor and Zwick [9] improved it to $(2 + \epsilon)n$ [9]. For a survey of previous work on lower bounds in the comparison tree model, see the paper by Dor and Zwick [9].

An (n, t)-selection network is a comparator network that selects the t smallest elements in a set of n elements. Alekseev [2] proved that an (n, t)-selection network has at least size $(n - t)\lceil \log(t + 1) \rceil$.[1] For $t = \Omega(n^\epsilon)$ and $0 < \epsilon < 1$, the existence of a work optimal selection network immediately follows by the sorting networks of Ajtai et al. However, since selection networks do not need to do as much as sorting networks, and due to the big constant hidden by the sorting networks in [1], selection networks with improved constant factors in both depth and size have been developed. In particular, Pippenger proposes a $(n, \lfloor n/2 \rfloor)$-selection network with size $2n \log n$ and depth $\mathcal{O}(\log^2 n)$ [15]. More recently, Jimbo and Marouka have constructed a $(n, \lfloor n/2 \rfloor)$-selection network of depth $\mathcal{O}(\log n)$ and of size at most $Cn \log n + \mathcal{O}(n)$, for any arbitrary $C > 3/\log 3 \approx 1.89$, which improves Pippenger's construction by a constant factor in size and at the same time by an order in depth [11].

The preceding summary shows that work optimal comparator networks have been studied for merging, sorting, and selection. Although the heap data structure has historically been strictly related to these problems, we are not aware of any comparator network for the heap construction problem. In this scenario, we show that heap construction can be done by comparator networks of size $\mathcal{O}(n \log \log n)$ and depth $\mathcal{O}(\log n)$, and that our networks reach optimal size by reducing the problem of selecting the smallest $\log n$ elements to heap construc-

---

[1] All logarithms throughout this paper have base 2

tion. Finally, since finding the minimum requires at least a network of size $n - 1$ and depth $\lceil \log n \rceil$, our heap construction networks also have optimal depth.

1 Preliminaries

Let us review some definitions, and agree on some notations used throughout the paper.

A *binary tree* of size n is a tree with n nodes, each of degree at most two. A node x of a binary tree belongs to level k if the longest simple path from the root to x has k edges. The height of the tree is the number of edges in the longest simple path starting at the root of the tree. The subtree T_x rooted at node x at *level* k is the tree induced by the descendants of x.

A *complete binary tree* is a binary tree in which all the leaves are at the same level and all the internal nodes have degree two. Clearly, it has *height* $\lfloor \log n \rfloor$.

A *heap shaped binary* tree of height h is a binary tree whose $h - 1$ uppermost levels are completed filled and the h-th level is filled from the left to the right.

In a *heap ordered* binary tree, each node contains one element which is greater or equal to the element at its parent.

Finally, a *binary heap* is defined as a heap-shaped and heap-ordered binary tree [17], which can be stored in an array H as an implicit tree of size n, as depicted in Fig. 1. The element of the root of the tree is at index 1 of the array, (i.e., root is stored in $H[1]$), and given an index i of a node x, the indices of its left and right children are $2i$ and $2i + 1$, respectively.

A *comparator network* with n inputs and size s is a collection of n horizontal lines, one for each input, and s comparators. A *comparator* between line i and j, briefly $i : j$, compares the current values on lines i and j and is drawn as a vertical line connecting lines i and j. After the comparison $i : j$, the minimum value is put on line i, while the maximum ends up on line j. Finally, a comparator network has *depth* d, if d is the largest number of comparators that any input element can pass through. Assuming that each comparator produces its output in constant time, the *depth* of a comparator network is the running time of such a network. From now on, let us refer to comparator networks simply as networks. For a comprehensive account of comparator networks, see [12, pp. 220-246].

2 Sequential Heap Construction

It is well known that an implicit representation of a binary heap H of size n can be built in linear sequential time by the heap construction algorithm of Floyd [10]. Because we base our heap construction networks on Floyd's algorithm, we rephrase it as follows:

Assuming that the two binary trees rooted at the children of a node i are heaps, the heap-order property in the subheap rooted at i can be reestablished simply by bubbling down the element $H[i]$. We let the bubbling down procedure be denoted Siftdown. At each step, Siftdown determines the smallest of the elements $H[i]$, $H[2i]$, and $H[2i+1]$. If $H[i]$ is the smallest, then the subtree rooted at node i is a heap and the Siftdown procedure terminates. Otherwise, the child with the smallest element and $H[i]$ are exchanged. The node exchanged with $H[i]$, however, may violate the heap order at this point. Therefore, the Siftdown procedure is recursively invoked on that subtree.

We can now apply Siftdown in a bottom-up manner to convert an array H storing n elements into a binary heap. Since the elements in the subarray $H[(\lfloor n/2 \rfloor + 1)..n]$ are all leaves, each is a 1-element heap to begin with. Then, the remaining nodes of the tree are visited to run the Siftdown procedure on each one. Since the nodes are processed level by level in a bottom up fashion, it is guaranteed that the subtrees rooted at the children of the node i are heaps before Siftdown runs at that node.

In conclusion, observe that the Siftdown routine invoked on a subheap of height i performs $2i$ comparisons in the worst case, and that the worst case running time of the heap construction algorithm of Floyd described above is $\sum_{i=0}^{\lfloor \log n \rfloor} \frac{n}{2^i} \cdot 2i = \mathcal{O}(n)$, which is optimal.

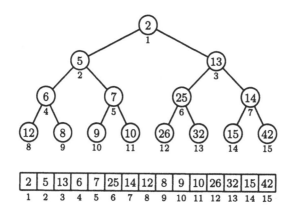

Fig. 1. A binary heap of size 15 and its implicit representation.

3 Heap Construction Networks of Size $n \log n$

In this section we present heap construction networks which have size at most $n \lfloor \log n \rfloor$ and depth $4 \lfloor \log n \rfloor - 2$. Notice that any sorting network could also be used as a heap construction network. The networks presented in this section are used in Sect. 4 to construct improved heap construction networks of size $\mathcal{O}(n \log \log n)$, and in Sect. 5 to give a reduction from selection to heap construction.

Lemma 1 gives a network implementation of the sifting down algorithm used in the heap construction algorithm by Floyd [10].

Lemma 1. *Let T be a binary tree of size n and height h. If the subtrees rooted at the children of the root satisfy heap order, then the elements of T can be rearranged to satisfy heap order with a network of size $n - 1$ and depth $2h$. At depth $2i + 1$ and $2i + 2$ of the network the comparators are only between nodes at level i and $i + 1$ in T. All comparators correspond to edges of T, and for each edge there is exactly one comparator.*

Proof. If the tree has height zero, no comparator is required. Otherwise let r be the root and u and v the children of r. If u or v is not present, the steps below which would involve v or u are skipped.

First we apply the comparators $r : u$ and $r : v$. Because T_u and T_v were assumed to be heap ordered subtrees, r now has the minimum. After the two comparators the heap order can be violated at the roots of *both* T_u and T_v. We therefore recursively apply the above to the subtrees T_u and T_v. Notice that the two recursively constructed networks involve disjoint nodes and therefore can be performed in parallel. If r only has one child we still charge the network depth two to compare r with its children to guarantee that all comparisons done in parallel by the network correspond to edges between nodes at the same levels in T.

The depth of the network is two plus the depth of the deepest recursively constructed network. By induction it follows that the depth of the network is $2h$, and that the network at depth $2i + 1$ and $2i + 2$ only performs comparisons between nodes at level i and $i + 1$ in T. Furthermore, the network contains exactly one comparator for each edge of T. $\square$

Notice that the network has $n - 1$ comparators while the corresponding algorithm of Floyd only needs h comparisons. By replacing the sifting down algorithm in Floyd's heap construction algorithm by the sifting down networks of Lemma 1, we get the following lemma.

Lemma 2. *Let T be a binary tree of size n and height h which does not satisfy heap order, and let n_i be the number of nodes at level i in T. Then a network exists of size $\sum_{i=0}^{h} i \cdot n_i$ and depth $4h - 2$ which rearranges the elements of T to satisfy heap order. All comparators correspond to edges of T.*

Proof. Initially all nodes at level h of T by definition are heap ordered binary trees of height zero. Iteratively for each level $i = h - 1, \dots, 0$ we apply the sifting down networks of Lemma 1 in parallel to the 2^i subtrees rooted at level i of T, to make these subtrees satisfy heap order. The resulting tree then satisfies heap order. By Lemma 1 all comparators correspond to edges of T.

The edge between a node v at level i and its parent corresponds to a set of comparators in the resulting network. These comparators are performed exactly when we apply the sifting down networks of Lemma 1 to an ancestor of v, i.e., there are exactly i comparators corresponding to this edge. The total number of comparators is $\sum_{i=0}^{h} i \cdot n_i$.

By Lemma 1 the depth of the network is $\sum_{i=0}^{h} 2i = h^2 + h$. But because the networks constructed by Lemma 1 proceeds top-down on T, having exactly depth two for each level of T, the applications of Lemma 1 can be *pipelined*. After the first two comparators of the applications of Lemma 1 to subtrees rooted at level i, the applications of Lemma 1 to subtrees rooted at level $i-1$ can be initiated. The application of Lemma 1 to the root of the tree can therefore be initiated at depth $2(h-1)+1$ of the network, i.e., the network has depth $2(h-1)+2h = 4h-2$. □

Theorem 1. *There exists a heap construction network of size at most $n\lfloor \log n \rfloor$ and depth $4\lfloor \log n \rfloor - 2$. All comparators correspond to edges of T.*

Proof. Let the n input lines represent a heap shaped binary tree of height $\lfloor \log n \rfloor$. The theorem then follows from Lemma 2. □

In Fig. 2 we show the network of Theorem 1 for $n = 15$. The network has size 34 and depth 10. Notice that the first two comparators of the application of Lemma 1 to the root of the tree ($1:2$ and $1:3$) are done in parallel with the third and fourth comparator of the applications of Lemma 1 to the subtrees rooted at nodes 2 and 3.

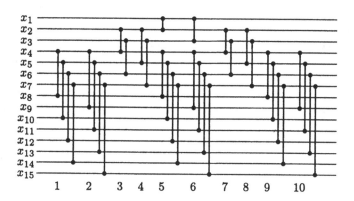

Fig. 2. A heap construction network for $n = 15$. All comparators are of the form $i : j$, where $i < j$.

4 Heap Construction Networks of Size $\mathcal{O}(n \log \log n)$

In the following we give improved heap construction networks of depth $\mathcal{O}(\log n)$ and size $\mathcal{O}(n \log \log n)$. The improved networks are obtained by combining the networks of Theorem 1 with efficient selection networks.

An arbitrary sorting network is obviously also an (n, t)-selection network, e.g., the sorting network of size $\mathcal{O}(n \log n)$ by Ajtai et al. [1]. Due to the large constants involved in the sorting network of Ajtai et al., Pippenger [15] and

Jimbo and Maruoka [11] have developed specialized $(n, \lfloor n/2 \rfloor)$-selection networks of size $\mathcal{O}(n \log n)$ where the involved constants are of reasonable size. The following lemma was developed by Jimbo and Maruoka [11].

Lemma 3 (Jimbo and Maruoka). *For an arbitrary constant $C > 3/\log 3 \approx 1.89$, there exist $(n, \lfloor n/2 \rfloor)$-selection networks of size at most $Cn \log n + \mathcal{O}(n)$ and depth $\mathcal{O}(\log n)$.*

Unfortunately, neither Pippenger [15] or Jimbo and Maruoka [11] give bounds for general (n, t)-selection networks. The following lemma is a consequence of Lemma 3, and is sufficient for our purposes.

Lemma 4. *For an arbitrary constant $C > 6/\log 3 \approx 3.79$, there exist (n, t)-selection networks of size $Cn \log t + \mathcal{O}(n)$ and depth $\mathcal{O}(\log n \cdot \log t)$.*

Proof. The n input lines are partitioned into $\lceil n/t \rceil$ blocks $B_1, \ldots, B_{\lceil n/t \rceil}$ of size t each. By applying the selection networks of Lemma 3 to $B_1 \cup B_2$ we find the t least elements of $B_1 \cup B_2$. By combining the $\lceil n/t \rceil$ blocks in a treewise fashion with $\lceil n/t \rceil - 1$ applications of Lemma 3 to $2t$ elements, we find the t least elements of the n inputs. The resulting network has size $(\lceil n/t \rceil - 1)(C \cdot 2t \log 2t + \mathcal{O}(2t)) = 2Cn \log t + \mathcal{O}(n)$ and depth $\mathcal{O}(\log n \cdot \log t)$, for $C > 3/\log 3$. □

We need the following definition. Let $\mathcal{P}$ be an arbitrary connected subset of nodes of a binary tree T which contains the root of T. Let $x_1 \leq x_2 \leq \cdots \leq x_{|\mathcal{P}|}$ be the set of elements in $\mathcal{P}$, and let $x_1' \leq x_2' \leq \cdots \leq x_{|\mathcal{P}|}'$ be the set of elements in $\mathcal{P}$ after applying a network $\mathcal{N}$ to T. We define a network $\mathcal{N}$ to be *heap-convergent*, if $\mathcal{N}$ for all possible inputs, all connected subsets $\mathcal{P}$ of nodes of T containing the root of T, and $i = 1, \ldots, |\mathcal{P}|$ satisfies $x_i' \leq x_i$. Notice that sorting networks are *not* heap-convergent. If $\mathcal{P}$ is the path to the rightmost node in the lowest level of a tree, then $\mathcal{P}$ always contains the maximum element after applying a sorting network, but the maximum element could initially be anywhere in the tree.

Lemma 5. *A comparator corresponding to an edge in a binary tree T is a heap-convergent network.*

Proof. Let the comparator be $u : v$, where v is a child of u in T. If $\mathcal{P}$ does not contain u it does not contain v either, implying that the elements in $\mathcal{P}$ are unchanged. If $\mathcal{P}$ contains both u and v, the set of elements is also unchanged. If $\mathcal{P}$ contains u but not v, the comparator $u : v$ can only replace the element at u with a smaller element from v in which case $x_i' \leq x_i$ for all $i = 1, \ldots, |\mathcal{P}|$. □

Because the networks constructed by Theorem 1 only contain comparators corresponding to tree edges and heap convergence is a transitive property we immediately have the following corollary:

Corollary 1. *The networks constructed by Theorem 1 are heap-convergent.*

Theorem 2. *If for some constants C and d, there exist (n, t)-selection networks of size $Cn \log t + \mathcal{O}(n)$ and depth $\mathcal{O}(\log^d n)$, then there exist heap construction networks of size $Cn \log \log n + \mathcal{O}(n \log \log \log n)$ and depth $4 \log n + \mathcal{O}(\log^d \log n)$.*

Proof. Assume without loss of generality that $n \geq 4$. Let the n input lines represent a heap shaped binary tree T of height $h = \lfloor \log n \rfloor$, and let $k = \lceil \log h \rceil \geq 1$. The heap construction network proceeds in three phases.

1. To each subtree T_v rooted at level $h - 2k + 1$, apply in parallel $(|T_v|, 2^k - 1)$-selection networks, such that all elements at the upper k levels of T_v become less than or equal to all elements at the remaining levels of T_v.
2. Apply the heap construction networks of Theorem 1 to the uppermost $h - k$ levels of T.
3. In parallel apply Theorem 1 to each subtree T_v rooted at level $h - 2k + 1$.

It follows immediately from Step 2 that the uppermost $h - 2k$ levels of the tree satisfy heap order and from Step 3 that each subtree rooted at level $h - 2k + 1$ satisfies heap order. What remains to be proven for the correctness of the algorithm is that for all nodes v at level $h - 2k + 1$, the subtree T_v only contains elements which are greater or equal to the elements on the path from the root to v.

After Step 1, the $2^k - 1$ least elements $e_0 \leq \cdots \leq e_{2^k-2}$ of T_v are at the uppermost k levels of T_v, which are exactly the levels of T_v which overlap with Step 2. Let $p_0 \leq \cdots \leq p_{h-2k}$ denote the elements on the path from the root to v (excluding v) after Step 2. Because the network applied in Step 2 is heap-convergent and $2^k - 2 \geq h - 2k$, we have $p_i \leq e_i$ for $i = 0, \ldots, h - 2k$ by letting $\mathcal{P}$ consist of the path from the root to v together with the upper k levels of T_v. We conclude that after Step 2 all elements on the path from the root to v are smaller than or equal to all the elements in T_v, and that after Step 3, T satisfies heap order.

From Theorem 1 we get the following upper bound on the size and depth of the resulting network. The size is bounded by

$$\left(Cn \log 2^k + \mathcal{O}(n)\right) + \mathcal{O}\left(\frac{n}{2^k} \log \frac{n}{2^k}\right) + \left(n \log 2^{2k} + \mathcal{O}(n)\right) ,$$

which is $(C + 2)n \log \log n + \mathcal{O}(n)$, and the depth is bounded by

$$\mathcal{O}\left(\log^d 2^{2k}\right) + (4(h - k) - 2) + (4(2k - 1) - 2) ,$$

which is $4 \log n + \mathcal{O}(\log^d \log n)$.

The "+2" in the size bound comes from the application of the heap construction networks of Theorem 1 in Step 3. If we instead apply the above construction recursively in Step 3, we get heap construction networks of size $Cn \log \log n + (C + 2)n \log \log \log n + \mathcal{O}(n)$ and depth $4 \log n + \mathcal{O}(\log^d \log n)$. $\quad\square$

Notice that in Steps 1 and 3 we could have used arbitrary sorting networks, but in Step 2 it is essential that the heap construction network used is heap-convergent. By applying the construction recursively $\mathcal{O}(\log^* n)$ times the asymptotic size could be slightly improved, but the constant in front of $n \log \log n$ would still be C. From Lemma 4 we get the following corollary:

Corollary 2. *For an arbitrary constant $C > 6/\log 3 \approx 3.79$, there exist heap construction networks of size $Cn \log \log n + \mathcal{O}(n \log \log \log n)$ and depth $4 \log n + \mathcal{O}(\log^2 \log n)$.*

5 A Lower Bound for the Size of Heap Construction Networks

We now prove that the construction of the previous section is optimal. Let $S(n,t)$ denote the minimal size of (n,t)-selection networks, and let $H(n)$ denote the minimal size of heap construction networks on n inputs. The following lower bound on $S(n,t)$ is due to Alekseev [2].

Lemma 6 (Alekseev). $S(n,t) \geq (n-t)\lceil \log(t+1) \rceil$.

Theorem 3. $H(n) \geq S(n, \lfloor \log n \rfloor) - \mathcal{O}(n)$.

Proof. The theorem is proven by giving a reduction from (n,t)-selection to heap construction. We prove that (n,t)-selection can be done by networks with size $H(n) + 2^{t+1} - 2t - 2$.

First we construct a heap over the n inputs with a network of size $H(n)$, and make the observation that the t least elements can only be at levels $0, \ldots, t-1$ of the heap.

The minimum is at the root, i.e., at output line one. To find the second least element we consider the implicit heap given by the lines $n, 2, 3, \ldots, 2^t - 1$. Notice that the root is now line n. By applying the sifting down network of Lemma 1 to the levels $0, \ldots, t-1$ of this tree the remaining $t-1$ least inputs are at levels $0, \ldots, t-2$ of this tree. The second least element is now at output line n. By iteratively letting the root be lines $n-1, n-2, \ldots, n-t-2$, and by applying Lemma 1 to trees of decreasing height, the t least elements will appear in sorted order at output lines $1, n, n-1, n-2, \ldots, n-t+2$. If the t smallest inputs are required to appear at the first t output lines, the network lines are permuted accordingly.

The total number of comparators for the $t-1$ applications of Lemma 1 is

$$\sum_{i=0}^{t-1} (2^{i+1} - 2) = 2^{t+1} - 2t - 2 .$$

We conclude that the resulting (n,t)-selection network has size $H(n) + 2^{t+1} - 2t - 2$, implying $H(n) \geq S(n,t) - 2^{t+1} + 2t + 2$. By letting $t = \lfloor \log n \rfloor$ the theorem follows. □

By combining Lemma 6 and Theorem 3, we get the following corollary.

Corollary 3. $H(n) \geq n \log \log n - \mathcal{O}(n)$.

6 Conclusion

The parallel construction of heaps has been addressed for several parallel models of computation: EREW-PRAM [14], CRCW-PRAM [6], the parallel comparison tree model and the randomized parallel comparison tree model [5]. These algorithms all achieve optimal $\mathcal{O}(n)$ work. In this paper we have addressed the problem for the most simple parallel model of computation, namely comparator networks.

As opposed to merging and selection, which both can be solved in sequential linear time but require networks of size $\Theta(n \log n)$, we have shown that heap construction can be done by networks of size $\mathcal{O}(n \log \log n)$ and depth $\mathcal{O}(\log n)$, and that this is optimal. By combining the results of Theorem 2 and Theorem 3, we get the following characterization of the leading constant in the size of heap construction networks compared to the leading constant in the size of (n, t)-selection networks.

Theorem 4. *If for constants C_1 and C_2,*

$$C_1 n \log t - \mathcal{O}(n) \le S(n, t) \le C_2 n \log t + \mathcal{O}(n) \ ,$$

then

$$C_1 n \log \log n - \mathcal{O}(n) \le H(n) \le C_2 n \log \log n + \mathcal{O}(n \log \log \log n) \ .$$

Acknowledgment

Thanks to Peter Sanders for his comments on an earlier draft of this paper.

References

1. Miklós Ajtai, János Komlós, and Endre Szemerédi. Sorting in $c \log n$ parallel steps. *Combinatorica*, 3:1–19, 1983.
2. Vladimir Evgen'evich Alekseev. Sorting algorithms with minimum memory. *Kibernetika*, 5(5):99–103, 1969.
3. Samuel W. Bent and John W. John. Finding the median requires $2n$ comparisons. In *Proc. 17th Ann. ACM Symp. on Theory of Computing (STOC)*, pages 213–216, 1985.
4. Manuel Blum, Robert W. Floyd, Vaughan Pratt, Ronald L. Rivest, and Robert Endre Tarjan. Time bounds for selection. *Journal of Computer and System Sciences*, 7:448–461, 1973.
5. Paul F. Dietz. Heap construction in the parallel comparison tree model. In *Proc. 3rd Scandinavian Workshop on Algorithm Theory (SWAT)*, volume 621 of *Lecture Notes in Computer Science*, pages 140–150. Springer Verlag, Berlin, 1992.
6. Paul F. Dietz and Rajeev Raman. Very fast optimal parallel algorithms for heap construction. In *Proc. 6th Symposium on Parallel and Distributed Processing*, pages 514–521, 1994.
7. Dorit Dor and Uri Zwick. Selecting the median. In *Proc. 6th ACM-SIAM Symposium on Discrete Algorithms (SODA)*, pages 28–37, 1995.

8. Dorit Dor and Uri Zwick. Finding the alpha n-th largest element. *Combinatorica*, 16:41–58, 1996.

9. Dorit Dor and Uri Zwick. Median selection requires $(2+\epsilon)n$ comparisons. In *Proc. 37th Ann. Symp. on Foundations of Computer Science (FOCS)*, pages 125–134, 1996.

10. Robert W. Floyd. Algorithm 245: Treesort3. *Communications of the ACM*, 7(12):701, 1964.

11. Shuji Jimbo and Akira Maruoka. A method of constructing selection networks with $O(\log n)$ depth. *SIAM Journal of Computing*, 25(4):709–739, 1996.

12. Donald E. Knuth. *The Art of Computer Programming, Volume III: Sorting and Searching*. Addison-Wesley, Reading, MA, 1973.

13. Peter Bro Miltersen, Mike Paterson, and Jun Tarui. The asymptotic complexity of merging networks. *Journal of the ACM*, 43(1):147–165, 1996.

14. Stephan Olariu and Zhaofang Wen. Optimal parallel initialization algorithms for a class of priority queues. *IEEE Transactions on Parallel and Distributed Systems*, 2:423–429, 1991.

15. Nicholas Pippenger. Selection networks. *SIAM Journal of Computing*, 20(5):878–887, 1991.

16. Arnold Schönhage, Michael S. Paterson, and Nicholas Pippenger. Finding the median. *Journal of Computer and System Sciences*, 13:184–199, 1976.

17. John William Joseph Williams. Algorithm 232: Heapsort. *Communications of the ACM*, 7(6):347–348, 1964.

18. Andrew C. Yao and Frances F. Yao. Lower bounds on merging networks. *Journal of the ACM*, 23:566–571, 1976.

Two-Variable Linear Programming in Parallel*

Danny Z. Chen and Jinhui Xu

Department of Computer Science and Engineering
University of Notre Dame
Notre Dame, IN 46556, USA
{chen,jxu}@cse.nd.edu

Abstract. Two-variable linear programming is a fundamental problem in computational geometry. Sequentially, this problem was solved optimally in linear time by Megiddo and Dyer using the elegant prune-and-search technique. In parallel, the previously best known deterministic algorithm on the EREW PRAM for this problem takes $O(\log n \log \log n)$ time and $O(n)$ work. In this paper, we present a faster parallel deterministic two-variable linear programming algorithm, which takes $O(\log n \log^* n)$ time and $O(n)$ work on the EREW PRAM. Our algorithm is based on an interesting parallel prune-and-search technique, and makes use of new geometric observations which can be viewed as generalizations of those used by Megiddo and Dyer's sequential algorithms. Our parallel prune-and-search technique also leads to efficient EREW PRAM algorithms for other problems, and is likely to be useful in solving more problems.

1 Introduction

Linear programming (LP) is a central problem in discrete algorithm study and plays a key role in solving numerous combinatorial optimization problems [27]. From the geometric viewpoint, the LP problem can be considered as one of finding a point p^* such that p^* is maximal or minimal in a given direction v and that p^* is contained in the convex polyhedron which is the common intersection of n half-spaces in a d-dimensional space. Due to its importance in many areas including computational geometry, the LP problem has attracted a great deal of attention from computational geometry researchers, especially for the fixed dimension case [1, 2, 5, 8, 9, 13–17, 19, 20, 23–26, 28–30]. In this paper, we are concerned with solving deterministically in parallel the 2-dimensional case of the LP problem, called two-variable linear programming [28].

The parallel computational model we use is the EREW PRAM. The PRAM is a synchronous parallel model in which all processors share a common memory and each processor can access any memory location in constant time [22]. The EREW PRAM does not allow more than one processor to simultaneously access the same memory address. We denote the *time × processor* product of a parallel

---

* This research was supported in part by the National Science Foundation under Grant CCR-9623585.

algorithm as its *work* (i.e., the total number of operations performed by the parallel algorithm). When analyzing the complexity bounds of our parallel algorithms, we often give only their time and work bounds; the desired processor bounds of all our algorithms in this paper can be easily derived from Brent's theorem [4].

For two and three-variable linear programming, Megiddo [25] and Dyer [14] discovered optimal sequential linear time algorithms. For LP in any fixed dimension, Megiddo [26] was the first to show that a sequential linear time algorithm is also possible. The sequential linear time algorithms of Dyer and Megiddo [14, 25] for two and three-variable LP are based on the elegant prune-and-search technique. There have also been parallel deterministic LP algorithms in PRAM models. Deng [13] presented a CRCW PRAM algorithm for two-variable LP that takes $O(\log n)$ time and $O(n)$ work; Deng's algorithm is a parallel version of Megiddo and Dyer's prune-and-search approach [14, 25]. Ajtai and Megiddo [1] developed an LP algorithm in a very powerful PRAM model (counting only "comparison" steps) for any fixed dimension d, in $O((\log \log n)^d)$ time and a sub-optimal $O(n(\log \log n)^d)$ work. Sen [30] found a work-optimal algorithm on the CRCW PRAM which takes $O((\log \log n)^{d+1})$ time. (Simulating these PRAM algorithms [1, 13, 30] on the EREW PRAM in a straightforward manner would incur at least one additional $\log n$ factor in time.) Dyer [16] gave an EREW PRAM algorithm for LP in any fixed dimension d, in $O(\log n(\log \log n)^{d-1})$ time and a sub-optimal $O(n\log n(\log \log n)^{d-1})$ work. A work-optimal EREW PRAM algorithm was recently obtained by Goodrich and Ramos [20], which takes $O(\log n(\log \log n)^{d-1})$ time.

We present in this paper a faster parallel deterministic algorithm for two-variable LP. Our algorithm takes $O(\log n \log^* n)$ time and $O(n)$ work on the EREW PRAM, thus improving the time bound of the previously best known work-optimal EREW PRAM solution for two-variable LP [20] by a factor of $\frac{\log \log n}{\log^* n}$. Our algorithm is based on an interesting parallel prune-and-search technique, and makes use of new geometric observations which can be viewed as generalizations of those used by Dyer and Megiddo's sequential algorithms [14, 25]. Our parallel prune-and-search technique relies on the parallel algorithm by Chen *et al.* [7] for partitioning sorted sets. Note that previous parallel prune-and-search techniques such as the one used in Cole's EREW PRAM selection algorithm [10] does not seem to yield an $O(\log n \log^* n)$ time solution for two-variable LP (more discussion on Cole's prune-and-search technique will be given in Section 2). On the other hand, the best known EREW PRAM algorithm [20], while it solves the LP problem in any fixed dimension, uses a sophisticated derandomization technique. In comparison, our two-variable LP algorithm is relatively simple.

Our parallel prune-and-search technique also leads to efficient EREW PRAM algorithms for other problems such as the weighted selection [12]. Previously, Cole [10] gave an EREW PRAM algorithm for the (unweighted) selection problem [3, 12], which takes $O(\log n \log^* n)$ time and $O(n)$ work. For the weighted version, an EREW PRAM algorithm in $O(\log n \log \log n)$ time and $O(n)$

work was presented in [7]. By using our parallel prune-and-search technique, an $O(\log n \log^* n)$ time, $O(n)$ work EREW PRAM algorithm can be obtained for weighted selection. Note that recently there is another EREW PRAM algorithm by Hayashi, Nakano, and Olariu [21] for weighted selection which has the same complexity bounds as ours. The algorithm in [21], also making use of the parallel partition approach in [7], is an independent discovery.[1] Our parallel prune-and-search technique is likely to be useful in solving other problems.

The rest of this paper is organized as follows: Section 2 discusses certain interesting features of our parallel prune-and-search technique, in comparison with Cole's technique. In Sections 3 and 4, we describe parallel algorithms for the two-variable linear programming and weighted selection problems, respectively. Section 5 concludes the paper with some remarks.

2 Parallel Prune-and-Search

Our EREW PRAM algorithms are based on a new parallel prune-and-search technique. Note that prune-and-search techniques have been used in a number of sequential geometric algorithms [28]. In parallel, Cole [10] gave an efficient prune-and-search technique for solving the selection problem [3, 12] on the EREW PRAM. Cole's parallel selection algorithm takes $O(\log n \log^* n)$ time and $O(n)$ work. When the k-th smallest element is to be selected from an n-element set S, Cole's algorithm performs the following main steps:

1. Use k to find in S two "good" approximations to the k-th smallest element with ranks k_1 and k_2 respectively, such that $k_1 \leq k \leq k_2$ and such that $k_2 - k_1$ is sufficiently small.
2. Eliminate from S all the elements whose ranks are not in the range of $[k_1, k_2]$, and recursively solve the problem on the remaining elements in S.
3. When $|S|$ is $O(n/\log n)$, use Cole's parallel merge sort [11] to finish the selection.

Cole's selection algorithm consists of $O(\log^* n)$ recursion levels, and the computation of finding a desirable approximation at each of these levels is done by a recursive procedure which also has $O(\log^* n)$ recursion levels.

A key feature of Cole's parallel selection algorithm is that the approximations are determined based on an already known rank k, and this feature could limit the applicability of Cole's parallel prune-and-search technique. For certain problems, finding in parallel good approximations based on rank information as in [10] can be difficult, since useful rank information is not as readily available as for the selection problem. In fact, it is not clear to us that how the parallel prune-and-search technique in [10] can be applied to obtain $O(\log n \log^* n)$ time, $O(n)$ work EREW PRAM algorithms for several more complicated problems for which linear time sequential prune-and-search algorithms are already known.

---

[1] In fact, we exchanged papers on parallel weighted selection with the authors of [21] before their work appeared at the ISAAC'97 conference.

Examples of such problems include weighted selection [12] and two- and three-variable linear programming [14, 25].

Actually, the previously best known EREW PRAM algorithms for weighted selection and two-variable linear programming can be obtained by a straightforward parallelization of the corresponding sequential prune-and-search algorithms: Select the median of the candidate elements, eliminate a constant fraction of these elements, and recurse on the remaining elements. By using Cole's EREW PRAM algorithm for approximate median selection [10], one may be able to simulate in parallel a sequential prune-and-search procedure, until the number of the remaining elements is $O(n/\log n)$ (at that point, one can simply apply some known parallel algorithms such as Cole's merge sort [11]). Parallel prune-and-search algorithms of this kind can be obtained for weighted selection and two-variable linear programming; such a parallel algorithm takes $O(\log n \log \log n)$ time and $O(n)$ work on the EREW PRAM, and consists of $O(\log \log n)$ recursion levels.

Our parallel prune-and-search technique is different from Cole's [10] in the following aspects: (1) Our technique does not depend on any given rank information, (2) although this technique also consists of $O(\log^* n)$ recursion levels, the computation at each level involves no recursive procedure, and (3) this technique can be used to design efficient parallel algorithms for more problems than [10]. A key procedure used in our parallel prune-and-search algorithms is the parallel partition algorithm of Chen *et al.* [7].

The following two sections show how this technique enables us to obtain $O(\log n \log^* n)$ time, $O(n)$ work EREW PRAM algorithms for two-variable linear programming and weighted selection.

3 Parallel Two-Variable Linear Programming

We use the parallel two-variable linear programming algorithm as an example to illustrate our parallel prune-and-search technique. For the two-variable LP problem, a set of n half-planes that are respectively bounded by lines $a_i x + b_i y = c_i$, where $i = 1, 2, \ldots, n$, and an *objective function* $g(x, y) = ax + by$ in the plane are given. The goal is to find a point $p^* = (x^*, y^*)$ in the plane such that p^* is contained in the common intersection of all the n half-planes and that the objective function $g(x, y)$ attains at p^* the minimum (or maximum) value $ax^* + by^*$ [14, 25]. The common intersection of the n half-planes is a convex region called the *feasible region*. The feasible region can be unbounded or empty for a certain set of half-planes. Without loss of generality (WLOG), we assume that the objective function $g(x, y)$ corresponds to the class of all horizontal lines, that is, $g(x, y) = y$ (the general case can be easily reduced to this case by properly rotating the axes of the coordinate system). Figure 1 illustrates this case of the two-variable LP problem.

As in the sequential algorithms [14, 25], we partition the set of the n input half-planes into two subsets H_u and H_l, such that H_u (resp., H_l) contains all the *upper* (resp., *lower*) half-planes whose points are above (resp., below) or on

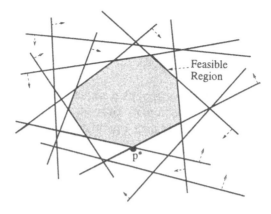

Fig. 1. Two-variable linear programming with $g(x, y) = y$ as the objective function.

their corresponding bounding lines. For the sake of simplicity in our exposition, we will illustrate the main idea and steps of our algorithm by considering the special case in which H_l is empty (the algorithm for the general case can be obtained from the one for this special case by only a few minor modifications [28]). We henceforth assume that $H_l = \emptyset$. WLOG, we also assume that no two upper half-planes in H_u are the same.

Some of the observations on which our parallel algorithm is based are actually generalizations of the geometric observations used in the sequential prune-and-search algorithms for two-variable linear programming [14, 25]. For the purpose of comparison, we first review the key observations used in [14, 25].

The sequential algorithms in [14, 25] work as follows. They group the upper half-planes of H_u into pairs in an arbitrary fashion. For each pair of half-planes, if their bounding lines are parallel with each other, then the upper half-plane whose bounding line is lower than that of the other upper half-plane is eliminated; otherwise, the algorithms compute the intersection point of the two bounding lines. For any vertical line L in the plane, the algorithms can decide in linear time whether the optimal point p^* is on, to the left of, or to the right of L. A key observation for these algorithms is that if p^* is known to lie to the left or right of the intersection point of the two bounding lines of a half-plane pair, then one of the two half-planes can be eliminated from further consideration without affecting the sought solution p^*. The algorithms in [14, 25] then find the median of the x-coordinates of the intersection points from the (at most) $\frac{n}{2}$ pairs of half-planes, and decide the position of p^* with respect to the vertical line L passing through the median. In this manner, it is ensured that $\frac{n}{4}$ intersection points are known to lie either to the left or the right of the line L, and hence at least $\frac{n}{4}$ half-planes can be eliminated or "pruned" from further consideration. The algorithms then recurse on the remaining (at most) $\frac{3n}{4}$ half-planes, for its subsequent search for the optimal solution p^*. Note that one can modify the algorithms in [14, 25] by stopping their recursion when the number of the remaining half-planes is $O(n/\log n)$, and then using an optimal algorithm for computing the intersection of half-planes [28] to find the point p^* from the remaining $O(n/\log n)$ half-planes. The number of recursion levels of such a modified algorithm is clearly $O(\log \log n)$.

One can parallelize the above modified sequential prune-and-search algorithm. It is easy to see that the number of recursion levels of this sequential algorithm affects the time bound of such a parallel algorithm. Actually, by using Cole's parallel algorithm for approximate median selection [10], it is possible to obtain a parallel version of the above sequential prune-and-search algorithm that takes $O(\log n \log \log n)$ time and $O(n)$ work on the EREW PRAM. In fact, Deng's CRCW PRAM algorithm [13] is based on this idea.

It might be tempting for one to modify the above sequential algorithm in the following manner, in the hope that the number of recursion levels of the resulted algorithm (and hence its parallelization) would be asymptotically reduced: Instead of using one vertical line (passing through the median) to determine the elimination of half-planes, use multiple vertical lines that evenly partition the intersection points into subsets. However, it can be shown that regardless of the number of vertical lines used, the number of half-planes that can be eliminated based on the observations in [14, 25] is at most $\frac{n}{2}$. Thus, such an algorithm would still have $O(\log \log n)$ recursion levels.

Our idea for achieving a faster parallel algorithm for this problem is nevertheless to use multiple vertical lines to decide the elimination of half-planes at each recursion level, so that more half-planes than a constant fraction would be removed and the number of recursion levels of the algorithm would be asymptotically reduced. To reach this goal, we must be able to do the following: (1) Use different geometric observations from those in [14, 25], (2) at each recursion level, use an efficient parallel procedure for computing such multiple vertical lines, and (3) perform linear work for the overall computation of the parallel algorithm. Actually, being able to do these will also lead to a sequential linear time prune-and-search algorithm that is quite different from [14, 25].

The following observation is a key to our parallel prune-and-search algorithm.

Lemma 1. *Let H' be a non-empty subset of the upper half-planes in H_u, and let $I(H')$ be the convex region for the common intersection of the half-planes in H'. Suppose that it is already known that the optimal point p^* for all the half-planes in H_u lies to the left (resp., right) of a vertical line L. Then the half-planes in H' of the following types can be eliminated from further consideration without affecting the search for the final solution p^*:*

1. *The half-planes in H' whose bounding lines do not contain any edge of the boundary of $I(H')$.*
2. *The half-planes in H' whose bounding lines contain an edge e of the boundary of $I(H')$, such that e does not intersect L and such that e is to the right (resp., left) of L (see Figure 2).*

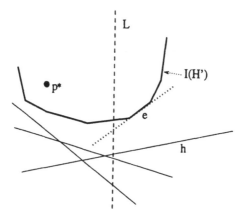

Fig. 2. Illustrating Lemma 1.

Proof. For half-planes in H' of type 1 (e.g., the half-plane with h as its bounding line in Figure 2), it is clear that they do not contribute to defining the optimal point p^*. For half-planes in H' of type 2 (e.g., the half-plane whose bounding line contains the edge e of $I(H')$ in Figure 2), their bounding lines intersect L at points that are strictly below the intersection of L and the boundary of $I(H')$ due to the convexity of $I(H')$. Hence these half-planes of H' also do not contribute to defining the point p^*.

The next lemma is for computing the multiple vertical lines used at each recursion level of our parallel algorithm.

Lemma 2. *Given a set S of n elements that is organized as n/m sorted subsets of size m each, it is possible to partition S into m subsets $D_1, D_2, \ldots, D_m$ of size $O(n/m)$ each, such that for any two indices i and j with $1 \leq i < j \leq m$, no element in D_i is bigger than any element in D_j. This partition can be obtained in parallel in $O(\log n)$ time and $O(n)$ work on the EREW PRAM.*

Proof. By using the EREW PRAM partition algorithm of Chen et al. [7].

We are now ready to present our EREW PRAM algorithm for two-variable LP.

Similar to [10], we define a class of functions $f(i)$ as follows: $f(1) = 2$ and $f(i+1) = 2^{f(i)}$ for any integer $i \geq 1$. Hence $\log(f(i+1)) = f(i)$ for $i \geq 1$, and $\log n \leq f(\log^* n) < n$ for $n \geq 2$.

Our parallel algorithm starts with the half-plane set $H(1) = H_u$ and the function $f(1)$ (i.e., with $i = 1$). Suppose that the algorithm is currently working with a subset $H(i)$ of H_u and the function $f(i)$ for some integer $i \geq 1$. Then the following steps are performed:

1. Partition, in an arbitrary fashion, the half-planes of $H(i)$ into $O(\frac{|H(i)|}{(f(i+1))^2})$ groups $G_1, G_2, \ldots,$ of size $O((f(i+1))^2)$ each.

2. In parallel, compute the common intersection of the half-planes in each group G_j, by using Chen's EREW PRAM algorithm [6]. Let the convex region for the intersection of the half-planes in G_j so obtained be denoted by $I(G_j)$. Then the boundary of $I(G_j)$ can be specified by a sequence of $O((f(i + 1))^2)$ vertices in a sorted order of their x-coordinates. As a result, there are $O(\frac{|H(i)|}{(f(i+1))^2})$ such sorted vertex sequences, one for each group G_j.

3. Partition the set of vertices which is the union of those in the $O(\frac{|H(i)|}{(f(i+1))^2})$ sorted vertex sequences (of size $O((f(i + 1))^2)$ each) into $O((f(i + 1))^2)$ subsets of size $O(\frac{|H(i)|}{(f(i+1))^2})$ each, based on Lemma 2. The EREW PRAM algorithm of Chen et al. [7] finds $O((f(i + 1))^2)$ values, in sorted order, for this partitioning. Let a vertical line pass through each of these values used for the partitioning (hence there are $O((f(i+1))^2)$ vertical lines L_1, L_2, ..., in sorted order).

4. Determine the two consecutive vertical lines L_k and L_{k+1} such that the optimal point p^* lies between L_k and L_{k+1}.

5. Eliminate half-planes from $H(i)$ based on Lemma 1. That is, for every group G_j, any half-plane of G_j whose bounding line does not contain a boundary edge e of $I(G_j)$ such that e intersects the planar region between L_k and L_{k+1} is eliminated. Let $H(i + 1)$ consist of the remaining half-planes after this elimination.

6. If $|H(i + 1)| > n/\log n$, then recurse on the half-plane set $H(i + 1)$ and the function $f(i + 1)$.

7. If $|H(i + 1)| \leq n/\log n$, then use Chen's EREW PRAM algorithm [6] to compute the intersection of the half-planes in $H(i+1)$, and find p^* from this intersection.

The correctness of the above algorithm is ensured by Lemma 1. In the rest of this section, we discuss the details of this parallel algorithm, and analyze its parallel complexity bounds.

It is easy to perform the partitioning of Step 1 in $O(\log |H(i)|)$ time and $O(|H(i)|)$ work on the EREW PRAM. For Step 2, the intersection of the $O((f(i+1))^2)$ half-planes in each group G_j can be computed by Chen's EREW PRAM algorithm [6] in $O(\log(f(i+1))) = O(f(i))$ time and $O((f(i+1))^2 \log(f(i+1))) = O((f(i+1))^2 f(i))$ work. Hence Step 2 takes $O(f(i))$ time and $O(|H(i)|f(i))$ work over all the $O(\frac{|H(i)|}{(f(i+1))^2})$ groups G_j. Since the vertex sequence for every group G_j is already sorted by their x-coordinates, the partitioning of Step 3, by Lemma 2, can be done in $O(\log |H(i)|)$ time and $O(|H(i)|)$ work on the EREW PRAM. The algorithm of Chen et al. [7] finds, for this partitioning, $O((f(i+1))^2)$ values r_1, r_2, ..., which are the x-coordinates of some vertices from the $O(\frac{|H(i)|}{(f(i+1))^2})$ vertex sequences. These values are obtained in sorted order. As a result, there are $O((f(i + 1))^2)$ vertical lines L_1, L_2, ..., with the line L_s passing through the value r_s. Note that for a vertical line L_s, one can decide the position of the optimal point p^* with respect to L_s, as follows:

Compute the intersection between L_s and the boundary of every convex region $I(G_j)$, and obtain the slope information of the boundary of $I(G_j)$

around this intersection (this can be done by performing a binary search for the value r_s on the sorted vertex sequence along the boundary of $I(G_j)$). Then use the intersection between L_s and the boundary of each $I(G_j)$ and the slope information of $I(G_j)$ around this intersection to decide the position of p^* with respect to L_s (the computation for this has been discussed in [14, 25], and it can be done by performing $O(1)$ parallel prefix operations [22]).

Since there are $O((f(i+1))^2)$ vertical lines L_s and since the size of the vertex sequence for each group G_j is also $O((f(i+1))^2)$, one can compute the intersections between all such vertical lines and the boundary of each $I(G_j)$ by a parallel merge [22] instead of binary searches. Altogether, Step 4 takes $O(\log |H(i)|)$ time and $O(|H(i)|)$ work on the EREW PRAM. The elimination of half-planes in Step 5 can be easily done in $O(\log |H(i)|)$ time and $O(|H(i)|)$ work on the EREW PRAM. Step 7 takes $O(\log n)$ time and $O(n)$ work on the EREW PRAM by applying Chen's algorithm [6] to a problem of size $O(n/\log n)$.

Summarizing over all the non-recursive steps of the above algorithm, it is clear that the non-recursive computation on the subset $H(i)$ of H_u takes altogether $O(\log |H(i)|)$ time and $O(|H(i)|f(i))$ work on the EREW PRAM.

We now analyze the size of the subset $H(i+1)$ that contains the remaining half-planes from $H(i)$ after the elimination of Step 5. Suppose that the optimal point p^* lies between the two consecutive vertical lines L_k and L_{k+1}. Then by Lemma 2, Step 3 ensures that only $O(\frac{|H(i)|}{(f(i+1))^2})$ vertices from the $O(\frac{|H(i)|}{(f(i+1))^2})$ vertex sequences are between L_k and L_{k+1}. If a convex region $I(G_j)$ has w boundary vertices between L_k and L_{k+1}, then $w+1$ edges on the boundary of $I(G_j)$ intersect the planar region between L_k and L_{k+1}, and hence $w+1$ half-planes of G_j belong to $H(i+1)$ by Lemma 1. Since there are $O(\frac{|H(i)|}{(f(i+1))^2})$ groups and each of them contributes at most one additional half-plane to $H(i+1)$ (other than those half-planes that are associated with the $O(\frac{|H(i)|}{(f(i+1))^2})$ vertices between L_k and L_{k+1}), the total number of half-planes in the set $H(i+1)$ is $O(\frac{|H(i)|}{(f(i+1))^2})$.

From the above analysis of the size of $H(i+1)$, it is easy to see that the number of recursion levels of our EREW PRAM prune-and-search algorithm for two-variable linear programming is $O(\log^* n)$. Hence the time bound of our algorithm is $O(\log n \log^* n)$. The work bound of the algorithm is $\sum_{i=1}^{O(\log^* n)} O(|H(i)|f(i)) = O(n)$.

Theorem 1. *The two-variable linear programming of size n can be solved in $O(\log n \log^* n)$ time and $O(n)$ work on the EREW PRAM.*

Proof. It follows from the above discussion.

4 Parallel Weighted Selection

Given a set S of n elements $a_1, a_2, \ldots, a_n$ such that each a_i is associated with a nonnegative weight w_i, and given a number w, the weighted selection problem is to find an element $a_k \in S$ such that $\sum_{a_i \in S, \; a_i < a_k} w_i < w$ and $\sum_{a_i \in S, \; a_i \leq a_k} w_i \geq w$ [12]. (If two elements a_i and a_j, $i \neq j$, are equal to each other, we assume that there is a consistent way to break the tie, e.g., by the values of their indices i and j.) Note that the selection problem [3, 12] is a special case of the weighted selection (by letting the weights be all equal). The previously best known EREW PRAM algorithm for weighted selection takes $O(\log n \log \log n)$ time and $O(n)$ work.

Our parallel weighted selection algorithm actually is quite similar to and is simpler than our parallel two-variable linear programming algorithm. We define the functions $f(i)$ for $i = 1, 2, \ldots$, as in the previous section.

Our parallel weighted selection algorithm starts with the element set $S(1) = S$, the input value w, and the function $f(1)$ (i.e., with $i = 1$). Suppose that the algorithm is currently working with a subset $S(i)$ of S, the value w, and the function $f(i)$ for some integer $i \geq 1$. Then the following steps are performed:

1. Partition, in an arbitrary fashion, the elements of $S(i)$ into $O(\frac{|S(i)|}{(f(i+1))^2})$ groups $G_1, G_2, \ldots$, of size $O((f(i+1))^2)$ each.
2. In parallel, sort the elements of each group G_j by using Cole's merge sort [11].
3. Partition the elements from the $O(\frac{|S(i)|}{(f(i+1))^2})$ sorted groups (of size $O((f(i+1))^2)$ each) into $O((f(i+1))^2)$ subsets $A_1, A_2, \ldots$, of size $O(\frac{|S(i)|}{(f(i+1))^2})$ each, based on Lemma 2.
4. Determine the subset A_s such that A_s contains the element a_k that we seek as the output for the weighted selection problem. This can be done by first computing the sum W_r of the weights of all the elements in each subset A_r, then computing the prefix sums of $W_1, W_2, \ldots$, and finally locating the subset A_s by using these prefix sums and the value w.
5. Let $S(i+1)$ be A_s, and appropriately adjust the value of w (for recursively solving the problem on $S(i+1)$).
6. If $|S(i+1)| > n/\log n$, then recurse on the element set $S(i+1)$, the modified value w, and the function $f(i+1)$.
7. If $|S(i+1)| \leq n/\log n$, then use Cole's merge sort [11] to sort the elements in $S(i+1)$, and use parallel prefix [22] on the weights of these sorted elements to find the sought element a_k from $S(i+1)$.

It is easy to see that the above parallel algorithm correctly solves the weighted selection problem. This algorithm takes $O(\log n \log^* n)$ time and $O(n)$ work on the EREW PRAM. The analysis is similar to that for our parallel two-variable linear programming algorithm.

Theorem 2. *The weighted selection problem of size n can be solved in $O(\log n \log^* n)$ time and $O(n)$ work on the EREW PRAM.*

Proof. It follows from the above discussion.

5 Conclusion

We have proposed in this paper a new parallel prune-and-search technique and used it to solve several problems such as two-variable linear programming and weighted selection. The algorithms we developed run in $O(\log n \log^* n)$ time and $O(n)$ work on the EREW PRAM, thus improving the time bound of the previously best known solutions. Our parallel technique could be applicable to solving other problems, such as three-variable linear programming [14, 25, 28].

It is not clear to us whether the $O(\log n \log^* n)$ parallel time bound for these two problems is optimal, provided that work-optimal EREW PRAM algorithms are desired. Of course, one could use work-suboptimal EREW PRAM algorithms [6, 11] to solve these problems in $O(\log n)$ time. But can one achieve $O(\log n)$ time, $O(n)$ work EREW PRAM algorithms for two-variable linear programming and weighted selection?

References

1. M. Ajtai and N. Megiddo, "A deterministic poly(log log N)-time N-processor algorithm for linear programming in fixed dimension," *SIAM J. Comput.*, 25 (6) (1996), pp. 1171-1195.
2. N. Alon and N. Megiddo, "Parallel linear programming in fixed dimension almost surely in constant time," *Proc. 31st Annual IEEE Symp. on Foundation of Computer Science*, 1990, pp. 574-582.
3. M. Blum, R.W. Floyd, V.R. Pratt, R.L. Rivest, and R.E. Tarjan, "Time bounds for selection," *Journal of Computer and System Sciences*, 7 (1973), pp. 448-461.
4. R.P. Brent, "The parallel evaluation of general arithmetic expressions," *Journal of the ACM*, 21 (1974), pp. 201-206.
5. B. Chazelle and J. Matoušek, "On linear-time deterministic algorithms for optimization problems in fixed dimension," *Proc. 4th Annual ACM-SIAM Symp. on Discrete Algorithms*, 1993, pp. 281-290.
6. D.Z. Chen, "Efficient geometric algorithms on the EREW PRAM," *IEEE Trans. on Parallel and Distributed Systems*, 6 (1) (1995), pp. 41-47.
7. D.Z. Chen, W. Chen, K. Wada, and K. Kawaguchi, "Parallel algorithms for partitioning sorted sets and related problems," *Proc. of 4th Annual European Symp. on Algorithms*, 1996, pp. 234-245.
8. K.L. Clarkson, "Linear programming in $O(n3^{d^2})$ time," *Information Processing Letters*, 22 (1986), pp. 21-24.
9. K.L. Clarkson, "A Las Vegas algorithm for linear programming when the dimension is small," *J. of the ACM*, 42 (1995), pp. 488-499.
10. R.J. Cole, "An optimally efficient selection algorithm," *Information Processing Letters*, 26 (1987/1988), pp. 295-299.
11. R.J. Cole, "Parallel merge sort," *SIAM J. Computing*, 17 (1988), pp. 770-785.
12. T.H. Cormen, C.E. Leiserson, and R.L. Rivest, *Introduction to Algorithms*, McGraw-Hill, 1990.
13. X. Deng, "An optimal parallel algorithm for linear programming in the plane," *Information Processing Letters*, 35 (1990), pp. 213-217.

14. M.E. Dyer, "Linear time algorithms for two- and three-variable linear programs," *SIAM J. Computing*, 13 (1984), pp. 31–45.

15. M.E. Dyer, "On a multidimensional search technique and its applications to the Euclidean one-center problem," *SIAM J. Computing*, 15 (1986), pp. 725–738.

16. M.E. Dyer, "A parallel algorithm for linear programming in fixed dimension," *Proc. 11th Annual Symp. on Computational Geometry*, 1995, pp. 345–349.

17. H. Edelsbrunner, *Algorithms in Combinatorial Geometry*, Springer-Verlag, New York, 1987.

18. M.T. Goodrich, "Geometric partitioning made easier, even in parallel," *Proc. 9th Annual ACM Symp. on Computational Geometry*, 1993, pp. 73–82.

19. M.T. Goodrich, "Fixed-dimensional parallel linear programming via relative epsilon-approximations," *Proc. 7th Annual ACM-SIAM Symp. on Discrete Algorithms*, 1996, pp. 132–141.

20. M.T. Goodrich and E.A. Ramos, "Bounded-independence derandomization of geometric partitioning with applications to parallel fixed-dimensional linear programming," to appear in *Discrete & Computational Geometry*.

21. T. Hayashi, K. Nakano, and S. Olariu, "Weighted and unweighted selection algorithms for k sorted sequences," *Proc. 8th International Symp. on Algorithms and Computation*, 1997, pp. 52–61.

22. J. JáJá, *An Introduction to Parallel Algorithms*, Addison-Wesley, Reading, Massachusetts, 1992.

23. G. Kalai, "A subexponential randomized simplex algorithm," *Proc. 24th Annual ACM Symp. on Theory of Computing*, 1992, pp. 475–482.

24. J. Matoušek, M. Sharir, and E. Welzl, "A subexponential bound for linear programming," *Algorithmica*, 16 (1996), pp. 498–516.

25. N. Megiddo, "Linear time algorithms for linear programming in R^3 and related problems," *SIAM J. Computing*, 12 (4) (1983), pp. 759–776.

26. N. Megiddo, "Linear programming in linear time when the dimension is fixed," *Journal of ACM*, 31 (1) (1984), pp. 114–127.

27. C.H. Papadimitriou and K. Steiglitz, *Combinatorial Optimization: Algorithms and Complexity*, Prentice-Hall, 1982.

28. F.P. Preparata and M.I. Shamos, *Computational Geometry: An Introduction*, Springer-Verlag, Berlin, 1985.

29. R. Seidel, "Small-dimensional linear programming and convex hulls made easy," *Discrete & Computational Geometry*, 6 (1991), pp. 423–434.

30. S. Sen, "A deterministic poly($\log \log n$) time optimal CRCW PRAM algorithm for linear programming in fixed dimension," Technical Report 95-08, Dept. of Computer Science, University of Newcastle, 1995.

Optimal Deterministic Protocols for Mobile Robots on a Grid*

Roberto Grossi[1], Andrea Pietracaprina[2], and Geppino Pucci[3]

[1] Dipartimento di Sistemi e Informatica, Università di Firenze, Italy
E-mail: grossi@dsi.unifi.it
[2] Dipartimento di Matematica Pura e Applicata, Università di Padova, Italy
E-mail: andrea@artemide.dei.unipd.it
[3] Dipartimento di Elettronica e Informatica, Università di Padova, Italy
E-mail: geppo@artemide.dei.unipd.it

Abstract. A *Multi Robot Grid System* consists of m robots that operate in a set of $n \geq m$ work locations connected by aisles in a $\sqrt{n} \times \sqrt{n}$ grid. From time to time the robots need move along the aisles, in order to visit disjoint sets of locations. The movement of the robots must comply with the following constraints: (1) no two robots can collide at a grid node or traverse an edge at the same time; (2) a robot's sensory capability is limited to detecting the presence of another robot at a neighboring node. We present an efficient deterministic protocol that allows $m = \Theta(n)$ robots to visit their target destinations in $O\left(\sqrt{dn}\right)$ time, where each robot visits at most $d \leq n$ targets in any order. We also prove a lower bound that shows that our protocol is optimal. Prior to this paper, no optimal protocols were known for $d > 1$. For $d = 1$ optimal protocols were known only for $m = O\left(\sqrt{n}\right)$, while for $m = O(n)$ only a randomized suboptimal protocol was known.

1 Introduction

A *Multi Robot Grid* (MRG) system consists of m *robots* that operate in a set of $n \geq m$ work locations connected by aisles in a $\sqrt{n} \times \sqrt{n}$ grid [ST95]. At any time, the robots are located at distinct grid nodes, and from time to time each robot is given a set of *targets*, i.e., some work locations it must visit. These sets are disjoint and no particular order is prescribed for visiting the targets in each set. Moreover, robots may end up at arbitrary locations, once their visits are completed. We may regard the system as representing a warehouse or a tertiary storage (tape) system, where robots are employed to gather or redistribute items. For simplicity, we assume that the system is synchronous, that is, all robots are provided with identical clocks. Our goal is to design an efficient on-line distributed protocol that every robot must follow to visit the assigned targets avoiding deadlocks and conflicts with other robots. More specifically, the protocol must comply with the following rules:

---

* Part of this work was done while the authors were attending the *Research Retreat on Sense of Direction and Compact Routing* held in June 1997 at Certosa di Pontignano, Siena, Italy. This work was supported, in part, by CNR and MURST of Italy.

1. At any time all the robots reside in the grid, i.e., no robot can leave or enter the grid. No two robots can collide at a grid node or traverse an edge at the same time.
2. A robot cannot exchange information with other robots directly. However, each robot is equipped with a short-range sensor that is able to detect the presence of other robots occupying nodes at (Manhattan) distance one from its current location.
3. In one time unit, a robot can perform a constant amount of internal computation, read the output of its short-range sensor, and decide whether to remain at the current grid node, or to move to a neighboring node not occupied by another robot.

In the MRG(n, m, d) *problem*, each of $m < n$ robots in an MRG system is required to visit (at most) $d \leq n$ targets, with no grid node being target for more than one robot.

Related Work The MRG(n, m, d) problem was originally introduced in [ST95] as a practical case study within the more general quest for *social laws* to coordinate the actions of agents in a common environment without resorting to central control. The problem is affine to a routing problem on a two-dimensional grid. However, as observed in [PU96], none of the many grid-routing algorithms known in the literature (see [Lei92] and [Sib95] for comprehensive accounts), appears to be directly applicable to solve the MRG(n, m, d) problem, because they do not comply with the stringent rules that govern the movement of robots in an MRG system. In particular, most known algorithms require buffering, that is, more than one packet (corresponding to a robot in our case) may be stored at one node at the same time. In fact, even *hot-potato* routing protocols, where the use of internal buffers is not permitted, work under the assumption that in one time unit a node may receive a packet from each neighbor, manipulate the information carried by the packet headers and redistribute the packets, suitably permuted, to the neighbors (e.g., see [NS95]). In addition, our model does not allow the robots to disappear from the grid or to be replicated, whereas packets can be destroyed and replicated.

More generally, typical on-line routing protocols require that the information carried by the packets be read and processed at grid nodes, which are computing units. In contrast, in an MRG system, grid nodes are passive entities, and the robots, which are the agents of the system, must orchestrate their movement according to a conflict-free motion plan, without being able to exchange information directly and under very limited sensory capabilities.

Any instance of the MRG(n, m, d) problem can be trivially completed in n steps by letting the robots circulate along a directed Hamiltonian cycle traversing all the grid nodes. In fact, Preminger and Upfal [PU96] proved that any deterministic protocol where robots are completely blind (i.e., a robot cannot detect the presence of other robots at any distance) requires $\Theta(n)$ time, thus implying the optimality of the trivial strategy in this case.

If the robots are not blind, $\Omega(\sqrt{n})$ time is necessary in the worst case, due to the grid diameter. A single robot with a single destination can achieve this bound by simply traversing the shortest path from its source to its target. For a larger number m of robots and a single destination per robot (the MRG$(n, m, 1)$ problem) two optimal, $\Theta(\sqrt{n})$-time protocols are presented in [ST92,ST95] for the following special cases. The first protocol is designed for $m \leq n^{1/4}$ robots, while the second one works for $m \leq \sqrt{n}$ robots, as long as they initially reside in distinct columns.

The only known protocol that deals with an arbitrary number of $m \leq n$ robots and a single destination is given in [PU96]. The algorithm is randomized and solves any MRG$(n, m, 1)$ problem in suboptimal $O(\sqrt{n} \log n)$ time, with high probability. However, the algorithm works under a relaxed set of rules. Specifically, it assumes that a robot's short-range sensor is able to detect the presence of other robots at distance at most two, that a robot may initially stay outside the grid for an arbitrary amount of time, and that robots disappear from the grid as soon as they have visited their target. No general deterministic protocols that take $o(n)$ time are known for the MRG$(n, m, 1)$ problem.

For the case of $d > 1$ targets, one could repeat the single-target protocol d times. However, as we will show in the paper, this strategy does not achieve optimal performance. To the best of our knowledge, no specific algorithm for the case of $d > 1$ targets has been developed so far.

Our Results We devise a simple and general protocol for the MRG$(n, m, 1)$ problem, with $m \leq n/4$, which attains optimal $\Theta(\sqrt{n})$ time. The algorithm implements a sorting-based routing strategy in a way that fully complies with the constraints imposed by an MRG system. Our protocol improves upon the work of [PU96] in several directions. First, the protocol is deterministic, hence it provides a worst-case guarantee on performance. Second, it achieves optimality in the general case, thus reducing the running time of [PU96] by an $O(\log n)$ factor. Third, it works in a weaker model in which the robots reside in the grid all the time and their sensor can detect other robots at distance one.

Next, we consider the case of $d > 1$ targets. If the order of the visits were fixed *a priori* for each robot, then it is immediate to prove that any protocol for the problem would require $\Omega(d\sqrt{n})$ time, hence applying our optimal MRG$(n, m, 1)$ protocol d times would yield an optimal $\Theta(d\sqrt{n})$-time general solution. However, since robots can arbitrarily rearrange the order of their targets, this approach becomes suboptimal. Indeed, we prove an $\Omega\left(\sqrt{dn}\right)$ lower bound to the MRG(n, m, d) problem and provide an optimal $\Theta\left(\sqrt{dn}\right)$-time protocol that matches the lower bound for arbitrary values of d and $m \leq (3/16)n$. Ours is the first nontrivial solution to the most general case of the MRG problem.

2 The Case of a Single Target

In this section we present a deterministic protocol that solves any instance of the MRG$(n, m, 1)$ problem in optimal $O(\sqrt{n})$ time for $m \leq n/4$ robots. The basic idea behind our protocol is to perform the routing through sorting, which is a typical strategy employed in the context packet routing. However, we need to develop specific primitives in order to implement such a strategy under the restrictive rules of an MRG system. In the following, we assume that at any time each robot knows the coordinates of its current location. Also, for the sake of simplicity, we assume that n is a power of 4. (Minor modifications are needed to deal with the general case.)

Let us consider the grid as partitioned into $n/4$ subgrids of size 2×2, which we call *tiles*. The protocol has a simple high-level structure consisting of the four phases outlined below.

Phase I — Balancing: The robots relocate on the grid so that each robot ends up in the top-left node of a distinct tile.

Phase II — Sorting-by-Row: The robots sort themselves by target row. The sorted sequence of robots is arranged on the grid (one robot per tile) according to the *Peano indexing* [Mo66] shown pictorially in Figure 1 and described mathematically later. In other words, at the end of the sorting, the i-th robot in the sorted order occupies the top-left corner of the tile of Peano index i.

Phase III — Permuting: The sorted sequence of robots permutes from the Peano indexing to the row-major indexing.

Phase IV — Routing: The robots visit their targets by exploiting their row-major order.

Before describing the four phases in more detail, we show how to perform some basic primitives in an MRG system which will be needed to implement the above phases.

Pack *Given $q \le t$ robots on a t-node linear array, pack them into q consecutive nodes at one end of the array.*
Solution: Each robot repeatedly crosses an edge towards the designated end whenever its short-range sensor detects that the node across the edge is empty. It is easy to see that no collisions can arise in this way, and after $2t$ time steps all the robots have completed the packing.

Count *Given $q \le t$ robots on a t-node linear array, make q known to each robot.*
Solution: The robots first pack at one end of the array and then at the other. A robot that ends up at the i-th location from one end and at the j-th location from the other, sets $q = i + j - 1$. Count requires no more than $4t$ steps.

Compare-Swap in a Tile *Given a tile with two robots in it, sort the two robots so that the one associated with the smaller target row goes to the top left corner, while the other goes to the bottom left corner.*
Solution: Suppose that the two robots start at the top and bottom left corners of the tile. The robots execute a number of rounds until they "learn" their relative order in the sorted sequence. Specifically, in the i-th round, the robots "compare" the i-th most significant bit of the binary representation of their respective target row as follows. A robot moves one step to the right if its bit is 1. Then each robot can determine the other robot's bit by simply checking

for its presence in the same column. The first time that the robots find different bits, the robot whose bit is 0 moves to the top left corner, while the other moves to the bottom left corner, and the algorithm ends. If the robots have the same target row (i.e., all bits are equal) they stay in the original positions. Overall, compare-swap takes no more than $\log n$ steps. A similar procedure can be used when the robots start (and end) in the top left and right corners.

In the following subsections, we describe the four phases of our protocol in more detail.

2.1   Phase I: Balancing

In this phase, the $m \leq n/4$ robots start at arbitrary positions in the grid and must distribute themselves among the $n/4$ tiles so that each tile contains at most one robot in its top-left node. This is accomplished in $\log n - 2$ *balancing steps*, numbered from 0 to $\log n - 3$, according to the following inductive scheme.[1] At the beginning of Step i, with i even, the robots are already distributed evenly among square subgrids of size $\sqrt{n/2^i} \times \sqrt{n/2^i}$ by induction. (This holds vacuously for $i = 0$). During the step, the robots work independently within each square subgrid, and partition themselves evenly among rectangular subgrids of size $\sqrt{n/2^i} \times \sqrt{n/2^{i+2}}$. Analogously, in Step i with i odd, the robots work independently within each rectangular subgrid of size $\sqrt{n/2^{i-1}} \times \sqrt{n/2^{i+1}}$, and partition themselves evenly among square subgrids of size $\sqrt{n/2^{i+1}} \times \sqrt{n/2^{i+1}}$. Since n is a power of 4, at the end of Step $\log n - 3$ the robots are evenly partitioned among the tiles, with at most one robot per tile. At this point, each robot moves to the top-left corner of its tile.

We now describe the implementation of balancing step i, with i odd (the implementation of a balancing step of even index requires only minor modifications). Consider an arbitrary $t \times t/2$ rectangular subgrid, with $t = \sqrt{n/2^{i-1}}$, and suppose that there are p robots in the subgrid. Let the rows (resp., columns) of the subgrid be numbered from 1 to t (resp., $t/2$). At the end of the step we want to have $\lfloor p/2 \rfloor$ robots in the upper half (top $t/2$ rows) and the remaining $\lceil p/2 \rceil$ in the lower half (bottom $t/2$ rows). This is done through the following substeps.

1. The robots in each row pack towards the left.
2. The robots in each column pack towards the bottom.
   Comment: After this step, the robots form a "staircase" ascending from the bottom-right corner to the top-left corner of the subgrid.
3. In each column $k < t/2$, each robot counts the number of robots in the column. If this number is odd, the topmost robot (referred to as *leftover*) is sent to the top of the column.
4. All leftovers pack towards the right of the topmost row. Then they move down along column $t/2$ towards the bottom. Then, in column $t/2$, each robot determines the number of robots.

---

[1] Unless otherwise specified, all logarithms are taken to the base 2.

Comment: It is easy to show that if $p < t^2/4$ (which is always the case) then there is enough room in column $t/2$ to hold all leftovers.

5. For every column k, let x be number of robots in the column after Step 4. (Note that x may be odd only for $k = t/2$.) If $k < t/2$, the robots pack around the column center, i.e., on rows $(t-x)/2+1, (t-x)/2+2, \ldots, (t+x)/2$. If $k = t/2$, the robots pack so that $\lfloor x/2 \rfloor$ of them end up in the upper half and the remaining $\lceil x/2 \rceil$ end up in the lower half.

Lemma 1. *The balancing phase for $m \leq n/4$ robots takes $O(\sqrt{n})$ time.*

Proof. The correctness of the above strategy is immediate. The time bound follows by observing that balancing Step i can be executed in $O\left(\sqrt{n/2^i}\right)$ time by using the pack and count primitives presented before.

2.2 Phase II: Sorting-by-Row

At the end of the balancing phase, the robots are spread among the grid nodes in such a way that there is at most one robot in each tile, parked in the tile's top-left corner. The robots will now sort themselves according to their target row, with ties broken arbitrarily. Let $M = m/4$. The sorting algorithm relies upon a grid implementation of Batcher's bitonic sorting algorithm [Ba68] for sequences of size M or smaller. We recall that Batcher's algorithm is structured as a cascade of $\log M$ *merging stages*. At the beginning of the i-th merging stage, $1 \leq i \leq \log M$, the robots are partitioned into $M/2^{i-1}$ sorted subsequences each of size 2^{i-1}. Then, pairs of subsequences are merged independently so that, at the end of the stage, there are $M/2^i$ sorted subsequences each of size 2^i. In turn, the i-th merging stage is made of a sequence of i (i,j)-*compare-swap* steps, for $j = i-1, i-2, \ldots 0$. More specifically, an (i,j)-compare-swap step compares and swaps pairs of elements in each subsequence at distance 2^j (the direction of the compare/swap operator is fixed *a priori* and depends on the values of i and j).

In order to efficiently implement Batcher's algorithm on the grid, we number the M tiles according to the so-called *Peano indexing* [Mo66], which is defined as follows (see Figure 1). Split the set of indices $I = \{0, \ldots, M-1\}$ into four equally sized subsets of consecutive indices $I_0 = \{0, \ldots, M/4-1\}, I_1 = \{M/4, \ldots, M/2-1\}, I_2 = \{M/2, \ldots, 3M/4-1\}, I_3 = \{3M/4, \ldots, M-1\}$. Similarly, split the grid into four quadrants of $M/4$ tiles each and assign the four subsets of indices to the four quadrants, namely, $H_{t\ell}, H_{b\ell}, H_{tr}$, and H_{br}, where t stands for "top," b for "bottom," ℓ for "left," and r for "right." Assign the set of indices I_0 to $H_{t\ell}, I_1$ to $H_{b\ell}, I_2$ to H_{tr} and I_3 to H_{br}. Then proceed recursively within the quadrants until subsets of size one and quadrants of one tile each are reached. Note that in the resulting linear ordering, a pair of tiles at distance $M/2$ occupy the same position within the two subgrids obtained by partitioning the grid vertically at column $n/2$. Analogously, tiles at distance $M/4$ occupy the same position within the two subgrids obtained by partitioning the grid horizontally at row $n/2$. This property carries on recursively in each of the four quadrants for geometrically smaller distances. In general, an easy induction shows that the

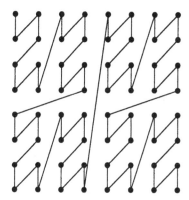

Fig. 1. 64 tiles arranged according to the Peano curve.

two tiles numbered h and $h + 2^j$ in the above linear ordering are connected by a path of at most $\sqrt{2^{j+1}}$ grid nodes and occupy the same position in two adjacent subgrids of 2^j nodes.

An (i, j)-compare-swap step can be performed as follows. Let k denote an arbitrary integer in $\{0, \ldots, m-1\}$ whose binary representation has a 0 in the j-th position. The following substeps are executed in parallel for all such values of k:

1. The robot residing in tile $k + 2^j$ in the Peano indexing moves to tile k.
2. The two robots in tile k execute the compare-swap operation according to their target row, with ties being broken arbitrarily.
3. The robot with the larger or smaller target moves to tile $k + 2^j$, depending on the direction of the (i, j)-compare-swap operator.

The routing implied by Steps 1 and 3 above is easily performed by the robots without collisions. In particular, when j is odd, the robots in tiles $k + 2^j$ first move to the bottom-left corner of the tile, and then move left until they reach the bottom-left corner of tile k (which, by our numbering, is on the the same row of tile $k+2^j$). When j is even, the robots in tiles $k+2^j$ first move to the top-right corner of the tile and then move upwards along the column, until they reach the top-right corner of tile k. Hence, Steps 1 and 3 require $O\left(\sqrt{2^j}\right)$ time altogether. By using the compare-swap primitive discussed before, Step 2 requires $O(\log n)$ time.

Lemma 2. *The sorting phase for $m \le n/4$ robots takes $O(\sqrt{n})$ time.*

Proof. The i-th merging stage of the sorting algorithm, $0 \le i \le \log n - 3$ consists of a sequence of (i, j)-compare-swap steps, for $j = i, i-1, \ldots, 1, 0$. An (i, j)-compare-swap step takes $O\left(\sqrt{2^j} + \log n\right)$ time altogether, hence the total running time of the algorithm is

$$T_{\text{sort}}(n) = \sum_{i=0}^{\log n - 3} \sum_{j=0}^{i} O\left(\sqrt{2^j} + \log n\right) = O\left(\sqrt{n}\right).$$

2.3 Phase III: Permuting

After the sorting phase, the robots are sorted by target row and are distributed one per tile according to the Peano indexing. In Phase III, the robots permute in such a way that the sorted sequence is rearranged in row-major indexing. Let us call *t-column* (resp., *t-row*) a column (resp., row) of tiles. The permutation is executed according to the following recursive protocol.

1. Each robot in H_{tr} swaps with the one occupying the same position in H_{bl}.
2. Within each quadrant, the sorted subsequence of robots recursively permutes from Peano to row-major indexing.
3. Within each quadrant, the robots permute so that those in odd rows pack to the top, while those in even rows pack to the bottom of the quadrant.
4. Each robot in the lower half of H_{tl} (resp., H_{bl}) swaps with the one occupying the same position in the top half of H_{tr}, (resp., H_{br}).

The correctness of the permutation protocol can be immediately established. Below, we give a pictorial illustration of the steps for $m = 64$ robots, where Step 4 is not shown as it gives the final permutation.

1	3	9	11	33	35	41	43
2	4	10	12	34	36	42	44
5	7	13	15	37	39	45	47
6	8	14	16	38	40	46	48
17	19	25	27	49	51	57	59
18	20	26	28	50	52	58	60
21	23	29	31	53	55	61	63
22	24	30	32	54	56	62	64

Initial configuration

1	3	9	11	17	19	25	27
2	4	10	12	18	20	26	28
5	7	13	15	21	23	29	31
6	8	14	16	22	24	30	32
33	35	41	43	49	51	57	59
34	36	42	44	50	52	58	60
37	39	45	47	53	55	61	63
38	40	46	48	54	56	62	64

After Step 1

1	2	3	4	17	18	19	20
5	6	7	8	21	22	23	24
9	10	11	12	25	26	27	28
13	14	15	16	29	30	31	32
33	34	35	36	49	50	51	52
37	38	39	40	53	54	55	56
41	42	43	44	57	58	59	60
45	46	47	48	61	62	63	64

After Step 2

1	2	3	4	17	18	19	20
9	10	11	12	25	26	27	28
5	6	7	8	21	22	23	24
13	14	15	16	29	30	31	32
33	34	35	36	49	50	51	52
41	42	43	44	57	58	59	60
37	38	39	40	53	54	55	56
45	46	47	48	61	62	63	64

After Step 3

1	2	3	4	5	6	7	8
9	10	11	12	13	14	15	16
17	18	19	20	21	22	23	24
25	26	27	28	29	30	31	32
33	34	35	36	37	38	39	40
41	42	43	44	45	46	47	48
49	50	51	52	53	54	55	56
57	58	59	60	61	62	63	64

Final configuration

Lemma 3. *The permuting phase for $m \leq n/4$ robots takes $O(\sqrt{n})$ time.*

Proof. It is immediate to see that the movements of robots implied by Step 1, Step 3 and Step 4 can be executed in a conflict-free fashion in $O(\sqrt{n})$ time. Since the recursive step is executed in parallel and independently within subgrids of geometrically decreasing side, we conclude that the overall permutation time is also $O(\sqrt{n})$.

2.4 Phase IV: Routing

The routing phase starts with the robots sorted by target row and arranged in a row major fashion with at most one robot per tile (in the tile's top-left corner). It is worth noting that there are $\sqrt{n}/2$ tiles in a t-column or in a t-row, and that due to the sorting, each t-column holds no more than two robots with targets in the same row. The routing is performed by first moving the robots to their target row and then to their final target. This is accomplished in parallel as follows.

1. The robot residing in t-column $2i$ and t-row j moves to the top-right corner of the tile in t-column $2i - 1$ and t-row j, for $1 \leq i \leq \sqrt{n}/4$ and $1 \leq j \leq \sqrt{n}/2$.
   Comment: After this step, in a nonempty t-column there can be up to four robots destined for the same row.

2. The robots in each odd-numbered t-column perform a complete rotation of the nodes of the t-column. When a robot traveling on the right side of the t-column reaches its target row, it attempts to shift right to the adjacent tile in the next t-column and the same t-row, and then moves to the rightmost unoccupied node in such tile.
   Comment: No more than two robots per row are able to move to the adjacent t-column.

3. The robots in each t-row perform a complete rotation of the nodes of the t-row, therefore visiting their target locations.

4. All the robots go back to the t-columns they occupied at the end of Step 1.

5. Steps 2–3 are repeated to deliver the remaining robots that have not reached their targets. To this end, the robots that have already completed their task will not attempt to shift to the next t-column during Step 2.
   Comment: The remaining robots are now able to move to the adjacent t-column.

We have:

Lemma 4. *The routing phase for $m \leq n/4$ robots takes $O(\sqrt{n})$ time.*

Proof. Steps 1–3 require $O(\sqrt{n})$ time altogether and are executed at most two times each. Step 4 can be executed as follows. In each odd-numbered t-column robots in each row pack to the left. Then, robots in each even-numbered t-column perform a rotation of the nodes of the t-column, and when a robot sees an empty spot in the adjacent t-column (to the left) it moves into such spot packing to the left. Thus, Step 4 requires $O(\sqrt{n})$ time. This implies that the whole routing phase also takes $O(\sqrt{n})$ time.

The following theorem is an immediate consequence of Lemmas 1, 2, 3 and 4.

Theorem 1. *Any instance of the MRG$(m, n, 1)$ problem, with $m \leq n/4$ can be solved in optimal $\Theta(\sqrt{n})$, in the worst-case.*

3   The Case of Multiple Targets

In this section we generalize the result of the previous section and present an optimal deterministic protocol that solves any instance of the MRG(n, m, d) problem with $m \leq 3n/16$ and $d \leq n$, where each of the m robots needs to visit $d' \leq d$ grid nodes and each grid node is visited by at most one robot. We first prove a more refined lower bound than the simple one based on the diameter argument, which will be needed to show the optimality of the protocol.

Lemma 5. *There exists an instance of the MRG(n, m, d) problem whose solution requires $\Omega\left(\sqrt{dn}\right)$ time.*

Proof. If $d < 4$ the bound trivially follows from the diameter argument. Consider the case $d \geq 4$ and let d' be the largest power of 4 smaller than d (hence $d' = \Theta(d)$). Let the grid be partitioned into d' square $\sqrt{n/d'} \times \sqrt{n/d'}$ subgrids and consider a MRG(n, m, d) problem where one robot has the d' centers of the subgrids among its targets. Clearly, in order to visit its targets the robot will have to traverse at least $\sqrt{n/d'}/2$ nodes in each of at least $d' - 1$ subgrids, which requires time $\Omega\left(d'\sqrt{n/d'}\right) = \Omega\left(\sqrt{dn}\right)$.

We now sketch the deterministic protocol for d targets by assuming that the robots know the value d. Such value can be easily learned by all robots in $o(\sqrt{dn})$ time (the details will be provided in the full version of the paper). Our protocol consists of $k = \lfloor \log_4 d \rfloor$ stages. In Stage 0, all robots with less than 16 targets visit their targets. In Stage i, with $1 \leq i \leq k-1$, all robots with $4^{i+1} \leq d' < 4^{i+2}$ targets visit their targets. We will show that Stage i is accomplished in $O\left(\sqrt{4^i n}\right)$ time, for every $0 \leq i < k - 1$, yielding an overall $O\left(\sqrt{dn}\right)$ running time for the entire protocol, which is optimal.

Note that Stage 0 can be easily executed within the desired time bound by running the single-target protocol presented in the previous section sixteen times. We now describe in detail the operations of Stage i for $i > 0$. Fix $i > 0$ and define $\delta_i = 4^{i-1}$. Consider the grid as partitioned into δ_i square i-*tiles* of size $\sqrt{n/\delta_i} \times \sqrt{n/\delta_i}$. Note that $\delta_i \leq n/4$, hence each i-tile contains at least four nodes. For $1 \leq u, v \leq \sqrt{\delta_i}$, let $T(u, v)$ denote the i-tile in row u and column v (according to the natural bidimensional indexing), and label the four quadrants of $T(u, v)$ with A, B, C and D as follows:

$$\begin{vmatrix} A & B \\ C & D \end{vmatrix}$$

u odd and $v < \sqrt{n/\delta_i}$

$$\begin{vmatrix} A & C \\ D & B \end{vmatrix}$$

u odd and $v = \sqrt{n/\delta_i}$

$$\begin{vmatrix} B & A \\ C & D \end{vmatrix}$$

u even and $v > 1$

$$\begin{vmatrix} C & A \\ B & D \end{vmatrix}$$

u even and $v = 1$

We also number the i-tiles from 1 to δ_i in a snake-like fashion, going left-to-right in odd rows and right-to-left in even rows. It is easy to see that the B quadrant of the ℓ-th i-tile has always one side in common with the A quadrant of the $(\ell+1)$-st i-tile, for $1 \le \ell < \delta_i$.

Let us call *active* those robots that visit their targets in Stage i, i.e., those with $4^{i+1} \le d' < 4^{i+2}$ targets to visit, and *inert* all other robots. Note that there are at most $n/4^{i+1} = n/(16\delta_i)$ active robots, hence all active robots fit in $1/16$ of the nodes of an i-tile. Stage i is executed as follows:

1. The robots relocate on the grid so that all active robots end up in $T(1,1)$, while the inert robots are evenly distributed among the i-tiles with at most $3n/(16\delta_i)$ robots in each i-tile;
2. Repeat the following sequence of steps $4^{i+2} + \delta_i$ times independently within each i-tile T.
   (a) Each active robot with unvisited targets in T visits one arbitrary such target;
   (b) All active robots that have no unvisited targets in T relocate to quadrant B of the tile, while all other robots (i.e., the inert ones plus those that still have unvisited targets in T) relocate to quadrant D;
   Comment: At the end of this step no robots occupy grid nodes in quadrants A and C.
   (c) All active robots move from quadrant B to quadrant A of the i-tile adjacent to T in the snake-like ordering.

Lemma 6. *Stage i is correct (i.e., all active robots visit their targets) and requires $O\left(\sqrt{4^i n}\right)$ time.*

Proof. In order to prove correctness we only need to show that $4^{i+2}+\delta_i$ iterations of Step 2 are sufficient for each active robot to visit its targets. Consider an active robot and let d_x be the number of its targets in the i-tile of index x, for $1 \le x \le \delta_i$. Such robot will stay in the x-th tile for $\max\{1, x\}$ iterations, hence the total number of iterations needed to visit all of its targets is

$$\sum_{x=1}^{\delta_i} \max\{1, x\} \le 4^{i+2} + \delta_i.$$

As for the running time we reason as follows. Step 1 can be executed in $\Omega\left(\sqrt{n}\right)$ time by a simple modification of the balancing and sorting phases employed in the protocol for a single target. Observe that at any time during Step 2, there are at most $3n/(16\delta_i)$ inert robots in each i-tile, since inert robots do not change i-tile after Step 1. Moreover, since there are at most $n/(16\delta_i)$ active robots overall, we can conclude that at the beginning of any iteration of Step 2, there are at most $n/(4\delta_i)$ robots in each tile. As a consequence, we can use the single-target protocol to execute Substep 2.(a) and a combination of the basic primitives described before to execute Substeps 2.(b) and 2.(c) in $O\left(\sqrt{n/\delta_i}\right)$ time altogether. Thus, the overall running time of Stage i is $O\left(\sqrt{n} + (4^i + \delta_i)\sqrt{n/\delta_i}\right) = O\left(\sqrt{4^i n}\right)$.

The following theorem is an immediate consequence of Lemma 6 and the previous discussion.

Theorem 2. *Any instance of the MRG(n, m, d) problem with $m \leq 3n/16$ and $d \leq n$ can be solved in optimal $\Theta\left(\sqrt{dn}\right)$ time, in the worst case.*

4 Conclusions

We studied the complexity of moving a set of m robots, with limited sensory capabilities, in a multi robot grid system of size $\sqrt{n} \times \sqrt{n}$. We provided an $O(\sqrt{n})$ deterministic protocol that governs the movement of up to $\Theta(n)$ robots, where each robot may visit an arbitrary number of disjoint locations, but not two robots visit the same location. We also proved a lower bound showing that the protocol is optimal. An interesting open problem concerns the extension of the protocol to allow distinct robots to visit the same location.

Acknowledgments. We thank Elena Lodi, Fabrizio Luccio and Linda Pagli for many interesting discussions at the early stages of this work.

References

[Ba68]   K.E. Batcher. Sorting networks and their applications. In *Proc. ot the AFIPS Spring Joint Computer Conference*, vol. 32, pages 307–314, 1968.

[Lei92] F.T. Leighton. *Introduction to Parallel Algorithms and Architectures: Arrays • Trees • Hypercubes.* Morgan Kaufmann, San Mateo, CA, 1992.

[Mo66]   G. Morton. A computer oriented geodetic data base and a new technique in file sequencing. IBM Ltd. Internal Report, 1966.

[NS95]   I. Newman and A. Schuster. Hot-potato algorithms for permutation routing. *IEEE Trans. on Parallel and Distributed Systems*, 6(11):1068–1176, November 1995.

[PU96]   S. Preminger and E. Upfal. Safe and efficient traffic laws for mobile robots. In *Proc. ot the 5th Scandinavian Workshop on Algorithm Theory*, LNCS 1097, pages 356–367, 1996.

[ST92]   Y. Shoham and M. Tennenholtz. On traffic laws for mobile robots. In *Proc. of the 1st Conference on AI Planning Systems*, 1992.

[ST95]   Y. Shoham and M. Tennenholtz. On social laws for artificial agent societies: Off-line design. *Artificial Intelligence*, 73(1–2):231–252, 1995.

[Sib95] J.F. Sibeyn. Overview of mesh results. Technical Report MPI-I-95-1018, Max-Planck Institut für Informatik, Saarbrücken, Germany, 1995.

Concurrent Multicast in Weighted Networks

Gianluca De Marco, Luisa Gargano, and Ugo Vaccaro

Dipartimento di Informatica, Università di Salerno, 84081 Baronissi (SA), Italy
{demarco,lg,uv}@dia.unisa.it

Abstract. Concurrent multicast is a problem of information dissemination from a set of source nodes to a set of destination nodes in a network with cost function: Each source node s needs to multicast a block of data $B(s)$ to the set of destinations. We are interested in protocols for this problem which have minimum communication cost. We consider both the classical case in which any transmitted message can consist of an arbitrary number of data blocks and the case in which each message must consist of exactly one block of data. We show that the problem of determining the minimum cost to perform concurrent multicast is NP-hard under both assumptions. We also give approximation algorithms to efficiently perform concurrent multicast in arbitrary networks.

1 Introduction

In this paper we consider the problem of concurrent multicast, that is, the problem of information dissemination from a set of source nodes to a set of destinations nodes in a weighted communication network.

Multicasting has been the focus of growing research interest, many future applications of computer networks such as distance education, remote collaboration, teleconferencing, and many others will rely on the capability of the network to provide multicasting services. Our model considers concurrent multicast in which a group of nodes in the network needs to multicast to a same set of destinations.

Processors in a network cooperate to solve a problem, in our case to perform concurrent multicast, by exchanging messages along communication channels. Networks are usually modelled as connected graphs with processors represented as vertices and communication channels represented as edges. For each channel, the cost of sending a message over that channel is measured by assigning a weight to the corresponding edges. Our goal is to give algorithms to efficiently perform concurrent multicast in the network.

The typical measure of the communication cost of an algorithm is the number of messages sent across the network during its execution. This measure assumes that the cost of sending a message along any channel is equal to one. However, it is more realistic to include the edge weights into the communication cost of an algorithm. More specifically, the cost of transmitting a message over a channel should be equal to the weight of the corresponding edges. This point of view was advocated in [2, 22] and several papers have followed this line of research since then.

Statement of the problem and summary of our results. We consider the communication network modelled by a graph $H = (V, E)$, where the node set V represents the set of processors and the set of edges E represents the set of communication channels. Each edge (i, j) in H is labelled by the communication cost $c(i, j) > 0$ of sending a message from node i to node j, where $c(i, j) = c(j, i)$. *Concurrent Multicast Problem (CM):* Let S and D be two arbitrary subset of V. Nodes in S are the sources and nodes in D are the destinations. Each node $a \in S$ holds a block of data $B(a)$. The goal is to disseminate all these blocks so that each destination node $b \in D$ gets all the block $B(a)$, for all $a \in S$.

We are interested in protocols for the Concurrent Multicast Problem which have minimum communication cost, where the communication cost of a protocol is the sum of the costs of all message transmissions performed during its execution.

We first study the concurrent multicast problem under the classical assumptions that all the blocks known to a node i at each time instant of the execution of the protocol can be freely concatenated and the resulting message can be transmitted to a node j with cost $c(i, j)$. This assumption is reasonable when the combination of blocks results in a new message of the same size (for example, blocks are boolean values and each node in D has to know the AND of all blocks of the nodes in S [22]). It is not too hard to see that a protocol for the CM problem in this scenario can be obtained as follows: construct in H a Steiner Tree T with terminal nodes equal to the set $S \cup \{v\}$, $v \in V$, by transmitting over the edges of T one can accumulate all the blocks of the source nodes into v; then, by using another Steiner Tree T' with terminal nodes equal to $D \cup \{v\}$, one can broadcast the information held by v to all nodes in D, thus completing the CM. It is somewhat surprising to see that this two phase protocol, accumulation plus broadcasting, is essentially the best one can do. The simple protocol outlined above is presented in Section 4, the non-simple proof of its optimality is given in Section 3. As a by product of our results characterizing the structure of a minimum cost protocol for the CM problem, we also get that determining the cost of an optimal protocol for CM problem is NP-hard. In Section 4.2 we give an approximate-cost polynomial time algorithm for the CM problem. Subsequently, in Section 5 we turn our attention to a different scenario, in which the assumption that the cost of the transmission of a message be independent of the number of blocks composing it no longer holds, therefore message transmissions must consist of one block of data at time. Communication protocols which works by exchanging messages of limited size have recently received considerable attention (see for example [3, 7, 8, 12]). The CM problem remains NP-hard in this case, therefore we also provide polynomial time approximate cost solutions. In Section 4.3 we consider the on-line version of the CM problem. The on-line version specifies that the source and destination nodes be supplied one at the time and the existing solution be extended to connect the current sources an destinations before receiving a request to add/delete a node from the current source/destination set. We will prove that the characterization we have given for the optimal cost solution to the CM problems allows us to derive an efficient solution also to the on-line version.

Related work. In case $S = D = V$ the CM problem reduces to the gossiping problem which arises in a large class of scientific computation problems [9]. In case $|S| = 1$ and $D \subseteq V$ the CM problem reduces to the multicasting problem [1, 5, 21] and in case $D = V$ to the broadcasting problem, both problems have been well investigated because of their relevance in the context of parallel/distributed systems [20]. In particular the broadcasting and gossiping problems have been investigated under a varieties of communication models and have accumulated a large literature, we refer the reader to the survey papers [10, 13, 14]. In this section we limit ourselves to briefly discuss some works whose results are either strictly related to ours or can be seen as corollaries of the result of the present paper.

In case $S = D = V$ we get the problem of Gossiping in weighted networks, a problem first considered in weighted complete networks by Wolfson and Segall [22]. One of the main results of [22] was to prove that the communication cost of an optimal gossiping algorithm is equal to 2× (cost of a minimum spanning tree of the weighted complete graph). As a consequence of more general results (i.e. our characterization of optimal cost instances of the CM problem given in Section 4), we are able to extend above quoted results of [22] to general weighted graphs, i.e., not necessarily complete. Gossiping in weighted complete networks in which blocks of data cannot be freely combined was studied in [12]. If messages must consist of exactly one block then our result of Section 5 implies one of the results of [12], that the minimum cost of an instance is equal to $|V| \times$ (cost of a minimum spanning tree of H); again, in the present case H does not need to be the complete graphs. Another problem strictly related to ours is the Set-to-Set Broadcasting problem [19], which asks for the minimum number of message transmissions $call(S, D)$ to perform concurrent multicast from a set S to a set D in a complete graph. Our results imply a solution equivalent to the one given in [17, 18] for the so called "telephone communication model", namely

$$call(S, D) = \begin{cases} |S| + |D| & \text{if } S \cap D = \emptyset, \\ |S| + |D| - 2 & \text{if } S \cap D \neq \emptyset. \end{cases}$$

Because of page limits, some proofs are omitted from this extended abstract.

2 Multi–digraphs associated to instances of concurrent multicast

We introduce here the notion of multi–digraph associated to an instance of a CM algorithm. We will consider the concurrent multicast problem on a communication graph $H = (V, E)$ from the source set S to the destination set D.

The sequence of message transmissions (calls) of an instance of a concurrent multicast algorithm will be represented by a labelled multi–digraph $I = (V, A(I))$ having as node set the same set of nodes of H and as arc set the multiset $A(I)$ in which each arc (i, j) represents a call from i to j; arc labels represent the temporal order in which calls are made.

A path in I from node i to node j is called *ascending* if the sequence of labels of the arcs on the path is strictly increasing when moving from i to j. Since a node b receives the block of a source node $a \in S$ iff the I contains an ascending path from a to b, the following property obviously holds

Fact 1 *A labelled multi–digraph* $I = (V, A(I))$ *is an instance of concurrent multicast from* S *to* D *if and only if* I *contains an ascending path from* a *to* b, *for each source* $a \in S$ *and destination* $b \in D$

An arc $(i, j) \in A(I)$ has cost $c(i, j)$, the cost of the corresponding call along the edge $\{i, j\}$ in H. The *cost* of an instance (that is, the cost of the associated multi–digraph) I is then the sum of the costs of all the arcs of I, each added as many times as its multiplicity.

Example 1 *Let* $H = (\{1, 2, 3, 4, 5, 6, 7\}, E)$ *be the weighted graph of Figure 1(a). Consider the source set* $S = \{1, 2\}$, *the destination set* $D = \{4, 5, 6\}$, *and the instance consisting of the following calls:*
At time 1: node 1 sends $B(1)$ to 3; At time 2: node 2 sends $B(2)$ to 3; At time 3: node 3 sends $(B(1), B(2))$ to 4; At time 4: node 3 sends $(B(1), B(2))$ to 6 and node 4 sends $(B(1), B(2))$ to 5.
The corresponding multi–digraph I is shown in Figure 1(b); each arc is labelled with the time in which the corresponding call is made. We have $cost(I) = 5$.

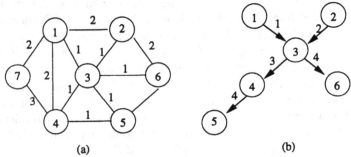

(a)   (b)

Figure 1

3   Characterization of a minimum cost instance

In this section we derive a lower bound on the cost of an optimal solution to the CM problem. Let I be an instance. A node $i \in V$ is called *complete at time* t if for each $a \in S$ there exists in I an ascending path α from a to i such that $t(e) \leq t$ for each arc e on α. In other words, a node is complete at time t if by time t it knows the blocks of all the source nodes in S. Notice that if a node i is complete at time t and i calls node j at time $t' > t$, then i can send to j any block $B(a)$, for $a \in S$, and thus make also j complete.

Given $I = (V, A(I))$ and $A' \subseteq A(I)$, call subgraph of I induced by the subset of arcs A' the graph (V', A') with
$$V' = \{i \in V \mid i \text{ has at least one arc of } A' \text{ incident on it}\}.$$

We will denote by $C_I = (V(C_I), A(C_I))$ the subgraph of I induced by the subset of arcs

$A(C_I) = \{(i,j) \in A(I) \mid$ there exists $t < t(i,j)$ s.t. i is complete at time $t\}$

and by $\overline{C}_I$ the subgraph induced by the subset of arcs $A(\overline{C}_I) = A(I) \setminus A(C_I)$.

Notice that C_I consists of all the arcs of I corresponding to calls made by a complete node.

Example 1 (continued). *For the instance I in Figure 1(b), the subgraphs C_I and $\overline{C}_I$ are given in Figure 2(a) and 2(b), respectively.*

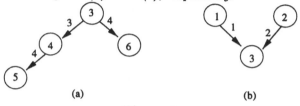

(a)   (b)

Figure 2

Lemma 1. *If I is a minimum cost instance, then*
1) *the multi-digraph C_I is a forest,*
2) *the node set of each tree in C_I contains at least one node in D.*

Proof. In order to prove that C_I is a forest, it is sufficient to show that at most one arc enters each node v in C_I. Suppose there exists a node v such that at least two arcs enter v in C_I. Then all arcs incoming on v but the one with smallest label, call it t, can be omitted. Indeed, since v is complete at time t, all successive calls to v are redundant. Hence, there exists an instance of smaller communication cost, contradicting the optimality of I.

We show now 2). It is obvious that if there exists a tree T in C_I containing no destination node in D, then the nodes of T are not in any ascending path from a to b, for each $a \in S$ and $b \in D$. Therefore, the calls corresponding to the arcs of T can be omitted. This contradicts the optimality of I.

We denote by $\mathcal{R}(I)$ the set of the roots of the trees in the forest C_I. Two roots $r_i, r_j \in \mathcal{R}(I)$, with $i \neq j$, are called *mergeable* if there exists an instance I' with $cost(I') \leq cost(I)$, $\mathcal{R}(I') \subseteq \mathcal{R}(I)$, and $|\mathcal{R}(I') \cap \{r_i, r_j\}| = 1$.

Given any multi-digraph G, let $indeg_G(x) = |\{y \mid (y,x) \in A(G)\}|$, that is, the number of *different* tails of arcs entering x in G. The following lemma is one of our main technical tools, its proof is omitted from this exended abstract.

Lemma 2. *Let I be a minimum cost instance with $|\mathcal{R}(I)| \geq 2$ and $indeg_{\overline{C}_I}(q) \geq 2$, for all $q \in \mathcal{R}(I)$. Then $\mathcal{R}(I)$ contains two mergeable roots.*

Theorem 1 *There exists a minimum cost instance I with $|\mathcal{R}(I)| = 1$.*

Proof. We show that given a minimum cost instance I, with $|\mathcal{R}(I)| > 1$, there exists another instance I^*, with $|\mathcal{R}(I^*)| = 1$, such that $cost(I^*) = cost(I)$. We distinguish two cases.

Case 1: there exists $r \in \mathcal{R}(I)$ with $indeg_{\overline{C}_I}(r) = 1$. We first notice that in such a case $r \in S$. Indeed assuming $r \in V \setminus S$ and that r has only one incoming arc $(x, r) \in E(\overline{C}_I)$, then we necessarily have that x is complete at a time smaller than $t(x, r)$. Thus r is not a root in $\mathcal{R}(I)$.

Let then $a \in \mathcal{R}(I) \cap S$ with $indeg_{\overline{C}_I}(a) = 1$. We show now that there exists an instance I^* such that $\mathcal{R}(I^*) = \{a\}$ and $cost(I^*) = cost(I)$. Let $(x, a) \in E(\overline{C}_I)$ with $t(x, a) = t$ such that a is complete at time t. Since a is complete at time $t = t(x, a)$, we get

1) for all $a' \in S \setminus \{a\}$ there exists an ascending path $\gamma(a') = (a', \cdots, x, a)$, from a' to a;

Moreover, we must have

2) there is no ascending path $(a, \cdots, y, x)$ with $t(y, x) < t(x, a)$,

otherwise considering all the paths in 1) and the path $(a, \cdots, y, x)$, we would have x complete at time $t(y, x) < t(x, a)$. Then $(x, a) \in E(\mathcal{C}_I)$ that would imply $x \in \mathcal{R}(I)$ and $a \notin \mathcal{R}(I)$.

To get the instance I^*, let us make the following modifications on the instance I:

- leave unchanged the labels of all the arcs on the paths $\gamma(a')$, for each $a' \in S \setminus \{a\}$,
- increase the label of all the other arcs in I of the value $t = t(x, a)$.

In such a way we obtain a multi–digraph I^* which is again an instance. Indeed, the paths $\gamma(a')$ in 1) make a complete at time t in I^*. Also, since $a \in S$, we have that for all $r \in \mathcal{R}(I) \setminus \{a\}$ there exists in I an ascending path $(a, \cdots, r)$. Because of the above modifications, all these paths have label greater than t. Hence, I^* contains an ascending path from a to every node $r \in \mathcal{R}(I)$ and therefore, to every node $b \in D$. This implies that I^* is an instance. Obviously, I^* has the same cost of I, moreover we have also $\mathcal{R}(I^*) = \{a\}$.

Case 2: $indeg_{\overline{C}_I}(r) \geq 2$, for each $r \in \mathcal{R}(I)$. Lemma 2 implies that $\mathcal{R}(I)$ contains two mergeable roots, that is, there exists an instance I' with $cost(I') \leq cost(I)$ and $|\mathcal{R}(I')| \leq |\mathcal{R}(I)| - 1$.

The theorem follows by iteratively applying Case 2 until we get an instance I^* such that either $|\mathcal{R}(I^*)| = 1$ or I^* satisfies Case 1.

The following Theorem 2 is the main result of this section. Given the graph $H = (V, E)$, let us denote by $ST(X)$ the Steiner tree in H on the terminal node set X, that is, the minimum cost tree T with $E(T) \subseteq E$ and $X \subseteq V(T) \subseteq V$.

Theorem 2 *Consider the communication graph $H = (V, E)$ and the sets $S, D \subseteq V$. Let $I_{\min}$ be a minimum cost instance for the CM problem from the source set S to the destination set D. Then*

$$cost(I_{\min}) \geq \min_{v \in V}\{cost(ST(S \cup \{v\})) + cost(ST(D \cup \{v\}))\}. \tag{1}$$

Proof. Consider the graph H and the sets $S, D \subseteq V$. Theorem 1 implies that there exists a minimum cost instance I such that its subgraph $\mathcal{C}_I$ is a tree. Let

r denote the root of C_I. By definition, r is the first node to become complete in I and each node $b \in D$ (a part r, if $r \in D$) becomes complete by receiving a message along a path from r to b in the tree C_I. Therefore, C_I is a tree whose node set includes each node in $D \cup \{r\}$ and it must hold

$$cost(C_I) \geq cost(ST(D \cup \{r\})). \tag{2}$$

Moreover, for each $a \in S$ we have that either $a = r$ or the complement $\overline{C}_I$ of C_I contains an ascending path from a to r. Therefore,

$$cost(\overline{C}_I) \geq cost(ST(S \cup \{r\})). \tag{3}$$

The above inequalities (2) and (3) imply
$$cost(I) = cost(C_I) + cost(\overline{C}_I) \geq cost(ST(D \cup \{r\})) + cost(ST(S \cup \{r\})).$$

4 Algorithms

In this section we present algorithms for the concurrent multicast problem. To this aim we first show that the inequality in Theorem 2 holds with equality.

Theorem 3 *Consider the communication graph $H = (V, E)$ and the sets $S, D \subseteq V$. Let I_{min} be a minimum cost instance for the concurrent multicast problem on S and D. Then*

$$cost(I_{min}) = \begin{cases} \min_{v \in V} \{cost(ST(S \cup \{v\})) + cost(ST(D \cup \{v\}))\} & \text{if } S \cap D = \emptyset, \\ \min_{v \in S \cap D} \{cost(ST(S \cup \{v\})) + cost(ST(D \cup \{v\}))\} & \text{if } S \cap D \neq \emptyset, \end{cases} \tag{4}$$

where $ST(X)$ represents a Steiner tree of the communication graph H spanning all nodes in X.

Proof. By Theorem 2, $cost(I_{min})$ is lower bounded by $\min_{v \in V} \{cost(ST(S \cup \{v\})) + cost(ST(D \cup \{v\}))\}$. If we denote by r the node in V for which the above minimum is reached, then

$$cost(I_{min}) \geq cost(ST(S \cup \{r\})) + cost(ST(D \cup \{r\})). \tag{5}$$

We show now that there exists an instance I of cost equal to $cost(ST(S \cup \{r\})) + cost(ST(D \cup \{r\}))$.

Consider the Steiner tree $ST(S \cup \{r\})$, and denote by T_1 the directed tree obtained from $ST(S \cup \{r\})$ by rooting it in r and directing all its edges toward the root r. Label each arc of T_1 so that each directed path in T_1 is ascending, let Δ denote the maximum label used in T_1.

Consider now the Steiner tree $ST(D \cup \{r\})$, and denote by T_2 the directed tree obtained from $ST(D \cup \{r\})$ by rooting it in r and directing all its edges away from the root r. Label each arc (i, j) of T_2 with a label $t(i, j) > \Delta$ so that each directed path in T_2 is ascending.

Consider then the multi-digraph I such that $V(I) = V$ and $E(I) = E(T_1) \cup$

$E(T_2)$. By definition of T_1 and T_2 we get that I contains an ascending path $(a, \ldots, r, \ldots, b)$, for each $a \in S$ and $b \in D$. Hence, I is an instance of the CM problem and its cost is

$$cost(I) = cost(T_1) + cost(T_2) = cost(ST(S \cup \{r\})) + cost(ST(D \cup \{r\})).$$

Since $cost(I) \geq cost(I_{min})$, by (5) we have

$$cost(I_{min}) = cost(ST(S \cup \{r\})) + cost(ST(D \cup \{r\}))$$
$$= \min_{v \in V} cost(ST(S \cup \{v\})) + cost(ST(D \cup \{v\})).$$

Finally, it is easy to see that in case $S \cap D \neq \emptyset$, at least one node for which the minimum is attained must be a node in $S \cap D$. Hence the theorem holds.

4.1 Complexity

If $|S| = 1$ and $D = V$ (or $S = V$ and $|D| = 1$) the CM problem easily reduces to the construction of a minimum spanning tree of the communication graph H. When H is the complete graph on V and $S = D = V$, Wolfson and Segal [22] proved that the problem is again equivalent to the construction of a minimum spanning tree of H. We notice that our Theorem 3 proves that for $S = D = V$ this result of [22] is true for any communication graph $H = (V, E)$. However, determining the minimum communication cost of a concurrent multicast instance is in general NP-hard. To this aim consider the following associated decision problem.

MIN–COST–CM

Instance: A weighted graph $H = (V, E)$, two sets $S, D \subseteq V$, and a bound C;
Question: Is there a concurrent multicast instance for S, D of cost $\leq C$?

By Theorem 3 and the NP-completeness of determining the cost of Steiner trees [21], we get

Theorem 4 *MIN–COST–CM is NP-hard.*

4.2 Approximate Algorithms

Since finding an algorithm of minimum cost is computationally infeasible, we must relax requirements for optimality and look for approximation algorithms. We present here a distributed algorithm for the CM problem. We assume, as in other papers (see [22, 11, 12]), that each node knows the identity of all the other nodes and the set of communication costs of the edges. The algorithm $CM(H, S, D)$ given in Figure 3 is executed by each node.

The trees T_S and T_D are subgraphs of the graph H and have node sets such that $S \cup \{r\} \subseteq V(T_S)$, $D \cup \{r\} \subseteq V(T_D)$, for some node r; a more detailed description will be given later. The trees T_S and T_D are identical at all the nodes given that the construction procedure is identical at all nodes.

The algorithm is asynchronous and does not require nodes to know when the blocks of the nodes in S are ready nor the time messages take to travel between pairs of nodes.

CM(H, S, D) /* executed at node x, given the graph H and the sets S and D

1. Construct the trees T_S and T_D, root them in the (common) node r;
2. [A node in $(V(T_S) \cap V(T_D)) \setminus \{r\}$ executes both 2.1 and (after) 2.3]
    2.1. If in $V(T_S) - \{r\}$, wait until received from all sons in T_S. Send to the parent in T_S a message containing all blocks received plus the block $B(x)$, if $x \in S$;
    2.2. If equal to r wait until received from all sons in T_S. Send to each child in T_D a message containing all blocks received plus the block $B(r)$, if $r \in S$, that is, send all blocks $B(a)$, for each $a \in S$;
    2.3. If in $V(T_D) - \{r\}$, wait until received from the parent in T_D. Send to each child in T_D a message containing all blocks $B(a)$, for each $a \in S$.

Figure 3

It is easy to see that the algorithm terminates and each destination $b \in D$ knows the blocks of all the sources in S. We consider now its communication cost. Since the algorithm uses only once each edge in T_S and T_D we immediately get that its cost is $cost(T_S) + cost(T_D)$.

Let $ST_{apx}(X)$ denote the tree obtained by using an approximation algorithm for the construction of the Steiner tree $ST(X)$.

Several efficient algorithms have been proposed in the literature. The simpler algorithm [21] is greedy, it has complexity $O(|V|^2)$ and approximation factor 2, that is, $cost(ST_{apx}(X))/cost(ST(X)) \le 2$, for any set X. The polynomial algorithm with the best known approximation factor for Steiner trees in graphs has been given in [15] and has approximation factor 1.644.

Fixed an approximation algorithm, we can then choose r as the node such that $cost(ST_{apx}(S \cup \{r\})) + cost(ST_{apx}(D \cup \{r\}))$
$$= \min_{v \in V} cost(ST_{apx}(S \cup \{v\})) + cost(ST_{apx}(D \cup \{v\}))$$
and then choose the trees T_S and T_D used in the algorithm CM(H, S, D) as $ST_{apx}(S \cup \{r\})$ and $ST_{apx}(D \cup \{r\})$, respectively. Therefore, by using the best approximation algorithm for the construction of the trees, we get that the cost of the algorithm CM(H, S, D) is
$$cost(T_S) + cost(T_D) = \min_{v \in V} cost(ST_{apx}(S \cup \{v\})) + cost(ST_{apx}(D \cup \{v\}))$$
$$\le cost(ST_{apx}(S \cup \{s\})) + cost(ST_{apx}(D \cup \{s\}))$$
$$\le 1.644(cost(ST(S \cup \{s\})) + cost(ST(D \cup \{s\}))),$$
for each $s \ne r$. for each $s \ne r$. By choosing s as the node that gets the minimum in the lower bound (1), we get

Theorem 5 *The ratio between the cost of CM(H, S, D) and the cost of a minimum cost algorithm is upper bounded by 1.644.*

4.3 On–line algorithms

In this section we consider the dynamic concurrent multicast problem, which allows the sets of nodes to be connected vary on the time. We will show that the characterization we gave for the optimal cost solution to the concurrent multicast problem allows to derive efficient algorithms also for the dynamic version.

A dynamic algorithm receives in input a sequence of requests $r_i = (x_i, s_i, \rho_i)$, for $i = 1, 2, \ldots$, where x_i is a node in H, the component $s_i \in \{S, D\}$ specify if x_i is a source or destination node, and $\rho_i \in \{\text{add}, \text{remove}\}$ specifies if the node x_i must be added or removed from the set s_i. As an example, (x, D, add) defines the operation of adding x to the current set of destinations. The sets

$S_i = \{a \mid \text{there exists } j \leq i \text{ with } r_j = (a, S, \text{add}), r_\ell \neq (a, S, \text{remove}),$
$\qquad \text{for each } j < \ell \leq i\},$
$D_i = \{a \mid \text{there exists } j \leq i \text{ with } r_j = (a, D, \text{add}), r_\ell \neq (a, D, \text{remove}),$
$\qquad \text{for each } j < \ell \leq i\}$

are the source and destination sets on which we are required to perform CM after the request r_i.

The algorithm will be the same as given in Figure 2, but we will make a different choice of the trees T_S and T_D in order to have the possibility of dynamically and efficiently modify them according to the sequence of requests.

We first consider the case of no remove requests. W.l.o.g., assume that the first two requests are $r_1 = (a, S, \text{add})$ and $r_2 = (b, D, \text{add})$; that is, $S_2 = \{a\}$ and $D_2 = \{b\}$. We simply connect a and b by a minimum weight path α in H; formally, we have $T_{S_2} = (\{a\}, \emptyset)$ and T_{D_2} coincide with the path α.
In general, for sake of efficiency, with the request r_i we want to add the new node without modifying the existing trees [16]. If r_i requires to add a' to S_{i-1}, we connect a' to a node in $T_{S_{i-1}}$ by a shortest path in H form a' to a node in $T_{S_{i-1}}$. Similarly for $r_i = (b, D, \text{add})$. Therefore, at each step we have the tree T_{S_i} rooted in a which spans all nodes in S_i and a tree T_{D_i} rooted in a which spans all nodes in $D_i \cup \{a\}$. Using the results proved in [16] we can get

$$\begin{cases} cost(T_{S_i}) = O(\log |S_i|)cost(ST(S_i)), \\ cost(T_{D_i}) = O(\log |D_i|)cost(ST(D_i)) + cost(\alpha). \end{cases} \qquad (6)$$

Denote by CM(i) the algorithm of Figure 3 when using the trees T_{S_i} and T_{D_i}. By (6) and (1)

Theorem 6 *Consider a sequence of add requests $r_1, \ldots, r_k$ and let $n_i = |S_i| + |D_i|$. The ratio between the cost of CM(i) and the cost of an optimal CM algorithm on S_i and D_i is $O(\log n_i)$.*

In case of remove requests, it is not possible to have a bounded performance ratio if no rearrangements (changes in the structure) of the trees are allowed after requests [16]. Several papers have recently considered the problem of efficiently maintaining dynamic Steiner trees, that is, with a limited number rearrangements [1, 5, 16]. It is clear from the above results that any algorithm for the dynamic Steiner tree problem can be applied to dynamically maintain the trees T_{S_i} and T_{D_i} obtaining equivalent results for our CM problem.

5 Concurrent multicast without block concatenation

In this section we consider the concurrent multicast problem under the hypothesis that each message transmission must consist of exactly one block $B(a)$, for some $a \in S$. Under the hypothesis of this section we have the following result whose proof is omitted.

Theorem 7 *For any instance I, it holds that*
$$cost(I) \geq \sum_{a \in S} cost(ST(D \cup \{a\})).$$

Consider now the following algorithm. Again we assume that each node knows the identity of all the other nodes and the sets S and D. The algorithm BLOCK-CM(H, S, D) given in Figure 4 is executed by each node. The trees T_a are identical at all the nodes given that the construction procedure is identical at all nodes.

BLOCK-CM(H, S, D) /* executed at each node, given the graph H and the sets S and D

1. For each $a \in S$, construct a tree T_a spanning a and all nodes in D.
2. For each T_a: if a then send $B(a)$ to all neighbours, otherwise, wait until received $B(a)$ from one neighbour in T_a and resend it to each of the other neighbours (if any) in T_a.

Figure 4

We immediately get that the above algorithm is correct and that its cost is

$$\sum_{a \in S} cost(T_a). \tag{7}$$

Assuming T_a be a Steiner tree on $D \cup \{a\}$, by Theorem 7, we would get an optimal cost algorithm. The NP-completeness of the Steiner tree problem [21] implies that the CM problem without block concatenation is NP-hard.

Constructing T_a by using the approximation algorithm for the Steiner tree $ST(D \cup \{a\})$ given in [15], we get

$$cost(T_a) < 1.664 \, cost(ST(D \cup \{a\})). \tag{8}$$

By (7) and (8) and from Theorem 7 we obtain

Theorem 8 *The ratio between the cost of BLOCK-CM(H, S, D) and the cost of a minimum cost algorithm is upper bounded by 1.664.*

Since the algorithm is based on the use of approximate Steiner trees, it is clear that all the discussion done in Section 4.3 on the dynamic implementation can be applied also in this case.

References

1. E. Aharoni, R. Cohen, " Restricted Dynamic Steiner Trees for Scalable Multicast in Datagram Networks", Proc. *INFOCOM'97*, Kobe, Japan, April 1997.
2. B. Awerbuch, A. Barowtz, D. Peleg, "Cost–Sensitive Analysis of Communication Protocols", Proceedings of *PODC'90*, 177–187.
3. A. Bagchi, E.F. Schmeichel, and S.L. Hakimi, "Parallel Information Dissemination by Packets", *SIAM J. on Computing*, 23 (1994), 355-372.
4. A. Bar-Noy, S. Guha, J. Naor, and B. Schieber, "Multicasting in Heterogeneous Networks", Proceedings of *STOC '98*.
5. F. Bauer, A. Varma, "Aries: a Rearrangeable Inexpensive Edge–Based On–Line Steiner Algorithm, Proc. *INFOCOM'96*, 361–368.
6. P. Berman, C. Coulston, "On–line algorithms for Steiner Tree Problems", Proc. *STOC'97*, 344–353.
7. J.–C. Bermond, L. Gargano, A. Rescigno, and U. Vaccaro, "Fast Gossiping by Short Messages", *SIAM J. on Computing*, to appear.
8. J.–C. Bermond, L. Gargano, S. Perennes, "Sequential Gossiping by Short Messages", *Discr. Appl. Math.*, to appear.
9. D. P. Bertsekas, and J. N. Tsitsiklis, *Parallel and Distributed Computation: Numerical Methods*, Prentice–Hall, Englewood Cliffs, NJ, 1989.
10. P. Fraignaud, E. Lazard, "Methods and Problems of Communication in Usual Networks", *Discrete Applied Math.*, 53 (1994), 79–134.
11. L. Gargano and A. A. Rescigno, "Communication Complexity of Fault–Tolerant Information Diffusion", *Theoretical Computer Science*, to appear.
12. L. Gargano, A. A. Rescigno, and U. Vaccaro, "Communication Complexity of Gossiping by Short Messages", *Journal of Parallel and Distributed Computing*, 45, (1997), pp. 73–81. An extended abstract appeared in Proceedings of *SWAT'96*, Rolf Karlsson and Andrzej Lingas (Eds.), Lectures Notes in Computer Science, vol. **1097**, pp. 234–245, Springer-Verlag, 1996.
13. S. Hedetniemi, S. Hedetniemi, and A. Liestman, "A Survey of Gossiping and Broadcasting in Communication Networks", *NETWORKS*, 18 (1988), 129–134.
14. J. Hromkovivc, R. Klasing, B. Monien, and R. Peine, "Dissemination of Information in Interconnection Networks (Broadcasting and Gossiping)", in: Ding-Zhu Du and D. Frank Hsu (Eds.) *Combinatorial Network Theory*, Kluwer Academic Publishers, 1995, pp. 125-212.
15. M. Karpinski, A. Zelikovsky, "New Approximation Algorithms for the Steiner Trees Problems", *Journal of Combinatorial Optimization* 1 (1997), 47-65.
16. M. Imase, B.M. Waxman, "Dynamic Steiner Tree Problem", *SIAM J. Discr. Math*, 4 (1991), 369–384.
17. H.–M. Lee, G.J. Chang, "Set-to-Set Broadcasting in Communication Networks", *Discr. Appl. Math.*, 40 (1992), 411–421.
18. Q. Li, Z. Zhang, J. Xu, "A Very Short Proof of a Conjecture Concerning Set-to-Set Broadcasting", *Networks*, 23 (1993), 449–450.
19. D. Richards, A. Liestman, "Generalizations of Broadcasting and Gossiping", *Networks*, 18 (1988), 125–138.
20. A. S. Tanenbaum, *Computer Networks*, Prentice Hall, Englewood Cliffs, N.J., 1981.
21. P. Winter, "Steiner Problems in Networks: a Survey", *Netw.*, 17 (1987), 129–167.
22. O. Wolfson and A. Segall, "The Communication Complexity of Atomic Commitment and of Gossiping", *SIAM J. on Computing*, 20 (1991), 423–450.
23. A. Z. Zelikovsky, "An 11/6-Approximation Algorithm for the Network Steiner Problem", *Algorithmica*, 9:463-470, 1993.

Some Recent Strong Inapproximability Results

Johan Håstad

Royal Institute of Technology
johanh@nada.kth.se

Abstract. The purpose of this talk is to give some idea of the recent progress in obtaining strong, and sometimes tight, inapproximability constants for NP-hard optimization problems. Tight results have been obtained for Max-Ek-Sat[1] for $k \geq 3$, maximizing the number of satisfied linear equations in an over-determined system of linear equations modulo a prime p and Set Splitting.

1 General problem area considered

We know that many natural optimization problems are NP-hard. This means that they are probably hard to solve exactly in the worst case. In practice, however, it is many times sufficient to get reasonable good solutions for all (or even most) instances. In this talk we discuss the existence of polynomial time approximation algorithms for some of the basic NP-complete problems. We say that an algorithm is a C-approximation algorithm if it, for each instance produces, an answer that is off by a factor at most C from the optimal answer. The fundamental question is, for a given NP-complete problem, for what value of C can we hope for a polynomial time C-approximation algorithm. Posed in this generality this is a large research area with many positive and negative results. We here concentrate on negative results, i.e. results of the form that for some $C > 1$ a certain problem cannot be approximated within C in polynomial time. Furthermore, we focus on problems in the class Max-SNP [12] where C is known to be an absolute constant and the main goal is to determine the exact threshold of efficient approximability. The inapproximability results are invariably based on plausible complexity theoretic assumptions, the weakest possible being NP$\neq$P since if NP=P, all considered problems can be solved exactly in polynomial time.

The most basic NP-complete problem is satisfiability of CNF-formulas and probably the most used variant of this is 3-SAT where each clause contains at most 3 variables. For simplicity, let us assume that each clause contains exactly 3 variables. The optimization variant of this problem is to satisfy as many clauses as possible. It is not hard to see that a random assignment satisfies each clause with probability 7/8 and hence if there are m clauses it is not hard (even deterministically) to find an assignment that satisfies $7m/8$ clauses. Since we can never satisfy more than all the clauses this gives a 8/7-approximation algorithm. This was one of the first approximation algorithms considered [11] and one of

---

[1] Max-Ek-Sat is the variant of CNF-Sat where each clause is of length exactly k.

the results we try to explain is the result of [10] that this is optimal to within an arbitrary additive constant ϵ.

A problem that in many respects is as basic as satisfiability is that of solving a system of linear equations over a field. One reason that not many papers are written on this subject in complexity theory is the simple and powerful procedure of Gaussian elimination which makes it solvable in polynomial time.

Gaussian elimination is, however, very sensitive to incorrect equations. In particular, if we are given an over-determined system of equations it is not clear how to efficiently find the "best solution", at least if we interpret this as the assignment satisfying the maximal number of equations. This problem is NP-hard even over the field of two elements, since the special case of having equations of the form $x_i + x_j = 1$ is equivalent to Max-Cut. As with 3-SAT there is an obvious approximation algorithm that just does as well as picking the variables at random and in this case a random assignment satisfies half the equations and thus this yields a 2-approximation algorithm. We discuss the results of [10] that proves that this is, again upto an arbitrary $\epsilon > 0$ and based on NP$\neq$P, the best possible for a polynomial time approximation algorithm. These results also extend to linear equations over an arbitrary Abelian group G. By using the good local reductions provided by [14] the results can then be extended to many other NP-hard approximation problems.

2 Very short history

The question of proving NP-hardness of some approximation problems was discussed at length already in the book by Garey and Johnson [9], but really strong inapproximability results were not obtained until the connection with multiprover interactive proofs, introduced for a different reason by [6], was discovered in the seminal paper of Feige et al. [7] in 1990. There are a number of variants of the multiprover interactive proofs and the two proof models that we discuss are that of two-prover interactive proofs and that of probabilistically checkable proofs.

The first model was introduced in the original paper [6] and here the verifier interacts with two provers who cannot communicate with each other. Probabilistically checkable proofs, which we from here on abbreviate PCPs, correspond to oracle proof systems of [8], but where introduced and studied in a way relevant to this paper in [2]. In a PCP the verifier does random spot-checks in a written proof.

The first result proving hardness for the problems we are discussing here using these methods was obtained in the fundamental paper by Arora et al [1] where it was established that NP has a PCP where the verifier reads only a constant number of bits and uses a logarithmic number of random coins. This result implies that there is some constant $C > 1$ such that Max-3-Sat could not be approximated within C unless NP=P. The first explicit constant was given by Bellare et al [4] and based on slightly stronger hypothesis they achieved the constant 94/93. Bellare and Sudan [5] improved this to 66/65 and the strongest result prior to our results here are by Bellare, Goldreich and Sudan [3] obtaining the bound 27/26. This paper [3] also studied the problem of linear equations mod 2 and obtained inapproximability constant 8/7. As mentioned above these results were improved to get essentially optimal results in [10] and let us give an outline of the overall construction of that paper.

3 Some points of the construction of [10]

The starting point is the inapproximability result for 3-SAT of [1] described above. This translates naturally into a constant error, two-prover multiprover interactive proof for satisfiability. The verifier V chooses a clause at random and then randomly a variable appearing in that clause and asks one prover, P_1, for the values of all variables in the clause while the other prover, P_2, is asked for chosen variable. V accepts if the answers are consistent and satisfies the clause. It is easy to see that this proof system has constant error rate for any non-satisfiable formula output by the construction of [1]. This protocol has two important properties that are exploited. The first property is that the total size of the answers of the provers is bounded by the constant and the second property is that the answers by P_2 is simply used to check the answer by P_1. More formally, it is used that the V acceptance criteria is that the bit(s) sent by P_2 is equal to some specific bit(s) sent by P_1 together with some condition that only depends on the answer sent by P_1.

The protocol is modified to bring down the acceptance probability for unsatisfiable formulas. This is done by running u copies of the protocol in parallel. The strong result by Raz [13], implies that the error probability decreases as c^u for an absolute constant c. This parallelized two-prover interactive proof is transferred to a PCP by instead of writing down answers, the prover is asked to write down the *long code* of the answers.

The long code, a great discovery of [3], of an input is simply $x \in \{0,1\}^u$ is a string of length 2^{2^u}. The coordinates correspond to all possible functions $f : \{0,1\}^u \mapsto \{0,1\}$ and the coordinate corresponding to f takes the value $f(x)$. It is a very wasteful encoding but in the application u is a constant. The long code is universal in that it contains every other binary code as a sub-code. Thus it never hurts to have this code available, but it is still surprising that it is beneficial to have such a wasteful code.

A correct written proof in the constructed PC looks, as outlined above, the same for all the different Boolean predicates mentioned in the abstract. The adaptation, as advocated already in [3], is to, in a very strong sense, modify the process of checking the proof to suit the target approximation problem.

Let us, by a simplistic example, give an idea how this is done. Suppose we want to check the property that a given string is a long code of some input and that we are trying to make a PCP that should prove inapproximability for Max-3-Sat.

The string A to check is thus of length 2^{2^u} and the coordinates correspond to all the different functions $\{0,1\}^u \mapsto \{0,1\}$ and such a coordinate is denoted by $A(f)$. The ideas is to think of the proof as a set of variables and to write down a 3-Sat formula in these variables which has the property that a variable assignment that corresponds to a correct proof satisfies all the clauses but that an assignment that is "very incorrect" only satisfies a fraction less than 1 of the clauses. To describe this second part is slightly complicated (even to discuss how "very incorrect" should be defined) but let us describe how to satisfy the first property.

We pick three functions f_1, f_2 and f_3 and now we want to make sure that the three bits $A(f_1), A(f_2)$ and $A(f_3)$ should satisfy the clause that says the one of them is true. Since for a correct proof $A(f_i) = f_i(x_0)$ for the x_0 for which A is the long code it is sufficient to make sure the for each x, at least one of $f_1(x), f_2(x)$ and $f_3(x)$ is true. We can thus choose (f_1, f_2, f_3) according to any probability distribution satisfies this property and apart from this condition optimize the distribution to make the analysis of what happens when A is very incorrect as strong and simple as possible. It turns out to be useful to make sure that f_i, when regarded in isolation, is random function with the uniform distribution.

The main tool for the analysis of the protocols is to look at the Fourier transform of the supposed long codes. In the simplest case, which is the case of linear equations mod 2, the calculations are not very complicated, but we refer to [10] for the details.

References

1. S. ARORA, C. LUND, R. MOTWANI, M. SUDAN AND M. SZEGEDY. Proof verification and intractability of approximation problems. Proceedings of 33rd Annual IEEE Symposium on Foundations of Computer Science, Pittsburgh, 1992, pp 14-23.

2. S. ARORA AND S. SAFRA. Probabilistic checking of proofs: a new characterization of NP. Proceedings of 33rd Annual IEEE Symposium on Foundations of Computer Science, Pittsburgh, 1992, 2-13.

3. M. BELLARE, O. GOLDREICH AND M. SUDAN. Free Bits, PCPs and Non-Approximability—Towards tight Results. Proceedings of 36th Annual IEEE Symposium on Foundations of Computer Science, 1995, Milwaukee, pp 422-431. See also a more complete version available from ECCC, Electronic Colloquium on Computational Complexity (http://www.eccc.uni-trier.de/eccc).

4. M. BELLARE, S. GOLDWASSER, C. LUND AND A. RUSSELL. Efficient probabilistically checkable proofs and applications to approximation. Proceedings of the 25th Annual ACM Symposium on Theory of Computation, San Diego, 1993, pp 294-304. (See also Errata sheet in Proceedings of the 26th Annual ACM Symposium on Theory of Computation, Montreal, 1994, pp 820).

5. M. BELLARE AND M. SUDAN. Improved non-approximability results. Proceedings of 26th Annual ACM Symposium on Theory of Computation, Montreal, 1994, pp 184-193.

6. M. BEN-OR, S. GOLDWASSER, J. KILIAN, AND A. WIGDERSON. Multiprover interactive proofs. How to remove intractability. Proceedings of the 20th Annual ACM Symposium on Theory of Computation, Chicago, 1988, pp 113-131.

7. U. FEIGE, S. GOLDWASSER, L. LOVÁSZ, S. SAFRA, AND M. SZEGEDY. Interactive proofs and the hardness of approximating cliques. Journal of the ACM, 1996, Vol 43:2, pp 268-292.

8. L. FORTNOW, J. ROMPEL, AND M. SIPSER. On the power of Multi-Prover Interactive Protocols. Proceedings 3rd IEEE Symposium on Structure in Complexity Theory, pp 156-161, 1988.

9. M.R. GAREY AND D.S. JOHNSSON. Computers and Intractability. W.H. Freeman and Company, 1979.

10. J. HÅSTAD. Some optimal inapproximability results. Proceedings of 29th Annual ACM Symposium on Theory of Computation, El Paso, 1997, pp 1-10. See also a more complete version available from ECCC, Electronic Colloquium on Computational Complexity (http://www.eccc.uni-trier.de/eccc).

11. D.S. JOHNSSON. Approximation algorithms for combinatorial problems. J. Computer and System Sciences, 1974, Vol 9, pp 256-278.

12. C. PAPADIMITRIOU AND M. YANNAKAKIS. Optimization, approximation and complexity classes. Journal of Computer and System Sciences, Vol 43, 1991, pp 425-440.

13. R. RAZ. A parallel repetition theorem. Proceedings of 27th Annual ACM Symposium on Theory of Computation, Las Vegas, 1995, pp 447-456.

14. L. TREVISAN, G. SORKIN, M. SUDAN, AND D. WILLIAMSON. Gadgets, approximation and linear programming. Proceedings of 37th Annual IEEE Symposium on Foundations of Computer Science, Burlington, 1996, pp 617-626.

Minimal Elimination of Planar Graphs

Elias Dahlhaus[1]

Department of Mathematics and Department of Computer Science,
University of Cologne
and
Dept. of Computer Science
University of Bonn,
Germany
e-mail: dahlhaus@cs.uni-bonn.de and dahlhaus@informatik.uni-koeln.de

Abstract. We prove that the problem to get an inclusion minimal elimination ordering can be solved in linear time for planar graphs. An essential tool is the use of breadth-first search.

1   Introduction

One of the major problems in computational linear algebra is that of sparse Gauss elimination. The problem is to find a pivoting, such that the number of zero entries of the original matrix that become non zero entries in the elimination process is minimized. In case of symmetric matrices, we would like to restrict pivoting along the diagonal. The problem translates to the following graph theory problem [12].

Minimum Elimination Ordering: For an ordering $<$ on the vertices, we consider the fill-in graph $G'_< = (V, E')$ of $G = (V, E)$. $G'_<$ contains first the edges in E and secondly two vertices x and y form an edge in $G'_<$ if they have a common smaller neighbor in $G'_<$. *The problem of Minimum Elimination ordering is, given a graph $G = (V, E)$, find an ordering $<$, such that $G'_<$ has a minimum number of fill-in edges.* Note that this problem is NP-complete [15].

For this reason, we relativize the problem.

Minimal Elimination Ordering: *Given a graph G, find an ordering $<$, such that the edge set of $G'_<$ is minimal with respect to inclusion.* Such an ordering $<$ is called a *minimal elimination ordering (MEO)*.

In case that $G = G'_<$ ($<$ has no fill-in edges) $<$ is a perfect elimination ordering, and graphs having a perfect elimination ordering are exactly the *chordal graph*, i.e. graphs with the property that every cycle of length than greater three has an edge that joins two non consecutive vertices of the cycle.

In general, this problem can be solved in $O(nm)$ time [13]. We do not expect that we can reduce the time complexity in general. It seems to be impossible to circumvent matrix multiplication.

Note that minimal elimination orderings always can be represented by a maximal selection of cuts that do not cross [11]. In planar graphs, cuts always can be represented by cycles (not necessarily of the graph, but by cycles of a planar triangulation). We use this fact to get a minimal elimination ordering for planar graphs in linear time.

According to the relevance of the problem to restrict the minimal elimination ordering problem to planar graphs, the author would like to mention that the problem of Gauss elimination on matrices with planar underlying graphs appears for example in finite elements [6]. A minimal elimination ordering might be far away from the optimum solution. But still the problem is an interesting one, because the algorithm in this paper might be helpful as a subprocedure to solve the problem to transform the fill-in of a nested dissection ordering of a planar graph to a minimal fill-in. For general graphs the problem has been considered in [2] and in [4]. The complexities of the algorithms are not linear. For planar graphs it might be interesting to get a linear time algorithm.

In section 2 we will introduce the basic notation. In section 3 we show that breadth-first search leads to a first approximation of a minimal elimination ordering (distance levels). In section 4 we discuss the structure of the distance levels. In section 5, we will show how to get the final minimal elimination ordering.

2 Notation

A *graph* $G = (V, E)$ consists of a *vertex set* V and an *edge set* E. Multiple edges and loops are not allowed. The edge joining x and y is denoted by xy.

We say that x is a *neighbor* of y iff $xy \in E$. The set of neighbors of x is denoted by $N(x)$ and is called the *neighborhood*. The set of neighbors of x and x is denoted by $N[x]$ and is called the *closed neighborhood* of x.

Trees are always directed to the root. The notion of the *parent, child, ancestor,* and *descendent* are defined as usual.

A *subgraph* of (V, E) is a graph (V', E') such that $V' \subseteq V$, $E' \subseteq E$.

We denote by n the number of vertices and by m the number of edges of G.

A graph is called *chordal* iff each cycle of length greater than three has a chord, i.e. an edge that joins two nonconsecutive vertices of the cycle. Note that chordal graphs are exactly those graphs having a *perfect elimination ordering* $<$, i.e. for each vertex v the neighbors $w > v$ induce a complete subgraph, i.e. they are pairwise joined by an edge [7].

A *planar graph* is a graph with the property that the vertices and the edges can be embedded into the plane such that no edges cross. The areas bounded by the edges are called *faces*. A planar graph that is embedded into the plane is called *triangulated* if all faces are bounded by cycles with three edges (or vertices). For an embedded planar graph G, G' is called a *triangulation* of G if G is a subgraph of G' with the same vertex set and G' is a triangulated planar graph.

3 A First Approximation of an MEO by Breadth-First Search

We select a vertex r as maximum element, and by breadth-first search, we determine the levels L_i that consist of the vertices that have distance i from r (compare also [10]). Denote the i with $x \in L_i$ by $l(x)$, for any vertex x of G and call $l(x)$ the *level* of x.

Theorem 1. *There is a minimal elimination ordering $<$, such that*

1. *r is the maximum element of $<$ and*
2. *with $x \in L_i$, $y \in L_j$, and $j < i$, $x < y$.*

Sketch of Proof. By [5], the closed neighborhood C of a connected subset of the vertex set of G is a final segment of a minimal elimination ordering, i.e. there is a minimal elimination ordering $<$, such that with $x \notin C$ and $y \in C$, $x < y$. Q.E.D.

We call a minimal elimination ordering with the properties as stated in previous theorem a *faithful* ordering.

Note that the levels L_i can be determined in linear time.

4 The Structure of Distance Levels

4.1 Adding Auxiliary Edges

We assume that the faces are oriented clockwise. We define the set E_1 of *level diagonals* as follows. For each face f of G and all vertices x and y of the same level $l(x) = l(y)$, we join x and y by an edge in E_1 if all vertices between x and y in the clockwise enumeration of the vertices of f are in some level $L(j)$ with $j > l(x) = l(y)$.

We can determine E_1 in linear time as follows. Suppose the cycle that surrounds f is $(x_0, \ldots, x_{k-1}, x_0)$. Note that $l(x_i)$ and $l(x_{i+1 \bmod k})$ differ at most by one. In f, we determine E_1 as follows. If $l(x_{i+1 \bmod k}) = l(x_i) + 1$ then join x_i and the next x_j with $l(x_i) = l(x_j)$ by an edge $e \in E_1$. This can be done by putting x_i into a stack if $l(x_{i+1 \bmod k}) = l(x_i) + 1$ and joining x_j with x_i with an edge and removing x_i from the stack if x_i is on the top of the stack and of the same level as x_j if we consider the vertex x_j.

Lemma 1. $G_1 := (V, E \cup E_1)$ *is planar.*
All edges in E_1 are fill-in edges of any faithful elimination ordering $<$.

Proof. Note that edges in E_1 do not cross edges in E, because two vertices joined by an edge in E_1 have a common face. Consider two edges xy and $x'y'$ in E_1 that join vertices of the same face f. Without loss of generality, we assume that

$l(x) = l(y) \leq l(x') = l(y')$. Then x and y cannot be between x' and y' in the clockwise enumeration of the face f. That means the edges xy and $x'y'$ do not cross. Therefore G_1 is planar.

It remains to show that all vertices of E_1 are fill-in edges. If an edge $xy \in E_1$ is also in E then nothing has to be proved. Otherwise all the vertices of the clockwise enumeration of a face f between x and y are in a level L_j with $j > l(x) = l(y)$ and therefore in the same connected component of $\bigcup_{j>l(x)} L_j$. Let $<$ be a faithful elimination ordering and let v be the smallest vertex with $l(v) = l(x) = l(y)$. Then all vertices of f between x and y in the clockwise enumeration of f are in the same connected component Z of $G[\{z|z < v\}]$. Note that x and y are neighbors of Z and $x, y > z$, for each $z \in Z$. Since Z is connected, xy is a fill-in edge.   Q.E.D.

4.2 The Structure of $G_1[L_i]$

We assume now that G and therefore also G_1 is 3-connected.

Then we can show the following.

Lemma 2. *Let C be a connected component of $G[\bigcup_{j>i} L_j]$. Then the neighborhood of C with respect to G in L_i is the vertex set of a face in $G_1[L_i]$.*

Proof. Note that the complement of C is also connected, because $\bigcup_{j\leq i} L_j$ is connected and all other connected components of $\bigcup_{j>i} L_j$ are joined by at least one edge with L_i. Therefore in the planar embedding of G, one gets a cyclic orientation $e_1, \ldots, e_k$ of the edges $e_\nu = v_\nu w_\nu$ that join vertices $v_\nu \in C$ and vertices $w_\nu \notin C$. e_ν and $e_{\nu+1 \bmod k}$ share a face. Note that $w_\nu \in L_i$. When we move from w_ν to v_ν and proceed in moving in a face f till $w_{\nu+1}$ then we only pass vertices in levels L_j, $j > i$, and we get an edge $w_\nu w_{\nu+1}$ in G_1. The edges $w_\nu w_{\nu+1}$ form a face in G_1 that consists of the neighborhood of C in L_i. Q.E.D.

Note that all vertices of $\bigcup_{j<i} L_j$ appear in the same face of $G_1[L_i]$, because $\bigcup_{j<i} L_j$ induces a connected subgraph of G. We call this face the *outer face* of $G_1[L_i]$.

Lemma 3. *Let C_1 and C_2 be two non outer faces of $G_1[L_i]$. Then C_1 and C_2 have at most two vertices in common. If C_1 and C_2 have two vertices in common then they are adjacent in G_1.*

Proof. This follows from the fact that each vertex in L_i has a neighbor in L_{i-1} and therefore a neighbor in the connected subset $\bigcup_{j<i} L_j$.   Q.E.D.

4.3 Fill-in Structure of L_i

Let $G_i^{fill} := (L_i, E_i)$ be defined as follows.

1. The edges of $G_1[L_i]$ are edges of G_i^{fill},

2. if C is a non outer face of $G_1[L_i]$ and the neighborhood of a connected component of $\bigcup_{j>i} L_j$ then C is made a clique (in that case, we call C *complete*),

3. if C is a non outer of $G_1[L_i]$ and not the neighborhood of a connected component of $\bigcup_{j<i} L_j$ (and therefore a face of G_1) then it is triangulated in the plane (in that case, we call C *triangulated*).

We can observe the following.

Lemma 4. G_i^{fill} *is a chordal graph.*

Proof: Note that $G_1[L_i]$ consists of non outer faces and bridges, i.e. edges that do not belong to non outer faces.

Let F be a connected component of $G_1[L_i]$. Let v be an *articulation vertex* of F, i.e. a vertex with the property that $G_1[F \setminus \{v\}]$ is not connected. No two vertices in different connected components of $G_1[F \setminus \{v\}]$ appear in a common non outer face and are therefore not joined by an edge in G_i^{fill}. Therefore v is also an articulation vertex of $G_i^{fill}[F]$. If, for each connected component H of $G_1[F \setminus \{v\}]$, $G_i^{fill}[H \cup \{v\}]$ is chordal then we know also that $G_i^{fill}[F]$ is chordal. We can reduce the proof of the lemma to the sets $H' = H \cup \{v\}$. Note that each non outer face is a face of some $G_1[H']$. Therefore we may assume that $G_1[F]$ has no articulation vertex and therefore also no bridge edge. Now assume that $G_1[F]$ has more than one face. Since there is no articulation vertex, there is an edge uv that separates two faces f_1 and f_2. Since u and v are at the border of $G_1[F]$ and the edge uv is not a border edge of $G_1[F]$, $G_1[F \setminus \{u,v\}]$ is not connected, i.e. consists of exactly two connected components H_1 and H_2. Note that any vertex $x \in H_1$ and any vertex $y \in H_2$ have no non outer face in common. We may assume that $G_i^{fill}[H_1 \cup \{u,v\}]$ and $G_i^{fill}[H_2 \cup \{u,v\}]$ are chordal. Since both subgraphs have a complete subgraph as their intersection, also $G_i^{fill}[F]$ is chordal. Therefore we may now assume that F consists of a face or of a bridge edge. If F is a face then either F is made complete or is triangulated. In both cases, $G_i^{fill}[F]$ is a chordal graph. If F is a bridge edge then it trivially induces a chordal graph.   Q.E.D.

To get a compact representation of G_i^{fill}, we triangulate triangulated non outer faces of $G_1[L_i]$, but we do not complete complete cycles. We call this graph G_i'.

Remark 1. The cliques of G_i^{fill} are exactly the non outer faces and the bridge edges (i.e. edges not appearing in any cycle) of G_i'. The size of G_i' is in the order of the size of $G_1[L_i]$.

Lemma 5. *There is a faithful elimination ordering $<$ that, for each G_i^{fill}, is a perfect elimination ordering, i.e. all fill-in edges of $<$ joining vertices of the same level L_i are edges of G_i^{fill}.*

Proof. This follows from the following fact in [5]: If C is a final segment of a minimal elimination ordering and if the connected components $D_1, \ldots, D_k$ have minimal elimination orderings $<_1, \ldots, <_k$ then there is a minimal elimination $<$ ordering with C as a final segment, and the fill-in of $<$ restricted to D_i is the same as the fill-in of $<_i$. Note that $<_i$ is not necessarily the restriction of $<$ to D_i.

$\hfill$ Q.E.D.

5 The Final Ordering

We know now the fill-in restricted to each level L_i. But still we did not take the fill-in edges between adjacent levels L_i and L_{i+1} into account. Here we follow the ideas in [5].

Let F be a connected component of G_i^{fill}. We define as a cut an intersection c of two cliques C_1 and C_2, such that $C_1 \backslash c$ and $C_2 \backslash c$ appear in different connected components of $F \backslash c$. We call a connected component C of $F \backslash c$ *full* if all vertices of c are neighbors of C. Let c be a cut. We sort the full connected components of $F \backslash c$ by

1. the number of neighbors in L_{i-1} and in second priority
2. by the number of elements

to a sequence $D_1, \ldots, D_l$. If D_l is the unique component that has a maximum number of neighbors in L_{i-1} and in the second priority a maximum size then we call D_l the *dominator* of c. Note that the dominator of a cut c might not exist. If $x \in D_\nu$, D_ν is not the dominator of c, and $y \in c$ then we set $x <' y$. We do this for every cut c.

Lemma 6. *[5] $<'$ is a partial ordering. There is a minimal elimination ordering $<$ such that with $x <' y$, $x < y$.*

It remains to order those x, y that are joined by an edge in G_i^{fill} that are not comparable with respect to $<'$. Let $x \equiv y$ if xy is an edge in G_i^{fill} and neither $x <' y$ nor $y <' x$.

Lemma 7. *[5] $\equiv$ is an equivalence relation.*

Let $N'(x)$ be the set of neighbors of x or of a $y <' x$ in L_{i-1}. We sort any $\equiv$-equivalence class A to a sequence $a_1, \ldots, a_l$, such that for each j, $N'(a_j) \backslash \bigcup_{j' < j} N'(a_{j'})$ minimal in the set system of $N'(a_{j''}) \backslash \bigcup_{j' < j} N'(a_{j''})$, $j'' \geq j$. Let $a_j <'' a_{j'}$ if $j < j'$.

Lemma 8. *[5] Any linear ordering that is an extension of $<'$ and of $<''$ is a minimal elimination ordering.*

We have to determine the (partial) orderings $<'$ and $<''$ in linear time. We also have to code these orderings appropriately.

5.1 Determining $<'$

To get a compact representation of $<'$, we develop an ordering $<'_1$ on the cliques and cuts of F.

Let c be a cut and C be a clique or cut containing c as a subset. We say that C *belongs to the dominator* of c is $C \setminus c$ is a subset of the dominator of c. If C does not belong to the dominator of c then we set $C <'_1 c$. Otherwise we set $c <'_1 C$.

Lemma 9. *If a clique C does not belong to the dominator of cuts c_1 and c_2 that are contained in C then c_1 and c_2 are comparable with respect to inclusion. If, moreover, $c_1 \subset c_2$ then $c_2 <'_1 c_1$.*

Proof. Assume c_1 and c_2 are not comparable with respect to inclusion and C contains c_1 and c_2. Let D be a full component of c_1 C does not belong to and D' be a full component of c_2 C does not belong to. Then D and D' are disjoint. Since C does not belong to the dominators of c_1 and c_2, we may assume that D and D' are full components of c_1 and c_2 with a maximum number of neighbors in L_{i-1} and in second priority of maximum size. It is easily seen that we get a contradiction. This proves the first sentence of the lemma.

Now suppose $c_1 \subset c_2$. Then the vertices of $c_2 \setminus c_1$ belong to $C \setminus c_1$ and therefore not to the dominator of c_1. Therefore $c_2 <'_1 c_1$. $\hspace{2cm}$ Q.E.D.

Now we create a tree T with on the cliques and cuts with parent function P.

1. If c is a cut that is contained in a clique C with $c <'_1 C$ then we select some clique C with this property as $P(c)$.
2. Otherwise if c is a clique or a cut with the property that for each clique C containing c, $C <'_1 c$ then by previous lemma, there is a unique maximal intersecting cut d with $c <'_1 d$. We set $P(c) := d$.

The following is easily checked.

Lemma 10. *The cliques that do not belong to the dominator of a cut c are the cliques that are descendents of c in T.*

As a consequence from this lemma we get the following.

Lemma 11. *For each vertex v, the set $\{c | v \in c\}$ forms a subtree of T.*

Proof. Suppose $v \in c$ but $v \notin P(c)$. Note that always $c \subset P(c)$ or vice versa. Here we only have the possibility that $P(c) \subset c$. A consequence is that v is not in the dominator of $P(c)$. Therefore all cliques containing v are descendents of $P(c)$. Consider first any cut $d \subset c$ that contains v. Then $c \setminus d$ belongs to the dominator of d and therefore c is an ancestor of d. Therefore all cuts that are subsets of c and contain v are descendents of c. Consider now any cut or clique d that contains v. Then $c \cap d$ is a subset of a cut d' contained in c that separates

c and d. Then $d \setminus d'$ is not in the dominator of d'. Therefore d is a descendent of d' and d' is a descendent of c and therefore d is a descendent of c.

This proves that the cuts and cliques containing v form a subtree of T. Q.E.D.

We extend $<_1'$ to a partial ordering.

Lemma 12. *For any vertex v of F, there is a unique clique or cut $c(v)$ containing v that is maximal with respect to $<_1'$.*

Proof. This follows from previous lemma. $c(v)$ is just the root of the subtree of cuts and cliques containing v.   Q.E.D.

Lemma 13. $x \equiv y$ *if and only if* $c(x) = c(y)$.

Proof. If one of the two statements is true then x and y are joined by an edge of the fill-in graph and are therefore in a common clique. Suppose there is a cut c containing x but not y. If $x \equiv y$ then y belongs to the dominator of c.

Assume $x \equiv y$. Suppose d is the $<_1'$-maximum cut or clique containing x. If d does not contain y then if d is a cut then y belongs to the dominator of d or if d is a clique then there is a cut d' contained in d that separates $d \setminus d'$ and y. Since $xy \in E$, d' contains x and y is in the dominator of d'. In the second case, d' is an ancestor of d. This is a contradiction. In the first case, since $xy \in E$, there is a clique that contains x and y and that belongs to the dominator of d. This clique is not a T-ancestor of d. This is a contradiction.

Vice versa, let $c(x) = c(y)$. Then for each cut c containing x but not y, y belongs to the dominator of c and vice versa. Therefore $x \equiv y$.   Q.E.D.

Algorithmically, we proceed as follows.

1. We determine for each cut c, the full components and the dominator (in a compact representation)
2. For each cut c and each clique or cut C containing c, we check whether $C <_1' c$ or vice versa. We will see that this can be done in linear time.
3. We do topological sorting on $<_1'$ and get an ordering $<_2'$ on the cliques and cuts.
4. We determine for each vertex v, the maximum element $c(v)$ with respect to $<_2'$ containing v.

First we have to determine the dominator of each cut of G_i^{fill}.
The following statement is an immediate consequence of lemma 3.

Lemma 14. *1. Each cut of G_i^{fill} has a size ≤ 2.*
2. If a cut c has two vertices then they are also adjacent in G_i' and c is the intersection of exactly two cliques of G_i^{fill}. These two cliques are exactly the faces of G_i' that share c as an edge.

To get a linear time algorithm, we use the fact that each vertex of a connected component F of G_i' is a border vertex of F. We therefore have a cyclic enumeration $w_0, \ldots w_{k-1}, w_0$ of F (w.l.o.g. clockwise). Note that a vertex of F might appear more than once in this enumeration.

Remember that a vertex v of F is called an *articulation vertex* if it is the unique element of a cut of size one.

We have the following characterization of full components of one element cuts.

Lemma 15. *A vertex v of F is an articulation vertex if and only if v appears more than once in $w_0, \ldots, w_{k-1}$. The full components of $c = \{v\}$ are maximal intervals of the cyclic enumeration $w_0, \ldots, w_{k-1}$ of F that do not contain v.*

One can easily determine all these intervals in linear time. We only have to know the border vertices of the intervals.

Next we determine the full components of two element cuts.

Lemma 16. *The two full components of a cut $\{v, w\}$ are the intervals*

$$\{w_{i_1}, \ldots w_{j_1}\}$$

and

$$\{w_{i_2}, \ldots, w_{j_2}\}$$

not containing v and w with $v = w_{i_1-1} = w_{j_2+1}$ and $w = w_{j_1+1} = w_{i_2-1}$.

The vertices $w_{i_1}, w_{i_2}, w_{j_1}, w_{j_2}$ can easily be determined using the cyclic orientation of the incident edges of v and w with respect to a planar embedding. This again can be done in linear time.

Therefore the full components of any cut are intervals in the cyclic clockwise enumeration of F and can be determined in linear time.

Next we have to determine the neighborhoods of full components in L_{i-1}. Note that the neighborhood of F in L_{i-1} is a cycle in G_{i-1}', say $(x_0, \ldots, x_{l-1}, x_0)$. For any $w_j \in F$, define $N_i(w_j)$ to be the set of neighbors of w_i in L_{i-1} that are clockwise between w_{j-1} and w_{j+1}. Let $left(w_j)$ be the first and $right(w_j)$ be the vertex in $N_i(w_j)$ with respect to the clockwise enumeration of the neighbors of w_j. If $N_i(w_j) = \emptyset$ then let $w_{j'}$ be the first successor of w_j in the clockwise enumeration of F with $N_i(w_{j'}) \neq \emptyset$ and let $w_{j''}$ be the closest predecessor of w_j with $N_i(w_{j''}) \neq \emptyset$. In that case, $left(w_j) := left(w_{j'})$ and $right(w_j) := right(w_{j''})$.

Lemma 17. *For a full component $D = \{w_j \ldots w_{j'}\}$ of c, the neighborhood of D in L_{i-1} is a cyclic interval of $x_1, \ldots, x_l$ with border vertices $left(w_j)$ and $right(w_{j'})$.*

Therefore we get a representation of the neighborhoods of full components that can be determined in linear time.

Therefore we get the sizes of full components and their neighborhoods in linear time and therefore also the dominators of any cut of G_i^{fill}.

Since the number of cliques containing a two element cut is restricted to two and the number of cliques containing an articulation vertex v is restricted to the number of faces of G'_i that contain v (and therefore by the degree of v in G'_i), we have a representation of $<'_1$ of linear size. We easily can determine $<'_1$ in linear time.

5.2 Determining $<''$

It remains to order the vertices that have the same $<'_2$-maximum cut or clique.

First we assume that the maximum $c(v)$ is a clique.

We have to determine the neighborhood of $\{x | x \leq' v\}$ in L_{i-1}. Note that the clique $c(v)$ of G_i^{fill} is a face of G'_i and has therefore a cyclic enumeration $c_1, \ldots, c_p$.

Remember that $w_0, \ldots, w_{k-1}$ is the cyclic enumeration of the connected component F of G'_i.

Lemma 18. *Assume that the predecessor of v in the cyclic enumeration of $c(v)$ is w_j, the successor of v in the cyclic enumeration of $c(v)$ is $w_{j'}$ and there is no repetition of w_j or $w_{j'}$ in between. Then $I_v := \{x | x \leq' v\}$ is the interval $\{w_{j+1}, \ldots, w_{j'-1}\}$ of the cyclic enumeration of F.*

Proof. First note that the only two vertex cuts containing v that are contained in $c(v)$ might be $\{v, w_j\}$ and $\{v, w_{j'}\}$. Not necessarily they are cuts. Note that $c(v)$ must belong to the dominators of both two element cuts (if they are cuts). Moreover, if $\{v\}$ is a cut then $c(v)$ belongs to the dominator of $\{v\}$. Note that the vertices that are between w_j and $w_{j'}$ and not identical v are exactly the vertices that are in a non dominator full component of some of the three cuts that contain v and are subsets of $c(v)$. Therefore these vertices are exactly the vertices x with $x <' v$. Q.E.D.

Corollary 1. *The neighborhood $N'(v)$ of $\{x | x \leq' v\}$ in L_{i-1} is an interval of the cyclic enumeration of the neighborhood of F. The neighborhoods $N'(v)$ can be determined (as intervals) in linear time.*

It remains to do the elimination process, i.e. to update the intervals $N'(v)$.

Lemma 19. *1. Suppose $c(v) = c(w) = c$. If I_v and I_w have a nonempty intersection then v and w are neighbors in the clockwise enumeration of c.*

2. Suppose $c(v) = c(w) = c$. If I_v and I_w have an empty intersection then $N'(v)$ and $N'(w)$ have only vertices in common that are at the border of $N'(v)$ and of $N'(w)$.

3. If

$$\emptyset \neq I_v \setminus \{v\} \subset I_w$$

and $c(v) = c(w)$ then v is a border vertex of I_v and the intervals $I_v \setminus \{v\}$ and I_w have one border vertex in common.

4. *If $c(v) = c(w)$ and $N'(v) \subset N'(w)$ then either $N'(v)$ and $N'(w)$ share a border vertex (as intervals) or $N'(v)$ consists exactly of the border vertices of $N'(w)$.*

Proof: Note that the clockwise enumeration of c is a subsequence of the clockwise enumeration $w_0, \ldots, w_{k-1}$ of the connected component F containing c. The first statement follows easily from previous lemma.

The second statement can be checked as follows. If the neighborhoods of the disjoint intervals I_v and I_w in L_{i-1} have a non border vertex in common then edges must cross. This is a contradiction.

The third statement is checked as follows. The hypothesis of the third statement can only be true if v and w appear consecutively in the clockwise enumeration of c. We assume w.l.o.g. that v appears before w in the clockwise enumeration of F (the connected component of G'_i containing c). Then there is no other alternative than that the successor of v in the clockwise enumeration of F is the start vertex of $I_v \setminus \{v\}$. Note that this vertex is also the first vertex of I_w.

The fourth statement is checked as follows. Again we assume that v appears before w in the clockwise enumeration of F. All the neighbors of v in L_{i-1} are neighbors of I_w. Since there are no edge crossings, there is no other choice than that all the neighbors of v are border vertices of $N'(w)$. If both border vertices of $N'(w)$ are in the neighborhood of v then there is no other choice than that $N'(v)$ consists exactly of the border vertices of $N'(w)$. Otherwise, we assume that v appears before w. Vertices left from v in I_v also have only the left border vertex of $N'(w)$ as a neighbor in L_{i-1}. Vertices right from v in I_v are either vertices in I_w or also have only the left border of $N'(w)$ as neighbor in L_{i-1}. Q.E.D.

The elimination process is done as follows.

1. We first eliminate the vertices v, such that $N'(v)$ has only one element.
2. When we did this, we have the following.
   If $N'(v)$ and $N'(w)$ intersect then v is the successor or the successor of the successor of w in the cyclic enumeration of $c(v) = c(w) = c$ or vice versa.
   That means the number of intervals $N'(w)$ that intersect $N'(v)$ is bounded.

This strategy allows us to do the update of the intervals $N'(v)$ in linear time. Last lemma gives also the guarantee that $N'(v)$ remains an interval (whenever we eliminate v, $N'(w)$ is updated by $N'(w) \setminus N'(v)$).

If $c(v) = c(w) = c$ is a cut then $c = \{v, w\}$. We proceed as if c is a face surrounded by two copies of the edge vw of G'_i.

6 Conclusions

One could ask as in in [2] or [4] whether one could combine the nested dissection algorithm for planar graphs (see [10] and [1]) with a planar minimal elimination algorithm, such that we get the performance of the nested dissection algorithm affecting the number of fill-in edges and an inclusion minimal fill-in. We would like to get it in linear time. The nested dissection algorithm for planar graphs itself needs $O(n^{3/2})$ time.

References

1. A. Agrawal, P. Klein, R. Ravi, Cutting Down on Fill-in Using Nested Dissection, in *Sparse Matrix Computations: Graph, Theory Issues and Algorithms*, A. George, J. Gilbert, J.W.-H. Liu ed., IMA Volumes in Mathematics and its Applications, Vol. 56, Springer Verlag, 1993, pp. 31-55.
2. J. Blair, P. Heggernes, J.A. Telle, *Making an Arbitrary Filled Graph Minimal by Removing Fill Edges*, Algorithm Theory-SWAT '96, R. Karlsson, A. Lingas ed., LLNCS 1097, pp. 173-184.
3. P. Bunemann, *A Characterization of Rigid Circuit Graphs*, Discrete Mathematics 9 (1974), pp. 205-212.
4. E. Dahlhaus, *Minimal Elimination Ordering inside a Given Chordal Graph*, WG 97 (R. Möhring ed.), LLNCS 1335, pp. 132-143.
5. Elias Dahlhaus, Marek Karpinski, *An Efficient Parallel Algorithm for the Minimal Elimination Ordering (MEO) of an Arbitrary Graph*, Theoretical Computer Science 134 (1994), pp. 493-528.
6. M. Eiermann, O. Ernst, W. Queck, *Finite Element Tutorial*, TU-Bergakademie Freiberg.
7. M. Farber, *Characterizations of Strongly Chordal Graphs*, Discrete Mathematics 43 (1983), pp. 173-189.
8. F. Gavril, *The Intersection Graphs of Subtrees in Trees Are Exactly the Chordal Graphs*, Journal of Combinatorial Theory Series B, vol. 16(1974), pp. 47-56.
9. J. Gilbert, R. Tarjan, *The Analysis of a Nested Dissection Algorithm*, Numerische Mathematik 50 (1987), pp. 427-449.
10. R. Lipton, R. Tarjan, *A Separator Theorem for Planar Graphs*, SIAM Journal on Applied Mathematics 36 (1979)¡ pp. 177-189.
11. Parra, A., Scheffler, P., How to use minimal separators for its chordal triangulation, *Proceedings of the 20th International Symposium on Automata, Languages and Programming (ICALP'95)*, Springer-Verlag Lecture Notes in Computer Science **944**, (1995), pp. 123–134.
12. D. Rose, *Triangulated Graphs and the Elimination Process*, Journal of Mathematical Analysis and Applications 32 (1970), pp. 597-609.
13. D. Rose, R. Tarjan, G. Lueker, *Algorithmic Aspects on Vertex Elimination on Graphs*, SIAM Journal on Computing 5 (1976), pp. 266-283.
14. R. Tarjan, M. Yannakakis, *Simple Linear Time Algorithms to Test Chordality of Graphs, Test Acyclicity of Hypergraphs, and Selectively Reduce Acyclic Hypergraphs*, SIAM Journal on Computing 13 (1984), pp. 566-579. Addendum: SIAM Journal on Computing 14 (1985), pp. 254-255.
15. M. Yannakakis, Computing the Minimum Fill-in is NP-complete, *SIAM Journal on Algebraic and Discrete Methods* 2 (1981), pp. 77-79.

Memory Requirements for Table Computations in Partial k-tree Algorithms

Bengt Aspvall[1], Andrzej Proskurowski[2], and Jan Arne Telle[1]

[1] Department of Informatics, University of Bergen, Norway
[2] CIS Department, University of Oregon, USA

Abstract. This paper addresses memory requirement issues arising in implementations of algorithms on graphs of bounded treewidth. Such dynamic programming algorithms require a large data table for each vertex of a tree-decomposition T of the input graph. We give a linear-time algorithm that finds the traversal order of T minimizing the number of tables stored simultaneously. We show that this minimum value is lower-bounded by the pathwidth of T plus one, and upper bounded by twice the pathwidth of T plus one. We also give a linear-time algorithm finding the depth-first traversal order minimizing the sum of the sizes of tables stored simultaneously.

1 Introduction

Many NP-hard graph problems have linear-time algorithms when restricted to graphs of treewidth bounded by k (equivalently, partial k-trees), for fixed values of k. These algorithms have two stages: the first stage finds a bounded width tree-decomposition of the input graph and is followed by a dynamic programming stage that solves the problem in a bottom-up traversal of that tree. Theoretically, the first stage has linear-time complexity (see [4]), but no general practical implementation for treewidth $k \geq 5$ exists. In this paper we focus on practical implementations of the second stage.

Dynamic programming algorithms on bounded treewidth graphs have been studied since the mid-1980s leading to several powerful approaches, see [5] for an overview. Recent efforts have been aimed at making these theoretically efficient algorithms amenable also to practical applications [14, 15]. Experimental results so far have not been negative: for example, using a 150 MHz alpha processor-based Digital computer, the maximum independent set problem is solved for treewidth-10 graphs of 1000 vertices in less than 10 seconds [8]. The given tree-decomposition for this particular example had about 1000 nodes and associated with each node is a data table of 2^{11} 32-bit words, so without reuse of data tables the combined storage would exceed 8 Megabytes of memory. With our algorithms we need only 80 Kilobytes. These initial investigations already show that avoidance of time-consuming I/O to external memory is an important issue (see [6] for a good overview of issues related to external memory use in graph algorithms and the expected increased significance of these issues in coming years).

In the bottom-up traversal of the tree-decomposition that accompanies the dynamic programming stage, we are free to choose both the root and the traversal order. Moreover, the information contained in the table at a child node is superfluous once the table of its parent has been updated, and the table of the parent need not be created until it is ready to be updated by the table of the first child. Given a tree, a natural question is how to find a good root and a good traversal order to minimize the table requirement, *i.e.*, the minimum number of tables that need to be stored simultaneously, under the assumption that all tables will be kept in the main memory. In the next section we answer this question and give a careful description of the resulting algorithm to ease the task of implementation. In Section 3, we show an interesting relationship between the table requirement of a tree T and the pathwidth of T, giving asymptotically tight bounds relating these two parameters. In the case when the size of the table is not the same over all nodes of the tree-decomposition, we may ask instead for a traversal order that minimizes the sum of sizes of tables that need to be stored simultaneously. In Section 4 we give a practical linear-time algorithm answering this question for depth-first search (dfs) traversal orders. The traversal order output by the algorithm in Section 2 minimizing table requirement is indeed a dfs order, and dfs orders have the advantage of being easy to implement. We conclude in Section 5 with a discussion of future research.

2   Table requirement

In the following, we will refer not to the bounded treewidth graph G itself but instead focus on the adjacencies of its tree-decomposition T. The interested reader may see, *e.g.*, [5] for definitions related to bounded treewidth graphs. We assume that in order to solve a discrete optimization problem on G, some function on T has to be computed. The value of this function is independent of the choice of root for the tree, and is obtained from the completed computation of the table associated with the vertex chosen as root. The computation of this final table is a bottom-up process which involves computation of tables for all other vertices of the tree. The table of a leaf vertex is defined by the base case of the function. The table of a non-leaf vertex is updated according to the contents of the completed table of its child vertex, after which that latter table can be discarded. This table update operation requires the simultaneous presence of the parent and child tables. The table at a vertex is completed once it has been updated using the contents of the tables of all its children. We must thus pay for storage of the table at a vertex u from the moment of its creation, allow the table to become completed by being updated by completed tables of all children of u, update the table of the parent of u, and only then discard the table of u. We summarize the above discussion in a formal definition:

Definition 1 *An ordering of edges $e_1, ..., e_{n-1}$ of a rooted tree with $n > 1$ vertices is a bottom-up traversal if for any edge $e_j = (u, parent(u))$, all edges in the subtree rooted at u have indices less than j. The lifetime of the table* of vertex u in this traversal is the interval $[i, j]$ where e_i is the lowest-indexed edge containing u and e_j is the highest-indexed edge containing u. The table requirement of this traversal order is the maximum number of tables alive at any time, i.e. $\max_{1 \le i < n} |\{v_k : i \in lifetime(v_k)\}|$.

For adjacent vertices u and v of T we define $tabreq(u, v)$ as the minimum table requirement of the subtree rooted at u with overall root chosen so that u has parent v, taken over all possible bottom-up traversals of this subtree.

Theorem 2 *In a given rooted tree, let a non-root vertex u have the parent v and let x and y be the children of u (if any) such that $tabreq(x, u) \ge tabreq(y, u) \ge tabreq(w, u)$ for any other child w of u (if u has only one child, define $tabreq(y, u)$ to be 0.) Then, the table requirement to compute the table associated with u is given by*

$$tabreq(u, v) = \begin{cases} 1 & \text{for a leaf } u \\ \max\{tabreq(x, u), 2, tabreq(y, u) + 1\} & \text{for a non-leaf } u. \end{cases} \quad (1)$$

Proof. The base case of a leaf u follows from the definition of $tabreq(u, v)$. We continue with the non-leaf case. That the value given by (1) suffices follows from noting the following: (i) First compute recursively the table of the child x with largest table requirement; this requires at most $tabreq(x, u)$. (ii) Then create the table of u and update it with the table of x before discarding table of x; this requires at most 2 tables. (iii) Finally compute recursively tables of remaining children of u one by one, in any order, updating the table of u before discarding the table of the child; this requires at most $tabreq(y, u) + 1$. That the value given by (1) is necessary follows by noting: (i) To compute the table of a non-leaf vertex u, it is necessary to have computed the tables of all its children; this requires at least $tabreq(x, u)$. (ii) To update the table of u with the table of a child, both these tables must be present in memory; this requires at least 2 tables. (iii) After a table has been created, it must be stored at least until the table of its parent has been created. Thus, only for a single child c of u will it be possible to compute its table while storing only tables that belong to the subtree rooted at c. This entails a requirement of at least $tabreq(y, u) + 1$ (achieved by choosing $c = x$) which is minimal since $tabreq(x, u) \ge tabreq(y, u)$. ∎

For a tree T and a vertex r in T, we denote by T_r the rooted version of T resulting from designating r as its root. Since r does not have a parent in T_r, we denote the table requirement of T_r by $tabreq(T_r)$. This value is the minimum number of tables necessary to complete the table at r, and hence to compute the function on T_r solving the original problem for G. When the tree is rooted at u, the value of $tabreq(T_u)$ is given by the right-hand side of (1).

Corollary 3 *The right-hand side of (1) gives the value of $tabreq(T_u)$; in this case, all neighbors of u are considered in defining x and y.*

Corollary 4 *For an n-vertex tree T rooted in an arbitrary vertex u we have*

$$tabreq(T_u) \leq \lfloor \log_2 \frac{4}{3}(n+1) \rfloor$$

Proof. Let $S(i)$ be the number of nodes in the smallest rooted tree with table requirement i. By Theorem 2 we have the recurrence $S(1) = 1$, $S(2) = 2$ and $S(i) = 2S(i-1)+1$ for $i \geq 3$. The solution to this recurrence is $S(i) = 3 \times 2^{i-2} - 1$ for $i \geq 3$. Rearranging and taking logarithms the result follows. ∎

Note that $tabreq(T_u)$ will usually differ from $tabreq(T_x)$ for $x \neq u$. We define the table requirement of a tree T as the minimum value of $tabreq(T_u)$ over all vertices u of T. This parameter is of interest also because of its similarity to graph searching games (see, for instance, [7]), to the minimum stack traversal problem [2] and to the pathwidth parameter, discussed in the next section. Indeed, the algorithm we now present to compute this parameter uses a traversal strategy resembling closely that of [2] as well as those used by [13, 10] to compute the pathwidth of a tree.

The algorithm will consist of a single bottom-up phase, which will start at the degree one vertices of the tree (leaves), and end in a root r with minimum value of $tabreq(T_r)$ (over all vertices of T). For each vertex v, we keep track of $larg(v)$ and $next.larg(v)$, the two largest table requirements reported by neighbors of v. We also maintain an array S of stacks to guide the order of processing vertices of T. The stack $S[i]$ contains those leaf vertices of the remaining tree of unprocessed vertices whose subtrees have table requirements of value i. Vertices are popped from the non-empty stack $S[i]$ with the smallest value of i. The algorithm, which assumes that T has at least two nodes, can now be described as follows (see Figure 1 for an example of algorithm execution):

1 Set $i = 1$. Push all degree one vertices of T on $S[1]$. For all vertices v, set $larg(v) = next.larg(v) = 0$.
2 While T has more than one vertex remaining
   2a While $S[i]$ is empty increment i.
   2b Pop a vertex w from $S[i]$. Let v be the single neighbor of w (its parent) in the tree T of unprocessed vertices.
   2c Report the value i to v, update the value of $larg(v)$ to $\max\{i, larg(v)\}$, and also update $next.larg(v)$ to $\max\{next.larg(v), \min(larg(v), i)\}$.
   2d If v has received reports from all but one neighbor then push v on the stack $S[j]$, $j = \max\{larg(v), 1 + next.larg(v), 2\}$. Remove w from T.
3 Report the only remaining vertex r in T as an optimal root, with table requirement j such that $r \in S[j]$.

Theorem 5 *The node r of T which remains unprocessed upon termination of the algorithm (when the conditional in the outer while loop is false) is an optimal root for minimizing table requirement of T.*

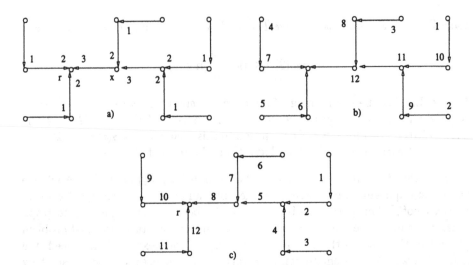

Fig. 1. a) A tree with its optimal root r and values of table requirement for subtrees, as computed by our algorithm. Optimal root r gives table requirement 3, while rooting in x would give table requirement 4. b) The order in which vertices were popped from the stack during computation of optimal root. c) An optimal traversal order, as defined by the order of execution of update-table$(u, parent(u))$ in call of DFS(r).

Proof. (By contradiction.) Assume that an optimal root $r^* \neq r$ has value $m^* = tabreq(T_{r^*}) < tabreq(T_r)$. Let r have the neighbors $w, w_1, ..., w_k$ with w on the path from r to r^* in T and let $tabreq(w_1, r) \geq tabreq(w_2, r) \geq ... \geq tabreq(w_k, r)$. Table requirements are defined by Theorem 2. We have $tabreq(w_1, r) \leq tabreq(r, w) \leq m^*$ and $tabreq(w_2, r) + 1 \leq tabreq(r, w) \leq m^*$. We also have $tabreq(w, r) \leq m^*$ since otherwise, considering the previous statement, r would be on a lower-indexed stack than w and r would not be processed last. Since we assumed $m^* < tabreq(T_r)$ we have w and w_1 the two neighbors of r with largest table requirements $tabreq(w, r) = tabreq(w_1, r) = m^* > tabreq(w_2, r)$. The two vertices, w and w_1, are thus the last neighbors of r processed by the algorithm, with the parameter $i = m^*$. But this implies that the penultimate neighbor of r processed by the algorithm causes r to be pushed on the stack $S[m^*]$ in step 2d. Thus, r would be popped next and not remain as the last unprocessed vertex. ∎

Using the observation that $tabreq(T_r)$ increases over its subtrees' requirement only when $larg(v) = next.larg(v)$, we arrive at the following result:

Corollary 6 $tabreq(T_r) - 1 \leq tabreq(T_{r^*}) = m^*$, for any vertex r and an optimal root r^* in a tree T.

Proof. Let P be the unique path $r^* = r_1, r_2, ..., r_d = r$ between r^* and r. Let C_i be the neighbors of r_i that are not on P. The subtree rooted in a given vertex x of C_i does not depend on the choice of overall root r or r^*. Thus in both cases x

reports the same memory requirement, of value at most m^*, to its parent r_i. By Theorem 2, a report of table requirement of value $m^* + 1$ or more occurs only if one child reports at least $m^* + 1$ or at least two children report m^*. Thus, in T_r such reports only occur on the path P, from some r_i to its parent $r_{i+1}, i \geq 1$, but in no other place. We conclude that $tabreq(T_r) \leq m^* + 1$. ∎

The values $\max\{larg(v), 1 + next.larg(v), 2\}$, computed in step 2c by the algorithm for each vertex v, gives the table requirement for the subtree of T_r rooted at v. To find a traversal order minimizing table requirement let a node v in T_r have children $child_1(v), ..., child_{c_v}(v)$, with $child_1(v)$ having the largest table requirement over all subtrees of T_r rooted at these vertices. A call of the procedure DFS(r) shown below will then perform the dynamic programming computation on T while minimizing the number of tables stored simultaneously:

```
DFS(v)
if v a leaf then create-table(v);
else {
    DFS(child₁(v)); /* The most expensive subtree first */
    create-table(v);
    update-table(child₁(v), v);
    discard-table(child₁(v));
    for i=2 to cᵥ{
        DFS(childᵢ(v));
        update-table(childᵢ(v), v);
        discard-table(childᵢ(v));}}
```

This strategy results in a traversal order minimizing table requirement as described in the proof of Theorem 2.

3 Pathwidth and table requirement

Pathwidth is a graph parameter whose definition is closely related to treewidth [12].

Definition 7 *A path-decomposition of a graph G is a sequence $X_1, ..., X_m$ of bags, which are subsets of the vertex set $V(G)$ of G, such that for each edge of G there is a bag containing both its end-vertices and the bags containing any given vertex form a connected subsequence. G has pathwidth at most k, $pw(G) \leq k$, if it has a path-decomposition where the cardinality of any bag is at most $k + 1$.*

As an aside, we remark that the definition of treewidth is similar, except that the bags are nodes of a tree, as opposed to being nodes of a path (sequence), and the bags containing a given vertex induce a connected subtree. There is a linear-time algorithm computing pathwidth and path-decomposition of a tree, [10], which uses a traversal similar to our minimum table requirement algorithm, based on the following Theorem:

Theorem 8 *[13] For a tree T, $pw(T) > k$ if and only if there exists a vertex v in T such that $T \setminus \{v\}$ has at least three components with pathwidth at least k.*

We will use the following corollary to Theorem 8:

Corollary 9 *In a tree T with $pw(T) = k + 1$, either there is a vertex x s.t. all components of $T \setminus \{x\}$ have pathwidth at most k, or the set of edges e, such that deleting e from T gives two trees both of pathwidth $k + 1$, induce a path $P = e_1, ..., e_p$, with $p \geq 1$.*

We will now show a close relation between the two parameters pathwidth and table requirement for a tree.

Theorem 10 *Any tree T has a node r such that*

$$pw(T) + 1 \leq tabreq(T_r) \leq 2pw(T) + 1$$

Proof. We show that the first inequality holds even for an arbitrary choice of r. Consider any bottom-up traversal order of T_r, $e_1, e_2, ..., e_{n-1}$ that defines the order of table updates. Let $e_i = (v_i, parent(v_i))$ and define $X_1 = \{v_1, parent(v_1)\}$ and $X_i = X_{i-1} \setminus \{v_{i-1}\} \cup \{v_i, parent(v_i)\}$ for $2 \leq i \leq n-1$. Then $X_1, ..., X_{n-1}$ is a path-decomposition of T as each edge is contained in a bag and, by construction, the bags containing a given vertex are consecutive. Moreover, the maximum cardinality of a bag is the memory requirement of this particular bottom-up traversal of T_r, as the table at a node is discarded precisely after the table of the parent of the node has been updated. Thus $pw(T) + 1 \leq tabreq(T_r)$.

We show the second inequality by induction on $pw(T)$. The base case is $pw(T) = 0$ with T being a single node r, and $tabreq(T_r) = 1$ as a single table suffices. For the inductive step, let us assume the second inequality for trees with pathwidth k and let $pw(T) = k + 1$. We must show that table requirement of T is at most $2k + 3$. Apply Corollary 9 to T and choose as root of T a vertex x as described by the Corollary, if it exists, or else consider the path $P = e_1, ..., e_p$ as described by the Corollary, with $e_i = \{v_{i-1}, v_i\}, 1 \leq i \leq p$ and choose v_p as the root. In the former case, the inductive assumption and Corollary 6, together with Theorem 2, gives us $tabreq(T_x) \leq 2k + 3 \leq 2pw(T) + 1$. In the latter case, note that the endpoint v_0 of P, the lowest node of P in the tree T_{v_p}, by Corollary 9 has all its subtrees of pathwidth at most k. Hence by the inductive assumption and Corollary 6 these children report table requirement at most $2k + 2$ to v_0, so that by Theorem 2 v_0 reports table requirement at most $2k + 3$ to its parent v_1. By induction on i, it follows that each node $v_i, i \geq 1$ on P receives report of at most $2k + 3$ from its child v_{i-1} and a report of at most $2k + 2$ from all its other children (since by Corollary 9 the subtrees rooted in these other children must have pathwidth at most k). By Theorem 2 we conclude that $tabreq(T_r) \leq 2k + 3 = 2(k + 1) + 1 = 2pw(T) + 1$. ∎

It is easy to check that for a complete ternary tree T with k levels and root r we have $pw(T) + 1 = tabreq(T_r) = k$. We can show that also the second bound in Theorem 10 is asymptotically tight, by giving a class of trees $G^k, k = 1, 2, ...$ with $pw(G^k) = k$ and minimum table requirement $2k$. However, we leave this out of this extended abstract, see [1].

4 Memory requirement

Minimal separators of a partial k-tree can have varying sizes. The size of a minimal separator, plus 1, is a lower bound on the size of the corresponding bag of a tree-decomposition T, which in turn determines the size of the table associated with the bag. These tables of varying size are updated during the dynamic programming computations on T.

In this section, we consider the case of variable size tables stored at nodes of the tree-decomposition T. In this case, minimizing table requirement is not enough. Instead, we would like to minimize the memory requirement, *i.e.*, the sum of the sizes of tables that must be stored simultaneously, over all bottom-up traversal orders of T. An efficient algorithm solving this problem still eludes us and we leave its design as an open question. Instead, we consider only traversals of T that follow a depth-first search (dfs) traversal order.

Definition 11 *An ordering of edges $e_1, ..., e_{n-1}$ of a tree T on $n > 1$ vertices is a dfs traversal if we can choose a root r of T and for each vertex v of T_r order its children $child_1(v), ..., child_{c_v}(v)$ so that a call of the procedure $DFS(r)$ of Section 2 will execute the instructions update-table(u, v) for edges $\{u, v\}$ in the order $e_1, ..., e_{n-1}$.*

For completeness, we also define a top-down traversal:

Definition 12 *An ordering of edges $e_1, ..., e_{n-1}$ of a rooted tree T_r on $n > 1$ vertices is a top-down traversal if for any edge $e_j = (u, parent(u))$ all edges in the subtree rooted in u have indices greater than j.*

Note that the traversal minimizing table requirement given in Section 2 is a dfs traversal. The implementation of a dfs traversal of a tree-decomposition is very simple, as illustrated by the DFS procedure given in Section 2. We will give a two-phase linear-time algorithm to find a dfs traversal order of a tree with weighted vertices minimizing its dfs memory requirement. Let $tab(u)$ be the size of the table at a node u of T. We define $mem(u, v)$ as the minimum dfs memory requirement of the subtree rooted in u with overall root chosen so that u has parent v, taken over all dfs traversals of this subtree.

Theorem 13 *In a given rooted tree, let a non-root vertex u have the parent v and let x and y be the children of u (if any) such that $mem(x, u) \geq mem(y, u) \geq mem(w, u)$ for any other child w of u (if u has only one child define $mem(y, u)$ to be 0.) Then, the minimum dfs memory requirement to compute the table associated with u is given by*

$$mem(u, v) = \begin{cases} tab(u) & \text{for a leaf } u \\ \max\{mem(x, u), \\ \quad tab(x) + tab(u), \\ \quad mem(y, u) + tab(u)\} & \text{for a non-leaf } u. \end{cases} \tag{2}$$

The proof of this theorem is analogous to the proof of Theorem 2, in which case table sizes are uniformly 1. We therefore leave the proof out, except for noting that the necessity of the value $mem(y, u) + tab(u)$ follows since we are restricting to dfs traversals. We define the dfs memory requirement of T_r, $mem(T_r)$, as the minimum dfs memory requirement necessary to complete the table at r.

Corollary 14 *The right hand side of Theorem 2 can also be used to compute* $mem(T_u)$; *in this case, all neighbors of u are considered in defining x and y.*

For any vertex r of tree T with neighbors $w_1, \ldots, w_c$, the values of dfs memory requirements for the principal subtrees of T_r, $mem(w_1, r), \ldots, mem(w_c, r)$, together with $tab(r)$, determine the value of $mem(T_r)$. For a pair of adjacent vertices u, v, the value of $mem(u, v)$ does not depend on the particular chosen root, only that the root is in the component of $T \setminus (u, v)$ that includes v. We therefore conclude that knowledge of table sizes and the values $mem(u, v)$ and $mem(v, u)$, for each edge (u, v) of T, provides all the information necessary to compute $mem(T_r)$, for any vertex r of T. The following two-phase algorithm utilizes this observation to find an optimal root and traversal order.

Let x be an arbitrary vertex of T.

1. **Bottom-up** Pick a root x and using Theorem 2, compute in an arbitrary bottom-up traversal of T_x the values $mem(u, v)$ for all edges uv of T where v is the parent of u in T_x.
2. **Top-down** Using Theorem 2, compute in an arbitrary top-down traversal of T_x the values $mem(u, v)$ for all edges uv of T where v is the child of u in T_x. When processing a vertex u, compute also $mem(T_u)$ using the right-hand side of Theorem 2 with children of u ranging over all its neighbors. During the traversal, select a vertex r that has the minimum value of $mem(T_r)$.

The vertex r above is an optimal root. To find a traversal order minimizing dfs memory requirement let a node v in T_r have children $child_1(v), \ldots, child_{c_v}(v)$, with $child_1(v)$ having the largest value of $mem(child_i(v), v)$ over $1 \leq i \leq c_v$, as computed by the algorithm. These values are the dfs memory requirements for subtrees of T_r rooted at these vertices. A call of DFS(r) of the procedure in section 2 will then perform the dynamic programming on T while minimizing the sum of the sizes of tables stored simultaneously, over all dfs traversals of the tree.

See Figure 2 for an example. At each node v of the tree we keep track of both the largest, second-largest and third-largest (these could be equal) dfs memory requirements $mem(w, v)$ reported by neighbors of v, and the identity of these neighbors. These values are incrementally updated. Assume that v has the neighbors r, x, y, z, w, with the bottom-up phase resulting in the values $mem(x, v) \geq mem(y, v) \geq mem(z, v) \geq mem(w, v)$. The three largest values are stored, and the two largest values are used in the bottom-up phase to compute $mem(v, r)$. In the top-down phase, the values used to compute, say, $mem(v, x)$ will be $mem(y, v)$ and the maximum of $mem(r, v)$ and $mem(z, v)$. These values and the identities of the corresponding neighbors are also used in the eventual call of DFS(r).

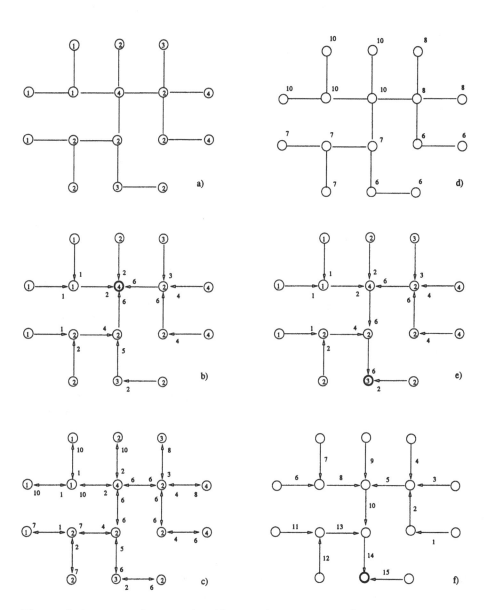

Fig. 2. a) An unrooted tree with table sizes for each node. b) Bottom-up phase of algorithm with arbitrary root in bold and dfs memory requirements for subtrees. c) Top-down phase of algorithm. d) dfs memory requirements by rooting at respective nodes. e) Minimum dfs memory requirement with optimal root in bold. f) An optimal dfs traversal order with root in bold.

Theorem 15 *The algorithm given above computes an optimal root r minimizing dfs memory requirement in linear time, and the call of DFS(r) gives an optimal traversal order.*

Proof. The correctness of the bottom-up phase of the algorithm follows by Theorem 13. By Corollary 14, the computation of $mem(T_u)$ for a vertex u in the top-down phase yields the correct value of the minimum dfs memory requirement with that vertex as the root. Using the data structures mentioned above (with values of the three largest memory requirements for subtrees of each node incrementally updated) each visit in the traversals takes constant time, hence the linear time complexity. The call of DFS(r) results in a traversal order minimizing table requirement as described in the proof of Theorem 13 and Theorem 2. ∎

5 Conclusions and Future Research

Recent efforts have been aimed at making the theoretically efficient dynamic programming algorithms on bounded treewidth graphs amenable also to practical applications. Thorup [15] has shown that the control-flow graphs of goto-free programs have treewidth bounded by 6, and that an optimal tree-decomposition of the graph can be found easily. This result has been used to give a linear-time algorithm that takes a goto-free program and for a fixed number of registers determines if register allocation can be done without spilling [3]. A key ingredient in this latter algorithm is dynamic programming on a bounded treewidth graph.

We have identified and described the problem of table and memory requirements of table computations in dynamic programming algorithms on bounded treewidth graphs and we have provided linear-time algorithms that can be applied in a pre-processing stage to minimize the effects of this problem. In Section 1 we mentioned the example of a width 10 tree-decomposition of 1000 nodes, with uniform table-size of 8 Kbytes. By applying Corollary 4 we see that the traversal order given by the algorithm of Section 2 for this example uses at most 10 tables, giving a memory requirement of 80 Kbytes.

Using our algorithms in a pre-processing stage it is possible to apply the dynamic programming stage of the partial k-tree algorithms to much larger input graphs, or graphs with larger treewidth, without paying the cost of external memory references. The overhead associated with this linear-time, small-constant pre-processing, is in most cases negligible in comparison with the cost of the dynamic programming itself, which for any NP-hard problem on an n-node treewidth k graph has a running time of $\Omega(nc^k)$ for some constant c (unless there are sub-exponential time algorithms for NP-hard problems). However, more programming work is needed to confirm this experimentally.

From Theorem 2 it is clear that table requirement two is achieved when there is a width k tree-decomposition T of the input graph G where T is a caterpillar, i.e. T is a tree obtained from a path by adding leaves adjacent to the path, or equivalently T is a connected graph of pathwidth 1. When T is a path, the treewidth and pathwidth of G coincide. This observation thus extends the "folklore" that dynamic programming on a path-decomposition (a path) is simpler than on a general tree-decomposition. A natural question to ask is for a characterization of graphs of treewidth k that have a width k tree-decomposition forming a caterpillar. A partial answer to this question is presented in [11]: every multitolerance graph has an optimal tree-decomposition forming a caterpillar. The more general question is how to find tree-decompositions having small value of memory requirement.

Finally, the design of an efficient algorithm finding the minimum memory requirement of a tree with varying table size, over all bottom-up traversal orders, remains open. Another interesting question is that of bounding the worst case performance ratio of the DFS-traversal algorithm over this minimum value.

References

1. B.Aspvall, A.Proskurowski and J.A.Telle, Memory requirements for table computations in partial k-tree algorithms, submitted special issue of *Algorithmica on Treewidth, Graph Minors and Algorithms*.
2. T. Beyer, S.M. Hedetniemi, S.T. Hedetniemi and A. Proskurowski, Graph traversal with minimum stack depth, *Congressus Numerantium, Vol. 32* , 121-130, 1981.
3. H. Bodlaender, J. Gustedt and J.A. Telle, Linear-time register allocation for a fixed number of registers and no stack variables, *Proceedings 9th ACM-SIAM Symposium on Discrete Algorithms (SODA'98)*, 574-583.
4. H. Bodlaender, A linear time algorithm for finding tree-decompositions of small treewidth, in *Proceedings 25th Symposium on the Theory of Computing* (STOC'93), 226-234.
5. H. Bodlaender, A tourist guide through treewidth, *Acta Cybernetica*, 11:1-21, 1993.
6. Y.-J. Chiang, M.T. Goodrich, E.F. Grove, R. Tamassia, D.E. Vengroff and J.S.Vitter, External-Memory Graph Algorithms, *Proc. ACM-SIAM Symp. on Discrete Algorithms* (SODA'95), pp. 139-149, 1995.
7. J.A. Ellis, I.H. Sudborough and J.S. Turner, The vertex separation number and search number of a graph, *Information and Computation* vol. 113, 50-79, 1994.
8. B. Hiim, Implementing and testing algorithms for tree-like graphs, Master's Thesis, November 1997, Dept. of Informatics, University of Bergen.
9. L. Kirousis and C. Papadimitriou, Searching and pebbling, *Theoretical Computer Science* 47, 205-218, 1986.
10. R. Möhring, Graph problems related to gate matrix layout and PLA folding, in *Computational Graph Theory, Computing Suppl. 7*, Springer-Verlag, 17-51, 1990.
11. A.Parra, Triangulating multitolerance graphs, Technical Report 392/1994, TU Berlin, Germany, 1994.
12. N. Robertson and P. Seymour, Graph Minors I. Excluding a forest, *Journal of Combinatorial Theory Series B* 35, 39-61, 1983.
13. P. Scheffler, A linear algorithm for the pathwidth of trees, *Topics in Combinatorics and Graph Theory, Physica-Verlag Heidelberg*, 613-620, 1990.
14. J.A. Telle and A. Proskurowski, Algorithms for vertex partitioning problems on partial k-trees, *SIAM Journal on Discrete Mathematics*, Vol. 10, No. 4, 529-550, November 1997.
15. M. Thorup, Structured Programs have Small Tree-Width and Good Register Allocation, *Proceedings 23rd Workshop on Graph-Theoretical Concepts in Computer Science* (WG'97), LNCS vol. 1335, 318-332.

Formal Language Constrained Path Problems

Chris Barrett, Riko Jacob, and Madhav Marathe*

Los Alamos National Laboratory, P.O. Box 1663, MS M997, Los Alamos, NM 87545.
{barrett,jacob,marathe}@lanl.gov

Abstract. Given an alphabet Σ, a (directed) graph G whose edges are weighted and Σ-labeled, and a formal language $L \subseteq \Sigma^*$, we consider the problem of finding a shortest (simple) path p in G complying with the additional constraint that $l(p) \in L$. Here $l(p)$ denotes the unique word given by concatenating the Σ-labels in G along the path p.

We consider the computational complexity of the problem for different classes of formal languages (finite, regular, context free and context sensitive), different classes of graphs (unrestricted, grids, treewidth bounded) and different type of path (shortest and shortest simple).

A number of variants of the problem are considered and both polynomial time algorithms as well as hardness results (**NP-**, **PSPACE**-hardness) are obtained. The hardness and the polynomial time algorithms presented here are a step towards finding such classes of graphs for which polynomial time query evaluation is possible.

1 Introduction

In many path finding problems arising in diverse areas, certain patterns of edge/vertex labels in the labeled graph being traversed are allowed/preferred, while others are disallowed. Thus, the feasibility of a path is determined by (i) its length (or cost) under well known measures on graphs such as distance, and (ii) its associated label. The acceptable label patterns can be specified as a formal language. For example, in transportation systems with mode options for a traveler to go from source to destination the mode selection and destination patterns of an itinerary that a route will seek to optimize can be specified by a formal language. The problem of finding label constrained paths arise in other application areas such as production distribution network, VLSI Design, Databases queries [13], etc. Here, we study the problem of finding shortest/simple paths in a network subject to certain formal language constraints on the labels of the paths obtained. We illustrate the type of problems studied here by discussing prototypical application areas:

1.1 Intermodal Route Planning

Our initial interest in the problem studied in this paper came from our work in the **TRANSIMS**[1] project at the Los Alamos National Laboratory. We refer the reader to [16] for a detailed description of this project.

---

* Research supported by the Department of Energy under Contract W-7405-ENG-36.

[1] TRANSIMS stands for the "TRansportation ANalysis and SIMulation System".

As a part of the intermodal route planning module of **TRANSIMS**, our goal is to find feasible (near optimal) paths for travelers in an intermodal network (a network with several mode choices, such as train, car, etc.) subject to certain mode choice constraints. The mode choices for each traveler are obtained by either processing the data from the microsimulation module or by certain statistical models built from real life survey data. We refer the reader to the book by Ben-Akiva and Lerman [3] for a detailed discussion and references on the theory of discrete choice analysis as applied to transportation science. The following example illustrates a prototypical problem arising in this context.

Example 1: We are given a directed labeled, weighted, graph G. The graph represents a transportation network with the labels on edges representing the various modal attributes (e.g. a label t might represent a rail line). Suppose, we wish to find a shortest route from s to d for a traveler. This is the ubiquitous shortest path problem. But now, we are also told that the traveler wants to go from s to d using the following modal choices: either he walks to the train station, then uses trains and then walks to his destination (office), or would like to go all the way from home to office in his car. Using t to represent trains, w to represent walking and c to represent car, the travelers mode choice can be specified as $w^+t^+w^+ \cup c^*$, where $\cup, +$ and $*$ denote the usual operators used to describe regular languages.

2 Problem Formulation

The problem to be discussed is formally described as follows: Let $G(V, E)$ be a (un)directed graph. Each edge $e \in E$ has two attributes — $l(e)$ and $w(e)$. $l(e)$ denotes the label of edge e. In most cases considered in this paper, the label is drawn from an (fixed, finite) alphabet Σ. The attribute $w(e)$ denotes the weight of an edge. Here, we assume that the weights are non-negative integers. Most of our positive results can in fact be extended to handle negative edge weights also (if there are no negative cycles). A path p of length k from u to v in G is a sequence of edges $\langle e_1, e_2, \ldots e_k \rangle$, such that $e_1 = (u, v_1)$, $e_k = (v_{k-1}, v)$ and $e_i = (v_{i-1}, v_i)$ for $1 < i < k$. A path is *simple* if all the vertices in the path are distinct. Given a path $p = \langle e_1, e_2, \ldots e_k \rangle$, the weight of the path is given by $\sum_{1 \le i \le k} w(e_i)$ and the label of p is defined as $l(e_1) \cdot l(e_2) \cdots l(e_k)$. In other words the label of a path is obtained by concatenating the labels of the edges on the path in their natural order. Let $w(p)$ and $l(p)$ denote the weight and the label of p respectively.

Definition 1. Formal Language Constrained Shortest Path:
Given a directed labeled, weighted, graph G, a source s, a destination d and a formal language (regular, context free, context sensitive, etc.) L, find a shortest (not necessarily simple) path p in G such that $l(p) \in L$.

Definition 2. Formal Language Constrained Simple Path:
Given a directed labeled, weighted, graph G, a source s, a destination pair d and a formal language (regular, context free, context sensitive, etc.) L, find a shortest simple path p in G such that $l(p) \in L$.

In general we consider the input for these problems to consist of a description of the graph (including labeling and weights) together with the description of the formal language as a grammar. By restricting the topology of the graph and/or the syntactic structure of the grammar we get modifications of the problems. If we claim a statement to be true 'for a fixed language' we refer to the variant of the problem, where the input consists of the graph only, whereas the language is considered to be part of the problem specification.

For the rest of the paper we denote the formal language constrained shortest path problem restricted to regular, context free and context sensitive languages by REG-SHP, CFG-SHP and CSG-SHP respectively. Similarly, we denote the formal language constrained simple path problem restricted to regular, context free and context sensitive languages by REG-SIP, CFG-SIP and CSG-SIP respectively.

Note that in unlabeled networks with non-negative edge weights, a shortest path between s and d is necessarily simple. This need not be true in the case of *label constraint* shortest paths. As a simple example, consider the graph $G(V, E)$ that is a simple cycle on 4 nodes. Let all the edges have weight 1 and label a. Now consider two adjacent vertices x and y. The shortest path from x to y consists of a single edge between them; in contrast a shortest path with label $aaaaa$ consists of a cycle starting at x and the additional edge (x, y).

3   Summary of Results

We investigate the problem of formal language constrained path problems. A number of variants of the problem are considered and both polynomial time efficient algorithms as well as hardness results (**NP-**, **PSPACE**-hardness) are proved. Two of the **NP**-hardness results are obtained by combining the simplicity of a path with the constraints imposed by a formal language. We believe that the techniques used to prove these results are of independent interest. The main results obtained in the paper are summarized in Figure 3 and include the following:

1. We show that CFG-SHP has a polynomial time algorithm. For REG-SHP with operators $(\cup, \cdot, *)$, we give polynomial time algorithms that are substantially more efficient in terms of time and space. The polynomial time solvability holds for the REG-SHP problem, when the underlying regular expressions are composed of $(\cup, \cdot, *, 2)^\dagger$ operators. We also observe that the extension to regular expressions over operators $(\cup, \cdot, *, -)$ is **PSPACE**-hard.

2. In contrast to the results for shortest paths, we show that the problem finding *simple* paths between a source and a given destination is **NP**-hard, even when restricted to a very simple, fixed regular language and very simple graphs (undirected unlabeled grid graphs).

3. In contrast to the results in (1) and (2) above, we show that for the class of treewidth bounded graphs, (i) the REG-SIP is solvable in polynomial

---

† operator $\Box^2$ or simple 2 stands for the square operator. R^2 denotes $R \cdot R$.

time, but (ii) CFG-SiP problem is **NP**-complete. The easiness proof uses a dynamic programming method; although the tables turn out to be quite intricate.

4. The results provide a tight bound on the complexity (**P** vs. **NP**) of the problems considered, given the following assumptions: The inclusions of classes of formal languages "finite $\subset$ locally testable" and "regular $\subset$ deterministic linear context free" are tight, i.e. there is no interesting class of languages "in between" these classes. Furthermore we consider grid-graphs to be the "easiest" graphs not having a bounded treewidth.

5. The theory developed here has several additional application, some of them already studied in the literature. See Section 7 for an example. More examples will appear in a complete version.

		RL	CFL	CSL
shortest	general graph	$SP(|R||G|)$	**FP**	undecidable
simple	directed chain	linear	cubic	**PSPACE**-c.
path	treewidth bounded	**FP**	NP-c.	**PSPACE**-c.
	planar/grid graph	NP-c.	NP-c.	**PSPACE**-c.

Fig. 1. Summary of results on formal language constrained simple/shortest paths. RL, CFL, CSL denote regular, context free and context sensitive languages respectively. For regular languages, the time bound hold for regular expressions with the operators $(\cup, \cdot, *)$. **FP** states that the sought path can be computed in deterministic polynomial time, even if the language specification is part of the input. On the other hand all of the hardness results are valid for a fixed language. A directed chain is a graph that forms exactly one directed simple path.

Preliminary versions of the algorithms outlined here have already been incorporated in the Route Planning Module of **TRANSIMS**. In [11] we conduct an extensive empirical analysis of these and other basic route finding algorithms on a realistic traffic network.

4 Related Work

As mentioned earlier, we refer the reader to [16] for a more detailed account of the **TRANSIMS** project. Regular expression constrained simple path problems were considered by Mendelzon and Wood [13]. The authors investigated this problem in the context of finding efficient algorithms for processing database queries (see [6, 13]). Yannakakis [20] in his keynote talk has independently outlined some of the polynomial time algorithms given here. The emphasis in [20] was on database theory. Online algorithms for regular path finding are given in [5]. Our work on finding formal language constrained shortest paths is also related to the work of Ramalingam and Reps [19]. The authors were interested in finding a minimum cost derivation of a terminal string from one or more non-terminals of a given context free grammar. The problem was first considered by Knuth [12] and is referred to as the *grammar problem*. [19] give an incremental algorithm for a version of the grammar problem and as corollaries obtain incremental algorithms for single source shortest path problems with positive edge weights. We close this section with the following additional remarks:

1. To our knowledge this is the first attempt to use formal language theory for modeling mode/route choices in transportation science.
2. The polynomial time algorithms and the hardness results presented here give a boundary on the classes of graphs and queries for which polynomial time query evaluation is possible. In [13] the authors state

   Additional classes of queries/dbgraphs for which polynomial time evaluation is possible should be identified ...

   Our results significantly extend the known hardness as well as easiness results in [13] on finding regular expression constrained simple paths. For example, the only graph theoretic restriction considered in [13] was acyclicity. On the positive side, our polynomial time algorithms for regular expression constrained simple path problem when restricted to graphs of bounded treewidth are a step towards characterizing graph classes on which the problem is easy. Specifically, it shows that for graphs with fixed size recursive separators the problems is easy. Examples of graphs that can be cast in this framework include chordal graphs with fixed clique size, outer planar graphs, etc.
3. The basic techniques extend quite easily (with appropriate time performance bounds) to solve other (regular expression constrained) variants of shortest path problems. Two notable examples that frequently arise in transportation science and can be solved are (i) multiple cost shortest paths [9] and (ii) time dependent shortest paths [18].

Due to the lack of space, the rest of the paper consists of selected proof sketches. Details will appear in a complete version.

5 Shortest Paths

5.1 Algorithm for REG-SHP and Extensions

In this subsection, we will describe our algorithms for regular expression constrained shortest path problems. We note that regular expressions over $(\cup, \cdot, *)$ can be transformed into equivalent NFAs[2] in $O(n)$ time [10], where n represents the size of the regular expression. Thus for the rest of this subsection we assume that the regular expressions are specified in terms of an equivalent NFA.

The basic idea behind finding shortest paths satisfying regular expressions is to construct an auxiliary graph (the product graph) combining the NFA denoting

---

[2] NFA stands for nondeterministic finite automata

the regular expression and the underlying graph. We formalize this notation in the following.

Definition 3. *Given a label directed graph G, a source s and a destination d define an NFA $M(G) = (S, \Sigma, \delta, s_0, F)$ as follows:*

1. $S = V$; $s_0 = s$; $F = \{d\}$;
2. Σ *is the set of all labels that are used to label the edges in G and*
3. $j \in \delta(i, a)$ *iff there is an edge (i, j) with label a.*

Note that this definition can as well be used to interpret a NFA as a labeled graph.

Definition 4. *Let $M_1 = (S_1, \Sigma, \delta_1, p_0, F_1)$, and $M_2 = (S_2, \Sigma, \delta_2, q_0, F_2)$, be two NFAs. The product NFA is defined as $M_1 \times M_2 = (S_1 \times S_2, \Sigma, \delta, (p_0, q_0), F_1 \times F_2)$, where $\forall a \in \Sigma$, $(p_2, q_2) \in \delta((p_1, q_1), a)$ if and only if $p_2 \in \delta_1(p_1, a)$ and $q_2 \in \delta_2(q_1, a)$.*

It is clear that $L(M_1 \times M_2) = L(M_1) \cap L(M_2)$.

ALGORITHM RE-CONSTRAINED-SHORT-PATHS outlines the basic steps for solving the problem.

ALGORITHM RE-CONSTRAINED-SHORT-PATHS:

- *Input:* A regular expression R, a directed labeled weighted graph G, a source s and a destination d.
- 1. Construct an NFA $M(R) = (S, \Sigma, \delta, s_0, F)$ from R.
  2. Construct the NFA $M(G)$ of G.
  3. Construct $M(G) \times M(R)$. The length of the edges in the product graph is chosen to be equal to the according edges in G.
  4. Starting from state (s_0, s), find a shortest path to all of the vertices (f, d), where $f \in F$. Denote these paths by p_i, $1 \le i \le w$. Also denote the cost of p_i by $w(p_i)$
  5. $C^* := \min_{p_i} w(p_i)$; $p^*: w(p^*) = C^*$.
     (If p^* is not uniquely determined, we choose an arbitrary one.)
- *Output:* The path p^* in G from s to d of minimum length subject to the constraint that $l(p) \in L(R)$.

We use $|I|$ to denote the size of an object I represented using Binary notation.

Theorem 1. ALGORITHM RE-CONSTRAINED-SHORT-PATHS *computes the exact solution for the problem REG-SHP with non-negative edge weights in time $O(|R||G|\log(|R||G|))$.*

Proof: The correctness of the algorithm follows by observing the following:

Consider a shortest path q^* in G of cost $w(q^*)$, that satisfies the regular expression R. Then there is a path q' of the same cost between (s_0, s) to (f, d) for some $f \in F$. So we know that the sought path is considered.

Conversely, for each path τ in in $M(G) \times M(R)$ of cost $w(\tau)$, that begins on a starting state and ends on a final state, there is a path of same cost in G from s to d that satisfies the regular expression R.

To calculate the running time of the algorithm, observe that the size of $M(R) \times M(G)$ is $O(|R||G|)$. Using efficient algorithms to solve shortest path problems in Step 4 the problem can be solved in time $O(|R||G|\log(|R||G|))$. $\square$

5.2 Algorithm for CFG-SHP

The algorithm for solving context free grammar constrained shortest paths is based on a dynamic programming. Hence we will first investigate the structure of an optimal shortest path from s to d in the graph G that obeys the context free Grammar R constraints. Assume that R is in Chomsky normal form (see [10] for definition). Consider any such shortest path p with $l(p) = a_1 a_2 \ldots a_m$. One important property of any CFG is that nonterminals are expanded independently. In the case of a Chomsky Normal form, the derivation forms a binary tree, which means that the label can be decomposed into two parts l_1 and l_2 such that $l(p) = l_1 l_2$, $S \to AB$, $A \xrightarrow{*} l_1$ and $B \xrightarrow{*} l_2$.

With this structure in mind let us define the quantity $D(i, j, A)$ as the shortest path distance from i to j subject to the constraint that the label on this path can be derived starting from the nonterminal A.

These values are well defined and fulfill the following recurrence:

$$D(i, j, A) = \min_{A \to BC \in R} \ \min_{k \in V} \{(D(i, k, B) + D(k, j, C))\} \tag{1}$$

$$D(i, j, a) = \begin{cases} w(i, j) & \text{if } l((i, j)) = a; \\ \infty & \text{otherwise.} \end{cases} \tag{2}$$

Observation 2 *These equations uniquely determine the function D and immediately imply a polynomial time dynamic programming algorithm.*

6   Simple Paths

Next, we investigate the complexity of finding formal language constrained simple paths. For the ease of exposition, we present our results for directed, multilabeled graphs. The following lemma shows how to extend these results to undirected and unilabeled graphs.

Lemma 1.   *1. CFG-SIP on directed graphs of treewidth k can be reduced to CFG-SIP on undirected graphs of treewidth k^2.*
   2. REG-SIP on directed, multilabeled grids can be reduced to REG-SIP on directed grids.
   3. REG-SIP on directed grids can be reduced to REG-SIP on undirected grids.

6.1 Finite languages

Consider a fixed finite constraining language L. Then there is a word of maximum length in L. Let k denote its length. Considering all k-tuples of nodes, and checking whether they form the sought path yields a polynomial time algorithm of running time roughly n^k. In contrast, we have the following hardness result:

Theorem 3. *Let C be a graph class (such as planar, grid, etc) such that the* HAMILTONIAN PATH *problem is **NP**-hard when restricted to C. Then the problem* REG-SIP *is **NP**-hard when restricted to C and finite languages.*

Proof: Consider a fixed class of graphs C for which the HAMILTONIAN PATH problem is **NP**-hard. Then given an instance G of the HAMILTONIAN PATH problem in which $G \in C$, with n nodes, we construct an instance G_1 of regular expression constrained simple path problem by labeling all the edges in G by a. We now claim that there is a Hamiltonian path in G if and only if there is a simple path in G_1 that satisfies a^n. $\qquad\square$

6.2 Hardness of REG-SIP and CFG-SIP

In this section we discuss hardness results for language constrained simple path problems.

Theorem 4. *The* REG-SIP *Problem is **NP**-hard for a complete multi-labeled directed grid and a fixed regular expression.*

This result extends to the following:

Corollary 1. *Let C be a class of graphs such that for all $k > 0$ there is an instance $\mathcal{I} \in C$ that contains as a subgraph a $k \times k$-mesh and both the graph and the subgraph are **FP**-computable, then the REG-SIP problem is **NP**-hard for C.*

*Thus the REG-SIP problem is **NP**-hard even for (i) complete cliques, (ii) interval graphs, (iii) chordal graphs, (iv) complete meshes, (v) complete hypercubes, (vi) permutation graphs.*

We show in more detail that CFG-SIP is **NP**-hard even on graphs with bounded treewidth. Before formally stating the proof, we give the overall idea. We present a reduction from 3-SAT (See e.g. [8] for definition). The basic idea is to have a path consisting of two subpaths. The first subpath uniquely chooses an assignment and creates several identical copies of it. The second subpath checks one clause at every copy of the assignment and can only reach the destination if the assignment satisfies the given formula.

Consider the language $L = \{w\#w^R\$w\#w^R \ldots w\#w^R | w \in \Sigma^*\}$. As is standard, w^R denotes the reverse of string w. At the heart of our reduction is the crucial observation that L can be expressed as the intersection of two context free languages L_1 and L_2. More precisely the resulting language is going to be $L = L_1 \cap L_2$, with $L_1 = \{w_0\#w_1\$w_1^R\#w_2\$w_2^Rt \ldots w_k\$w_k^R\#w_{k+1} | w_i \in \Sigma^*\}$ and $L_2 = \{v_1\#v_1^R\$v_2\#v_2^R \ldots v_k\#v_k^R | v_i \in \Sigma^*\}$. To see that $L = L_1 \cap L_2$, observe that $w_0 = v_1$, $v_1^R = w_1$, and for all i, $w_i^R = v_{i+1}, v_{i+1}^R = w_{i+1}$, establishing that $v_i = v_{i+1}$ holds for all i, and thus $L = L_1 \cap L_2$.

Now, imagine w representing an assignment to the variables of a 3-SAT formula with a fixed ordering of the variables. For every clause of the formula, we create a copy of this assignment. L is used to ensure that all these copies are consistent, i.e. identical.

Note that we have two basic objects for performing the reduction – a CFG and a labeled graph. We will specify L_1 as a CFG and use the labeled graph and simple paths through the graph to implicitly simulate L_2. Recall that there is a straightforward deterministic pushdown automaton M for accepting L_2. Our graph will consist of an "upward chain" of vertices and a "downward chain" of vertices along with a few additional vertices. The upward chain will simulate the behavior of M when it pushes w on the stack. The "downward chain" will then simulate the act of popping the contents of the stack and verifying that they match w^R. We will call such a gadget a "tower" as an analogy to the stack of M.

We now describe the proof in detail. For the purposes of simplicity, we prove the results for directed graphs, the extension to undirected graphs follows with Lemma 1.

Theorem 5. *The CFG-SIP problem is NP-hard, even for Graphs of constant treewidth and a fixed Context Free Language.*

Proof: Reduction from 3-SAT. Let $F(X, C)$ be a 3-SAT formula, where $X = \{x_1, \ldots, x_n\}$ denotes the set of variables and $C = \{c_1, \ldots, c_m\}$ denotes the set of clauses. Corresponding to F, we create an instance $G(V, E_1 \cup E_2)$ of CFG-SIP as follows. We will describe the reduction in two parts – first the subgraph (V, E_1) and then the subgraph (V, E_2).

The subgraph (V, E_1) is constructed as follows: Corresponding to each clause c_j, we have a Tower T^j. It consists of n "simple-path stack-cell" gadgets H, for each variable one. This basic gadget is depicted in Figure 2.

Consider a simple path p from one of the bottom nodes (marked by a square in the Figure) to one of the top nodes. Because of the labels of this gadget, we can define the *signature* y of this path by $l(p) = cyc$ with $y \in \{a, b\}$. Let q be another simple path, that has no vertex in common with p, starts at one of the top nodes and ends in one of the bottom nodes. Then we can again properly define the signature z of this path by $l(q) = czc$.

Because of the nodes α and β, the signatures are identical, i.e. $y = z$. Furthermore, if p uses the node x_i', the node x_i is not used at all, and q has to use the node $\neg x_i$. Similarly, if p uses the node $\neg x_i'$, the node $\neg x_i$ is not used at all, and q has to use the node x_i.

These gadgets are now composed to form towers T^j, by identifying the top terminal nodes of H_i^j with the bottom terminal nodes of H_{i+1}^j. The tower has four levels corresponding to every variable. We call this a *floor* of the tower. The bottom of the tower T^j is connected to the bottom of the tower T^{j+1}. The start vertex is connected to the bottom of the tower T^1. These connections are depicted in Figure 3. Before we describe the remaining edges, we discuss the properties of a tower.

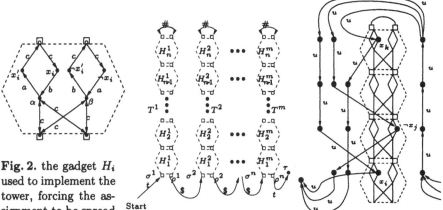

Fig. 2. the gadget H_i used to implement the tower, forcing the assignment to be spread consistently over the graph.

Fig. 3. Assembling the gadgets, building the first part of the graph.

Fig. 4. One tower of the second part of the graph, according to the clause $(\neg x_i \vee x_j \vee \neg x_k)$.

Consider a Tower T^j and a simple path r labeled according to the expression $(c(a \cup b)c)^* \# (c(a \cup b)c)^*$ that starts at one bottom vertex and reaches the other bottom vertex. Such a path has the following important properties:

1. The path r consists of two simple subpaths p and q separated by an edge labeled with $\#$. p starts at the bottom of the tower and is labeled according to the regular expression $(c(a \cup b)c)^*$. On every floor it uses exactly one of the nodes $\{x_i', \neg x_i'\}$, by this realizing an assignment to the variables of X. This assignment uniquely corresponds to the labeling of this path.
2. The constraints of *simplicity* and the direction of edges implies that q has the following structure: it starts at the unused top vertex, is labeled with $l(q) = l(p)^R$, avoids already used nodes, and reaches the unvisited bottom vertex. Furthermore q is *uniquely* determined given p.
3. Note that for each variable x_i, the simple path t visits exactly one of the nodes x_i and $\neg x_i$, If the path reflects the assignment that x_i is true, then x_i is unused and $\neg x_i$ is used and vice versa. These free nodes will be used in the second part of the reduction to verify that this is indeed a satisfying assignment for F.

Proposition 1. *Any simple path starting at the start node s, and reaching the intermediate node τ, with the constraint that the labeling belongs to $t((c(a \cup b)c)^* \# (c(a \cup b)c)^* \$)^* t$ generates the language*

$$L_2 = \{\, tw_1 \# w_1^R \$ w_2 \# w_2^R \$ \ldots w_n \# w_n^R t \mid w_i \in (c(a \cup b)c)^* \,\}.$$

Proof: The statement follows from the above. □

We now choose the constraining CFL for the path from s to τ to be
$$L_1 = \{ tw_1 \# w_2 \$ w_2^R \# w_3 \$ w_3^R \# \ldots w_{k-1} \$ w_{k-1}^R \# w_k t \mid k \in \mathbb{N}, w_i \in (ccc(a \cup b))^* \}.$$
The following important lemma follows from Proposition 1 and the definition of L_1.

Lemma 2. *1. Proposition 1 enforces that in every tower the used and unused nodes can be uniquely interpreted as an assignment to the variables of X.*

2. L_1 enforces that these assignment are consistent across two consecutive towers.

We now describe the subgraph (V, E_2). The label of each edge in this subgraph is u. The subgraph is composed of m subgraphs $D_1 \ldots D_m$, the subgraph D_i corresponding to clause c_i, depicted in Figure 4. Every D_i basically consists of four simple chains. The first one goes up the tower. There it splits into three downward directed paths. Each of them corresponds to a literal of c_i. The node in the tower T_i corresponding to that literal is used as part of the path. At the very bottom the three paths are joined and connected to D_{i-1}. At the boundaries D_0 is replaced by d and D_{m+1} by τ. This completes the description of the graph $G(V, E_1 \cup E_2)$.

The instance of CFG-SiP consists of the graph $G(V, E_1 \cup E_2)$ and the constraining CFL $L_1 \cdot u^*$. This enforces the path to go through every tower using edges in $G(V, E_1)$, visit vertex τ and then use the edges of $G(V, E_2)$ to reach d.

It is easy to see that G has bounded treewidth (can be shown to be < 16). The proof of correctness is omitted due to the lack of space. □

Corollary 2. CFG-SiP *is* **NP***-hard even when restricted to graphs of constant treewidth and a fixed linear, deterministic CFG.*

7 Extension: Finding Alternatives to Shortest Paths

To illustrate the applicability of our methods we briefly discuss one application. More applications will be outlined in a complete version.

There has been a considerable interest in finding algorithms for variations of shortest paths[1]. For example, in a recent paper, by Scott, Pabon-Jimenez and Bernstein [14], the authors consider the following problem — given a graph G, a shortest path SP in G and an integer parameter, find the shortest path $SP1$ in G that has at most k links in common with SP. Call this the BEST k-SIMILAR PATH. Our approach for solving the problem is based on using our algorithm for regular expression based shortest paths. The approach uses the fact that for fixed k the words, that have less or equal to k symbols a in them, is regular.

Acknowledgments: We want to thank Kevin Compton, Harry Hunt III, Sven Krumke, Alberto Mendelzon, S.S. Ravi, R. Ravi, Dan Rosenkrantz, M. Yannakakis, Hans-Christoph Wirth Kai Nagel, Terence Kelly, Robert White, Brian Bush and an anonymous referee for several useful discussions, pointers to related literature and suggested improvements in the course of writing this paper.

References

1. R. K. Ahuja, T. L. Magnanti and J. B. Orlin, *Network Flows: Theory, Algorithms and Applications*, Prentice-Hall, Englewood Cliffs, NJ, 1993.
2. S. Arnborg, J. Lagergren and D. Seese, "Easy Problems for Tree-Decomposable Graphs," *Journal of Algorithms*, vol. 12, pp. 308-340 (1991).
3. M. Ben-Akiva and S.R. Lerman, *Discrete Choice Analysis*, MIT Press Series in Transportation Studies, Cambridge, MA, 1985.
4. V. Blue, J. Adler and G. List, "Real-Time Multiple Objective Path Search for In-Vehicle Route Guidance Systems," *Proc. 76th Annual Meeting of The Transportation Research Board*, Washington, D.C. Paper No. 970944, January 1997.
5. A. Buchsbaum, P. Kanellakis and J. Vitter, "A Data Structure for Arc Insertion and regular Path Finding," *Proc. 1st ACM-SIAM Symposium on Discrete Algorithms*, 1990, pp. 22-31.
6. M. Cruz, A. Mendelzon and P. Wood, "A Graphical Query Language Supporting Recursion," *Proc. 9th ACM SIGMOD Conference on Management of Data* San Francisco, CA, 1990, 1987, pp. 323-330.
7. T.H. Cormen, C.E. Leiserson, and R.L. Rivest, *Introduction to Algorithms*, McGraw-Hill Book Co., 1990.
8. M. R. Garey and D. S. Johnson, *Computers and intractability: A guide to the theory of NP-completeness*, W. H. Freeman, San Francisco (1979).
9. R. Hassin, "Approximation schemes for the restricted shortest path problem", *Mathematics of Operations Research 17*, 1 (1992), 36–42.
10. J. E. Hopcroft and J. D. Ullman, *Introduction to Automata Theory, Languages and Computation*, Addison Wesley, Reading MA., 1979.
11. R. Jacob, M. Marathe, and K. Nagel, *Computational Experiences with Routing Algorithms for Realistic Traffic Networks*, in preperation.
12. D.E. Knuth, "A Generalization of Dijkstra's Algorithm," *Information Proc. Lett.*, 6(1), pp. 1-5, 1977.
13. A. Mendelzon and P. Wood, "Finding Regular Simple Paths in Graph Databases," *SIAM J. Computing*, vol. 24, No. 6, 1995, pp. 1235-1258.
14. K. Scott, G. Pabon-Jimenez and D. Bernstein, "Finding Alternatives to the Best Path," *Proc. 76th Annual Meeting of The Transportation Research Board*, Washington, D.C. Paper No. 970682, Jan. '97. Also available as Draft Report *Intelligent Transport Systems Program*, Princeton University, '97.
15. H. Straubing, *Finite Automata, Formal Logic, and Circuit Complexity*, Birkhäuser, 1994.
16. C. Barrett, K. Birkbigler, L. Smith, V. Loose, R. Beckman, J. Davis, D. Roberts and M. Williams, *An Operational Description of TRANSIMS*, Technical Report, LA-UR-95-2393, Los Alamos National Laboratory, 1995.
17. R. Tarjan, "A Unified Approach to Path Problems," *J. ACM* Vol. 28, No. 3, 1981, pp. 577-593.
18. A. Orda and R. Rom, "Shortest Path and Minimium Delay Algorithms in Networks with Time Dependent Edge Lengths," *J. ACM* Vol. 37, No. 3, 1990, pp. 607-625.
19. G. Ramalingam and T. Reps, "An incremental Algorithm for a Generalization of the Shortest-Path Problem," *J. Algorithms*, 21(2):267-305, September 1996.
20. M. Yannakakis "Graph Theoretic Methods in DataBase Theory," *Proc. 9th ACM SIGACT-SIGMOD-SIGART Symposium on Database Systems (ACM-PODS)*, Nashville TN, 1990, pp. 230-242.

Local Search Algorithms for SAT: Worst-Case Analysis

Edward A. Hirsch*

Steklov Institute of Mathematics at St.Petersburg
27 Fontanka, St.Petersburg 191011, Russia
hirsch@pdmi.ras.ru
http://logic.pdmi.ras.ru/~hirsch

Abstract. Recent experiments demonstrated that local search algorithms (e.g. GSAT) are able to find satisfying assignments for many "hard" Boolean formulas. However, no non-trivial worst-case upper bounds were proved, although many such bounds of the form $2^{\alpha n}$ ($\alpha < 1$ is a constant) are known for other SAT algorithms, e.g. resolution-like algorithms. In the present paper we prove such a bound for a local search algorithm, namely for CSAT. The class of formulas we consider covers most of DIMACS benchmarks, the satisfiability problem for this class of formulas is $\mathcal{NP}$-complete.

1 Introduction

Recently there has been an increased interest to local search algorithms for the Boolean satisfiability problem. Though this problem is $\mathcal{NP}$-complete (see e.g. [4]), B. Selman, H. Levesque and D. Mitchell have shown in [20] that an algorithm that uses local search can easily handle some of "hard" instances of SAT. They proposed a randomized greedy local search procedure GSAT (see Fig. 1) for the Boolean satisfiability problem and presented reassuring experimental results for graph coloring and n-queens problem. Starting with a random assignment, GSAT changes at each step the value of one variable so that the number of satisfied clauses increases as much as possible (the increase can be non-positive). If the satisfying assignment is not found in a certain number of steps, GSAT starts with another random assignment. Later many experimental results concerning various modifications of GSAT were obtained [17, 19, 5–7]. B. Selman and H. Kautz demonstrated in [18] that GSAT variants can solve large circuit synthesis and planning problems. However, no worst-case upper bounds were proved for these algorithms (even for a concrete class of formulas). The only theoretical results known to the author concern the average-case upper bounds [10, 8, 3].

In the present paper we study a variant of GSAT that does not use greediness. It chooses a variable to change at random from the set of variables that increase the number of satisfied clauses. Such algorithm was studied by I. Gent and T. Walsh [5, 6] who provided experimental evidence that greediness is not very

---

* Supported by INTAS-RFBR project No.95-0095

Algorithm GSAT.
Input:
    A formula F in CNF, integers MAXFLIPS *and* MAXTRIES.
Output:
    A truth assignment for the variables of F, or "No".

Method.

    Repeat MAXTRIES times:
      — Pick an assignment I at random.
      — Repeat MAXFLIPS times:
        — If I satisfies F, then output I.
        — Choose a variable v in F such that I with flipped (negated)
          value of v satisfies the maximal number of clauses of F.
        — Change the value of v in I.
    Output "No".

Fig. 1. Algorithm GSAT

important for GSAT efficiency. They called the corresponding algorithm CSAT (cautios SAT) (see Fig. 2). CSAT chooses a variable at random from the set of variables increasing the number of satisfied clauses, if there are no such variables, it chooses one that does not decrease the number of satisfied clauses, if there are none of those, it chooses any variable. For simplicity, we reduce the power of CSAT even more by disabling downward moves (moves that decrease the number of satisfied clauses) and moves on plateaus (moves that do not change the number of satisfied clauses). We show that even in this form CSAT solves the satisfiability problem for a natural class of Boolean formulas significantly better than a simple algorithm that just picks an assignment at random.

We consider CSAT as a Monte Carlo randomized algorithm that always outputs "no" if the input formula is unsatisfiable and outputs "yes" with probability greater than 50% if the input formula is satisfiable. In fact, in the latter case it finds a satisfying assignment. Namely, we study formulas in k–CNF–d, i.e. formulas in which each variable occurs at most d times and all clauses are of length at most k. The satisfiability problem is $\mathcal{NP}$-complete even for formulas in 3–CNF–3 [12]. We prove that CSAT can solve the satisfiability problem for formulas in k–CNF–d in the time $2^{(1-1/(kd))N}p(N)$ with error probability $1/e$ independent on the input formula (N is the number of variables in the input formula, p is a polynomial, $e = 2.718281828459\ldots$).

Starting from [13] (see also [14]) and [1] (see also [2]), there is a lot of theoretical results proving "less-than-2^N" bounds for SAT and its subproblems w.r.t. the number of variables, the number of clauses and the length of a formula. However, most of them use splitting algorithms. Splitting algorithms give the upper bounds $2^{0.309K}p(L)$ and $O(2^{0.106L})$ [9] for SAT (L is the length of the

---

Algorithm CSAT.
Input:
    A formula F in CNF, integers MAXFLIPS *and* MAXTRIES.
Output:
    A truth assignment for the variables of F, or "No".

Method.

    Repeat MAXTRIES times:
      — Pick an assignment I at random.
      — Repeat MAXFLIPS times:
        — If I satisfies F, then output I.
        — Choose at random a variable v in F such that I with flipped
          (negated) value of v satisfies more clauses in F than I does.
          If there are no such variables, choose at random a variable
          v that does not change the number of satisfied clauses.
          If there are none of those, choose a variable v at random
          from the set of all variables occuring in F.
        — Change the value of v in I.
    Output "No".

Fig. 2. Algorithm CSAT

input formula, K is the number of clauses in it, p is a polynomial), the bounds near $O(2^{0.59N})$ [11, 16] for the satisfiability problem for formulas in 3-CNF (N is the number of variables in the input formula) and the bound $O(L\phi_{k-1}^N)$ [14, 2] for formulas in k-CNF (ϕ_k is the only positive root of $x^{k+1} - 2x^k + 1 = 0$).

Though the bound presented in this paper is worse than the recent bound $O(2^{(1-1/k)N})$ for k-CNF which is due to R. Paturi, P. Pudlak and F. Zane [15] (their algorithm uses neither splitting nor local search), the result is nevertheless of particular interest because we present the first theoretically proved upper bound for experimentally well-studied local search algorithms for SAT.

In Sect. 2 we give definitions and formulate precisely the algorithm we study. Section 3 contains the proof of its worst-case upper bound for formulas in k–CNF–d.

2   Preliminaries

In this section F is a formula in CNF, I and J are truth assignments for the variables in F.

We denote by $\mathrm{sat}(F, I)$ the number of clauses that I satisfies in F.

The value (*True* or *False*) of a variable v in I is denoted by $v[I]$.

We denote by $\mathrm{dist}(I, J)$ the number of variables v such that $v[I] \neq v[J]$.

The values of all variables in assignment $I^{(v)}$ coincide with the values in I except the varaible v which has the opposite value. We say that we flip v in I and obtain $I^{(v)}$.

Let k and d be two natural numbers. We say that F is in k–CNF–d if every its clause contains at most k literals, and every variable occurs in F at most d times.

We now describe the algorithm we study (see Fig. 3). Algorithm SimpleCSAT either outputs a satisfying assignment after having considered at most K assignments (K is the number of clauses in the input formula), or does not stop at all (when it cannot choose a variable to change). It is based on I. Gent and T. Walsh's CSAT. For simplicity, we reduce the power of CSAT by disabling downward moves (moves that decrease the number of satisfied clauses) and moves on plateaus (moves that do not change the number of satisfied clauses). However, our result holds for the original CSAT as well. In the next section we prove that the probability that for a formula in k–CNF–d Algorithm SimpleCSAT outputs a satisfying assignment, is at least $2^{-\alpha N}/p(N)$, where $\alpha < 1$ is a constant, p is a polynomial. This implies the upper bound $2^{\alpha N}q(N)$ (q is another polynomial) on the worst-case running time of the Monte Carlo analog of Algorithm SimpleCSAT obtained by repeating Algorithm SimpleCSAT $2^{\alpha N}p(N)$ times (interrupting it each time it cannot choose the next variable).

Algorithm SimpleCSAT.
Input:
    A formula F in CNF.
Output:
    A truth assignment for the variables of F.

Method.

    Pick an assignment I at random.
    While I is not a satisfying assignment:
      — Choose a variable v in F such that $\mathrm{sat}(F, I) < \mathrm{sat}(F, I^{(v)})$. If there are several possible choices, choose it at random.
      — Assume $I = I^{(v)}$.
    Output I.

Fig. 3. Algorithm SimpleCSAT

3 Worst-Case Upper Bound for CSAT

In this section k and d are natural numbers, F is a satisfiable formula in k–CNF–d given as input to Algorithm SimpleCSAT, N is the number of variables in it.

Let I be a truth assignment for the variables of F. Let S be a satisfying assignment for F. We say that a variable v has the S-*correct* value in I, if $v[I] = v[S]$. Otherwise we say that v has the S-*incorrect* value in I. The variable v is *in a good position* w.r.t. formula F and assignment I, if it does not occur in clauses containing variables that have the S-incorrect values in I (maybe except the variable v itself). Otherwise we say that v is *in a bad position*. We say that I is *changed locally* w.r.t. S if all variables that have the S-incorrect values in it, are in a good position. We omit S in the above notation if its meaning is clear from the context.

Let i be an integer. We say that I satisfies the property $P_S(i)$ if $\text{dist}(I, S) = i$ and I is changed locally w.r.t. S. Informally speaking, I satisfies the property $P_S(i)$ if it is obtained from S by changing the values of i variables such that there is no clause in F containing more than one of these variables.

Below we estimate the running time of Algorithm SimpleCSAT in two steps.

1. We prove that there are many assignments satisfying $P_S(i)$ and thus it is likely that Algorithm SimpleCSAT picks one.
2. We prove that it is likely that starting with an assignment satisfying $P_S(i)$ Algorithm SimpleCSAT stops and outputs a satisfying assignment.

Remark 1. S is the only assignment satisfying $P_S(0)$.

Lemma 1. *Let S be a satisfying assignment for F. If an assignment I satisfies the property $P_S(i)$, then there are at most $id(k-1)$ variables v in a bad position that have the correct values in I.*

Proof. There are at most id clauses in F that contain variables having the incorrect values in I. These clauses contain at most $id(k-1)$ variables having the correct values in I. □

Lemma 2. *Let S be a satisfying assignment for F. If an assignment I satisfies the property $P_S(i-1)$, then there are at least $N - (i-1)(kd - d + 1)$ variables v such that $I^{(v)}$ satisfies the property $P_S(i)$.*

Proof. There are $N-(i-1)$ variables having the correct values in I. By Lemma 1 at most $(i-1)d(k-1)$ of them are in a bad position. The remaining $N - (i-1) - (i-1)d(k-1)$ variables are exactly what we want. □

Lemma 3. *Let S be a satisfying assignment for F. There are at least*

$$\frac{N(N - (kd - d + 1))(N - 2(kd - d + 1))\ldots(N - (i-1)(kd - d + 1))}{i!}$$

assignments satisfying the property $P_S(i)$ for $i \geq 1$.

Proof. We prove this statement by the induction on i.
Base ($i = 1$). Trivial.

Step. Suppose the statement holds for $i - 1$. By Lemma 2 we can map each assignment I satisfying $P_S(i-1)$ to at least $N - (i-1)(kd-d+1)$ assignments $I^{(v)}$ satisfying $P_S(i)$. Thus we have mapped each of

$$\frac{N(N-(kd-d+1))(N-2(kd-d+1))\ldots(N-(i-2)(kd-d+1))}{(i-1)!}$$

assignments guaranteed by the induction hypothesis, to at least $N - (i-1)(kd-d+1)$ assignments. Since we have mapped at most i assignments into one assignment, the number of assignments satisfying $P_S(i)$ is at least

$$\frac{(N-(i-1)(kd-d+1))}{i} \cdot$$

$$\cdot \frac{N(N-(kd-d+1))\ldots(N-(i-2)(kd-d+1))}{(i-1)!} =$$

$$= \frac{N(N-(kd-d+1))\ldots(N-(i-1)(kd-d+1)-1)}{i!}.$$

$\square$

Lemma 4. *Let S be a satisfying assignment for F. If an assignment I satisfies the property $P_S(i)$ and a variable v has the incorrect value in S, then $I^{(v)}$ satisfies the property $P_S(i-1)$.*

Proof. All variables having the correct values in I, have the correct values in $I^{(v)}$. $\square$

Lemma 5. *Let S be a satisfying assignment for F. If an assignment I is changed locally w.r.t. S, a variable v has the S-incorrect value in I, and $S^{(v)}$ is not a satisfying assignment for F, then $\operatorname{sat}(F, I^{(v)}) > \operatorname{sat}(F, I)$.*

Proof. W.l.o.g. we suppose $v[S] = True$. In the assignment I, the values of all variables in each of the clauses containing the literal $\bar{v}$, are correct. Thus, $I^{(v)}$ satisfies all these clauses, i.e. $\operatorname{sat}(F, I^{(v)}) \geq \operatorname{sat}(F, I)$.

Suppose $\operatorname{sat}(F, I^{(v)}) = \operatorname{sat}(F, I)$. Then I satisfies all clauses containing the literal v. Hence, the assignment $S^{(v)}$ satisfies the formula F, because I is changed locally w.r.t. S. $\square$

Lemma 6. *Let S be a satisfying assignment for F. If Algorithm SimpleCSAT starts with an assignment I that satisfies the property $P_S(i)$, then the algorithm stops in a polynomial time and outputs a satisfying assignment with probability at least $(kd)^{-i}$.*

Proof. There are i variables v having the S-incorrect values in I. By Lemma 5 for each such variable v, either $S^{(v)}$ is a satisfying assignment for F (suppose there are j such variables) or $\operatorname{sat}(F, I^{(v)}) > \operatorname{sat}(F, I)$. Now consider an assignment J that differs from S exactly in the values of the above mentioned j variables. Since I is changed locally, J is a satisfying assignment. For all $(i-j)$ variables v having the J-incorrect values in I, $\operatorname{sat}(F, I^{(v)}) > \operatorname{sat}(F, I)$.

Since I satisfies $P_S(i-j)$, by Lemmas 5 and 1 there are at most $(i-j)d(k-1)$ variables u having the J-correct values in I, such that $\text{sat}(F, I^{(u)}) > \text{sat}(F, I)$. Thus, the algorithm chooses a variable having the J-incorrect value in I with probability at least $(i-j)/(i-j+(i-j)d(k-1)) \geq 1/(kd)$. By Lemma 4 in this case the algorithm proceeds with processing an assignment having the property $P_J(i-j-1)$, etc. Thus, the algorithm stops and outputs a satisfying assignment with probability at least $(kd)^{-i}$. $\qquad\square$

Lemma 7. *If a satisfiable formula F in k–CNF–d is given as input to Algorithm SimpleCSAT, then Algorithm SimpleCSAT stops in a polynomial time and outputs a satisfying assignment with probability at least $2^{-(1-1/(kd))N}/p(N)$, where p is a polynomial.* $\qquad\square$

Proof. Let S be a satisfying assignment for F. By Lemmas 3 and 6, the probability of outputing S after i flips is at least

$$\frac{N(N-(kd-d+1))\dots(N-(i-1)(kd-d+1))}{i! \cdot 2^N}(kd)^{-i}$$

$$= \frac{1}{2^N} \cdot \frac{\frac{N}{kd} \cdot \frac{N-(kd-d+1)}{kd} \cdot \dots \cdot \frac{N-(i-1)(kd-d+1)}{kd}}{i!}$$

$$\geq \frac{\lfloor N/(kd)\rfloor!}{2^N i! \lfloor N/(kd)-i-1\rfloor!}.$$

We assume $i = \lfloor N/(2kd) \rfloor$. Then, using the Stirling formula, the above probability is at least $2^{-(1-1/(kd))N}/p(N)$ for some polynomial p.

By Lemma 7 the algorithm obtained by repeating $2^{(1-1/(kd))N}p(N)$ times Algorithm SimpleCSAT is a Monte Carlo algorithm for SAT for formulas in k–CNF–d with error probability

$$(1 - 2^{-(1-1/(kd))N}/p(N))^{2^{(1-1/(kd))N}p(N)} < 1/e,$$

where $e = 2.718281828459\dots$ Thus we have proved the following theorem.

Theorem 1. *Given a formula in k–CNF–d, the Monte Carlo analog of Algorithm SimpleCSAT performs correctly with error probability at most $1/e$. Its worst-case running time is at most $2^{(1-1/(kd))N}q(N)$, where N is the number of variables in the input formula, q is a polynomial, $e = 2.718281828459\dots$*

The result also holds for the original CSAT algorithm, since we can consider only the cases when it performs exactly as the Monte Carlo analog of Algorithm SimpleCSAT.

Corollary 1. *There exist polynomials p and q such that given a formula in k–CNF–d with K clauses and N variables, and values MAXFLIPS $= K$ and MAXTRIES $= 2^{(1-1/(kd))N}p(N)$, Algorithm CSAT returns a correct answer with probability at least $1 - 1/e$ (where $e = 2.718281828459\dots$) independent on the input formula. Its worst-case running time is at most $2^{(1-1/(kd))N}q(N)$.*

References

1. Dantsin, E.: Tautology proof systems based on the splitting method (in Russian). PhD thesis. Leningrad Division of Steklov Institute of Mathematics (LOMI) (1983)
2. Dantsin, E., Kreinovich, V.: Exponential upper bounds for the propositional satisfiability problem (in Russian). In: Proceedings of the 9th National Conference on Mathematical Logic. Leningrad (1988) p. 47
3. Franco, J., Swaminathan, R.: Average case results for satisfiability algorithms under the random clause model. Annals of Mathematics and Artificial Intelligence (to appear)
4. Garey, M. R., Johnson, D. S.: Computers and intractability. A guide to the theory of NP-completeness. W. H. Freeman and Company, San Francisco (1979)
5. Gent, I., Walsh, T.: The Enigma of Hill-climbing Procedures for SAT. Research Paper **605**, Department of Artificial Intelligence, University of Edinburgh (1992)
6. Gent, I., Walsh, T.: Towards an Understanding of Hill-climbing Procedures for SAT. In: Proceedings of 11th National Conference on Artificial Intelligence. Washington, DC, AAAI Press (1993) 28–33
7. Gent, I., Walsh, T.: Unsatisfied Variables in Local Search. In: Hallam, J. (ed.): Hybrid Problems, Hybrid Solutions (AISB-95). IOS Press, Amsterdam (1995)
8. Gu, J., Gu, Q.-P.: Average Time Complexity of The SAT1.2 Algorithm. In: Proceedings of the 5th Annual International Symposium on Algorithms and Computation (1994) 147–155
9. Hirsch, E. A.: Two new upper bounds for SAT. In: Proceedings of the 9th Annual ACM-SIAM Symposium on Discrete Algorithms (1998) 521–530
10. Koutsoupias, E., Papadimitriou, C. H.: On the greedy algorithm for satisfiability. Information Processing Letters **43(1)** (1992) 53–55
11. Kullmann, O.: Worst-case Analysis, 3-SAT Decision and Lower Bounds: Approaches for Improved SAT Algorithms. In: Du, D., Gu, J., Pardalos P. M. (eds.): DIMACS Proceedings SAT Workshop 1996. American Mathematical Society (1997)
12. Luckhardt, H.: Obere Komplexitätsschranken für TAUT-Entscheidungen. In: Proceedings of Frege Conference 1984. Schwerine, Akademie-Verlag Berline (1984) 331–337
13. Monien, B., Speckenmeyer, E.: 3-satisfiability is testable in $O(1.62^r)$ steps. Bericht Nr. **3/1979**, Reihe Theoretische Informatik, Universität-Gesamthochschule-Paderborn
14. Monien, B., Speckenmeyer, E.: Solving satisfiability in less than 2^n steps. Discrete Applied Mathematics **10** (1985) 287–295
15. Paturi, R., Pudlak, P., Zane, F.: Satisfiability Coding Lemma. In: Proceedings of the 38th Annual Symposium on Foundations of Computer Science (1997) 566–574
16. Schiermeyer, I.: Pure literal look ahead: An $O(1.497^n)$ 3-Satisfiability algorithm. In: Franco, J., Gallo, G., Kleine Büning, H., Speckenmeyer, E., Spera, C. (eds.): Workshop on the Satisfiability Problem, Technical Report. Siena, April, 29 – May, 3, 1996. University Köln, Report No. **96-230**
17. Selman, B., Kautz, H.: An Empirical Study of Greedy Local Search for Satisfiability Testing. In: Proceedings of the 11th Conference on Artificial Intelligence. Washington, DC, AAAI Press (1993)
18. Selman, B., Kautz, H.: Pushing the Envelope: Planning, Propositional Logic, and Stochastic Search. In: Proceedings of the 14th Conference on Artificial Intelligence (1996)

19. Selman, B., Kautz, H., Cohen, B.: Noise strategies for improving local search. In: Proceedings of the 12th Conference on Artificial Intelligence. Seattle, Washington, AAAI Press (1993) 1:337–343
20. Selman, B., Levesque, H., Mitchell, D.: A new method for solving hard satisfiability problems. In: Swartout, W. (ed.): Proceedings of the 10th Conference on Artificial Intelligence. San Jose, California, MIT Press (1992) 440–446

Speed is More Powerful than Clairvoyance

Piotr Berman   Chris Coulston

Department of Computer Science and Engineering
The Pennsylvania State University, University Park, PA 16802, USA

Abstract. We consider the problem of preemptive non-clairvoyant scheduling on a single machine. In this model a scheduler receives a number of jobs at different times without prior knowledge of the future jobs or the required processing time of jobs that are not yet completed. We want to minimize the total response time, i.e. the sum of times each job takes from its release to completion.

One particular algorithm, Balance, always schedules the job that was least processed so far. A comparison of an on-line scheduler running Balance against the optimal off-line shows a very large competitive ratio if both algorithms use machines of the same speed. However, it has been shown if Balance is run on a v times faster machine then the competitive ratio drops to at most $1 + 1/(v - 1)$. This result showed that speed can almost be as good as clairvoyance.

We show for $v \geq 2$ the competitive ratio of Balance is $2/v$. In other words, sufficiently high speed is more powerful than clairvoyance.

1   Introduction

The problem considered in this paper is similar to the situation that a uniprocessor computer encounters every day. Typically, the waiting queue contains a number of jobs and the operating system must determine which job should receive an additional time slice of the CPU. Unfortunately, we have no *a-priori* knowledge of how much processing time a job requires and when future jobs will enter the system. A reasonable goal for our computer would be to minimize the time that a user has to wait to see their job completed, i.e. the response time of the job. Since, there may be many jobs waiting for the CPU the CPU must decide which jobs to run at the expense of the response time of the non-active jobs.

If our computer were clairvoyant it would know the required execution times of all the jobs. It is well known [7] that with such information one can easily compute a schedule that minimizes the total response time.

In this paper we will use the following notation and terminology. A *schedule S* is a sequence $((l_1, s_1), (l_2, s_2), \cdots, (l_n, s_n))$ where the numbers $1, 2, \cdots, n$ identify individual jobs. The time when a job j is presented to a scheduler for processing is referred to as l_j, its *release time*. In a **proper schedule** we will require that the l_j's form a non-decreasing sequence. The amount of CPU time a job j will require is its *size*, s_j. The time when job j terminates is its *completion time*, c_j.

The *response time* $w_j = c_j - t_j$ of a job measures how much time has elapsed between its release and its completion. Note, for a particular job j, s_j is unknown to a non-clairvoyant scheduler.

This paper will compare two well known algorithms for the scheduling problem. The optimal algorithm [7, 8] at each point in time runs the job j which currently has the minimum remaining processing time. Obviously, this algorithm requires foreknowledge about the sizes of the jobs. In our analysis we will assume that our Adversary uses this algorithm, denoted A.

The second algorithm is Balance [3] denoted B. Balance runs the job was has the minimum p_j. It is interesting to notice that B uses only information about the past (i.e. p_i's) to formulate its decisions about which job(s) to run whereas A uses only information about the future (i.e. r_i's).

The quality of the solutions generated by an on-line algorithm are generally measured using the *competitive ratio* [2] i.e. the worst case ratio of the cost of the solution generated by the on-line algorithm to the cost of an optimal off-line solution The lower bound on the competitive ratio of any deterministic scheduler is $\mathcal{O}(n^{1/3})$ [5].

In spite of this dismal performance, scheduling algorithms perform quite well in practice. This observation has motivated Kalyanasundaram and Pruhs [3] to find an alternative means to measure the quality of on-line solutions. Under their paradigm the on-line algorithm is "enhanced" by being run on a machine that is faster by a factor of $1 + \epsilon$ than the adversaries machine. The resulting competitive ratio is called the ϵ-*weak competitive ratio*.

Phillips *et. al.* [6] generalized this approach calling it *resource augmentation*; in their point of view the system designer is given some specific performance requirements and must compute the necessary amount of resource, like processor speed.

Kalyanasundaram and Pruhs [3] showed the ϵ-weak competitive ratio of the balance algorithm is $\mathcal{O}(\frac{1+\epsilon}{\epsilon})$. This result, while offering a valuable insight concerning the quality of the Balance algorithm, is anomalous in the sense that even if the speed of the machine used by Balance is very large, the upper bound of its total response time is still larger than the total response time of the optimal algorithm. We remove this anomaly by showing that the ϵ-weak competitive ratio of the balance algorithm is $\frac{2}{1+\epsilon}$ for $\epsilon \geq 1$.

2 The Main Result

In all further discussion it will be assumed that B's processor has speed $v \geq 2$ and A's processor has speed 1. Define $C_X(S)$ as the cost of a schedule S run by algorithm X, i.e. the sum of the response times of the jobs in S run by algorithm X. The following theorem captures the main result of this paper:

Theorem 1. *If* $v \geq 2$ *then* $\frac{v}{2}C_B(S) < C_A(S)$

Theorem 1 states that if B is run on a machine which is v times faster than A's machine, then B's cost is $v/2$ times smaller than A's. Thus, the ϵ-weak competitive ratio for B is $\frac{2}{v}$ for $v \geq 2$. To prove this we will determine how the cost functions for A and B change when an additional job is introduced into the schedule. A schedule S can be appended by a new job (l, s) to yield a new schedule $((l_1, s_1), (l_2, s_2), \cdots, (l_n, s_n), (l, s))$ denoted $S \cdot (l, s)$. Let $\Delta_X(S, l, s) = C_X(S \cdot (l, s)) - C_X(S)$. This difference describes the change in cost associated with introducing a new job (l, s) into the schedule S on a machine running algorithm X.

Lemma 1. *If $S \cdot (l, s)$ is a proper schedule and $v \geq 2$ then*

$$\frac{v}{2} \Delta_B(S, l, s) < \Delta_A(S, l, s)$$

This lemma asserts that when we append a schedule with a new job the response time of A increases by a factor of $v/2$ more than the response time of B. Theorem 1 is a direct corollary of Lemma 1, because any schedule can obtained from the empty one by a sequence of append operations.

We will prove Lemma 1 using a sequence of auxiliary lemmas. The first two will provide us a way to compute $\Delta_X(S, l, s)$ for $X = A, B$.

Let $Q_X(S, t)$ be the set of jobs that have been released before time t (inclusive) but have not yet been completed. Each non-completed job j has its *remaining time* r_j, which is the amount of additional CPU time j requires to terminate. We also define $p_j = s_j - r_j$ as the *processing time* that j has received up to the present time. When we consider some $j \in Q_X(S, t)$, p_j and r_j, will mean the processing time and remaining time of j as a function of time t and algorithm X. Generally, t will be implicitly defined from the context.

The next two lemmas will characterize $\Delta_A(S, l, s)$ and $\Delta_B(S, l, s)$ in terms of the sets $Q_A(S, l)$ and $Q_B(S, l)$.

Lemma 2. *If $S \cdot (l, s)$ is a proper schedule then*

$$v\Delta_B(S, l, s) = s + \sum_{j \in Q_B(S,l)} (\min(s, s_j) \dotminus p_j) + \sum_{j \in Q_B(S,l)} \min(s, s_j)$$

where $a \dotminus b = \max(a - b, 0)$.

Proof. (Sketch) Clearly, the response times of jobs which have been completed prior to the arrival of (l, s) will be unaffected Thus, we need only examine the jobs present when (l, s) is released, i.e. jobs in $Q_B(S, l)$. We determine $\Delta_B(S, l, s)$ by examining:

1. The increase in the response times for each job in $Q_B(S, l)$
2. The response time of (l, s) itself.

The response time of a job $j \in Q_B(S, l)$ with $s_j > s$ is increased by s/v as j has to wait for (l, s) to be completed before it receives more than s processing time. The response time for a job $j \in Q_B(S, l)$ with $s_j \leq s$ is increased by s_j/v because before s_j will be completed, (l, s) receives s units of processing time. Thus, the increase in the response times of all the jobs in $Q_B(S, l)$ is $\frac{1}{v}\sum_{j \in Q_B}\min(s, s_j)$.

The response time of (l, s) is not increased by a job $j \in Q_B(S, l)$ with $p_j > s$, because (l, s) will complete before job j sees the processor again. The response time of (l, s) increases by $(s - p_j)/v$ for jobs $j \in Q_B(S, l)$ with $p_j \le s \le s_j$, because j will only start to compete for the processor when (l, s) has p_j processing time devoted to it and j will stop competing for the processor when (l, s) is completed. The response time of (l, s) increases by $(s_j - p_j)/v$ for jobs $j \in Q_B(S, l)$ with $s_j < s$, because j will only compete for the processor between the time that (r, l) starts to share the processor with j and when s_j completes. Considering all these cases together we have that the response time of (l, s) is $\frac{1}{v}\sum_{j \in Q_B(S,l)}(\min(s, s_j) \dotminus p_j)$. By adding both components of $\Delta_B(S, l)$ we conclude that:

$$v\Delta_B(S, l, s) = s + \sum_{j \in Q_B(S,l)} \min(s, s_j) \dotminus p_j + \sum_{j \in Q_B(S,l)} \min(s, s_j)$$

Lemma 3. *If* $S \cdot (l, s)$ *is a proper schedule then*

$$\Delta_A(S, l, s) = s + \sum_{j \in Q_A(S,l)} \min(s, r_j)$$

Proof. Consider a job j from $Q_A(S, l)$ as in Lemma 2. If $s \ge r_j$ then j will not have to wait for (l, s) and consequently the response time of j will not change, but (l, r) itself has to wait for j, thus j contributes r_j to the sum of the response times. Notice in this case $r_j = \min(s, r_j)$. On the other hand, if $s < r_j$ then (l, s) does not wait for j, instead j must wait for (l, s). In this way, the response time of j (and the overall sum) increases by s. Notice in this case $s = \min(s, r_j)$. Taking the sum over all jobs yields:

$$\Delta_A(S, l, s) = s + \sum_{j \in Q_A(S,l)} \min(s, r_j)$$

In the next Lemma we will focus on the difference described in the following formula:

$$\Delta\Delta(S, l, s) = 2\Delta_A(S, l, s) - v\Delta_B(S, l, s)$$

Clearly, one can restate Lemma 1 as follows: If $S \cdot (l, s)$ is a proper schedule, then $\Delta\Delta(S, l, s) > 0$. To prove Lemma 1 we will show how $\Delta\Delta(S, l, s)$ changes over time (i.e. as l increases).

If we call the empty schedule λ then Lemma 2 and 3 imply $v\Delta_B(\lambda, l, s) = s$ and $2\Delta_A(\lambda, l, s) = 2s$, thus $\Delta\Delta(\lambda, l, s) = s > 0$. To show that we always have $\Delta\Delta(S, l, s) \ge 0$ we will study the changes in $\Delta\Delta(S, l, s)$ as we move forward in time and append S with new jobs.

Lemma 4. $\Delta\Delta(S \cdot (l, s'), l, s) = \Delta\Delta(S, l, s)$

Proof. It suffices to show that after appending S with (l, s') both $v\Delta_B(S, l, s)$ and $2\Delta_A(S, l, s)$ change identically. Label the new job which has been appended $j' = (l, s')$. Obviously, at time l, just when j' has been released, $r_{j'} = s'$ and $p_{j'} = 0$. By Lemma 2, $v\Delta_B(S, l, s)$ gets two new terms:

$$\min(s, s') + \min(s, s') \div 0 = 2\min(s, s')$$

By Lemma 3, $\Delta_A(S, l)$ gains one extra term $\min(s, s')$, but this term is multiplied by 2 because of the form of $\Delta\Delta(S, l, s)$, yielding $2\min(s, r_{j'})$. Therefore, we have $2\Delta_A = v\Delta_B$.

The next step in our proof is to characterize the evolution of $\Delta_A(S, l, s)$. We say that $Q_A(S, t)$ is *s-sensitive* if for some $j \in Q_A(S, t)$ we have $r_j < s$. Moreover, $Q_B(S, t)$ is *s*-sensitive if for some $j \in Q_B(S, t)$ we have $p_j < s$.

Lemma 5.

a) If $Q_A(S, t)$ is *s*-sensitive then for some $\varepsilon_0 > 0$ we have:
$\Delta_A(S, t + \varepsilon, s) - \Delta_A(S, t, s) = -\varepsilon$ for $0 \le \varepsilon \le \varepsilon_0$.
b) If $Q_A(S, t)$ is not *s*-sensitive then for some $\varepsilon_0 > 0$ we have:
$\Delta_A(S, t + \varepsilon, s) - \Delta_A(S, t, s) = 0$ for $0 \le \varepsilon \le \varepsilon_0$.
c) If for $j \in Q_B(S, t)$ we have $p_j = s_j$ then:
$\lim_{\varepsilon \to 0+} v\Delta_B(S, t, s) - v\Delta_B(S, t - \varepsilon, s) \le -\min(s, s_j)$.
d) If $Q_B(S, t)$ is *s*-sensitive then for some $\varepsilon_0 > 0$ we have:
$v\Delta_B(S, t + \varepsilon, s) - v\Delta_B(S, t, s) = -v\varepsilon$ for $0 \le \varepsilon \le \varepsilon_0$.
e) If $Q_B(S, t)$ is not *s*-sensitive then for some $\varepsilon_0 > 0$ we have:
$v\Delta_B(S, t + \varepsilon, s) - v\Delta_B(S, t, s) = 0$ for $0 \le \varepsilon \le \varepsilon_0$.

Proof.

a,b) Assume that A is processing job j at time t. Clearly, j has a minimal r_j; any further processing given to j will make it unequivocally minimal. Therefore, j will be processed until it terminates, or until S is appended with a job smaller than r_j. In either case, for some $\varepsilon_0 > 0$, j is processed without interruption from t to $t+\varepsilon$. Thus, for $0 \le \varepsilon \le \varepsilon_0$ $\Delta_A(S, t+\varepsilon, s) - \Delta_A(S, t, s) = \min(s, r_j - \varepsilon) - \min(s, r_j)$
If $Q_A(S, t)$ is *s*-sensitive (case a) $\min(s, r_j - \varepsilon) - \min(s, r_j) = r_j - \varepsilon - r_j = -\varepsilon$.
Otherwise (case b), $\min(s, r_j - \varepsilon) - \min(s, r_j) = s - s = 0$.
c) We need to observe how the value of $v\Delta_B(S, t - \varepsilon, s)$ changes as ε decreases to 0. Lemma 3 provides a summation that is equal to $v\Delta_B(S, t - \varepsilon, s)$; each term in this sum corresponds to a job in $Q_B(S, t - \varepsilon)$. One can easily see that each of the terms is a continuous function of time, the only source of discontinuity is the fact that new terms are included when jobs are released and a term of job j is dropped when it is completed.
Under our assumptions, no jobs are released at time t while j is terminating. The drop of $v\Delta_B$ is the value of the term corresponding to j for some small θ this term equals $\min(s, s_j) + \min(s, s_j) \div (s_j - \theta\varepsilon)$ so in the limit this term equals $\min(s, s_j)$.

d) If $Q_B(S,l)$ is s-sensitive then we can define a non-empty set of jobs $I_t = \{j \in Q_B(S,t) : p_j < s\}$. By its very nature, B will only work on those jobs which are in I_t at time t. Obviously, for some sufficiently small $\varepsilon_0 > 0$ we have $Q_B(S,t) = Q_B(S,t+\varepsilon_0)$ and $I_t = I_{t+\varepsilon_0} = I$.

Consider now $0 < \varepsilon < \varepsilon_0$ and compare the summations that define $v\Delta_B(S,t,s)$ and $v\Delta_B(S,t+\varepsilon,s)$ according to Lemma 3. It is easy to see that only the terms containing p_j's for $j \in I$ change. Since jobs in I are being worked on, their p_j's are increasing, causing $\sum_{j\in I} p_j$ to increase by $v\varepsilon$, which in turn decreases $v\sum_{j\in I} \min(s,s_j) - p_j$ by $v\varepsilon$ (note that for $j \in I$, $\dot{-}$ is equivalent to $-$).

e) Since $s < p_j < s_j$ for all $j \in Q_B(S,t)$ then $\min(s,s_j) \dot{-} p_j = 0$. Lemma 2 now yields: $\Delta_B(S,t+\varepsilon,s) - \Delta_B(S,t,s) = 0$.

¿From Lemma 4 and 5 we can infer that the only time that we are in danger of $\Delta\Delta(S,l,s)$ dropping below 0 (and consequently violating lemma 1) is when B is not s-sensitive. This can occur when $p_j \geq s$ for all $j \in Q_B(S,t)$.

The next lemma will show that in this situation $\Delta\Delta(S,l,s) > 0$, thus proving Lemma 1 and consequently proving Theorem 1.

Lemma 6. *If $Q_B(S,t)$ is not s-sensitive, $S\cdot(l,r)$ is a proper schedule and $v \geq 2$ then*

$$\Delta\Delta(S,l,s) > 0$$

Proof. We will prove the lemma by normalizing any schedule so that $\Delta_B(S,t)$ is unchanged and $\Delta_A(S,t)$ is not increased. The resulting schedule will have a very specific form, making the claim easy to verify.

We define the *snapshot* of B as the set $\{(s_j,p_j) : j \in Q_B(S,t)\}$ Note that the snapshot fully determines $\Delta_B(S,t)$.

Let $P = (p^1, \cdots p^k)$ be the vector of p_j's such that (s_j,p_j) belongs to the snapshot of B and $p^i < p^{i+1}$, for $i = 1 \cdots k$ (notice that p^1 is the processing time of the current job). We will define $t_0 \ldots t_k$ as follows: $t_0 = t$ (the current time). For $i > 0$ t_i is the latest (largest) time $t_i \leq t$ when (or up to that time) the processor was working on a job that had more processing time than it had at time t_{i-1}.

Observe first that for $1 \leq i \leq k$ the time t_i is properly defined - the snapshot of B witnesses the fact that the event defining t_i indeed happened. If t_k is undefined we set $t_k = 0$, finally for completeness we set $t_0 = t$. Moreover, the very definition of the t_i's assures that $t_{i+1} > t_i$ for $1 \leq i \leq k$.

Now observe that a job that has processing time p^i in the snapshot of B had to be released between times t_i and t_{i-1}: were such a job released before time t_i, by time t_i it would receive more than p^i processing time, and were it released at time t_{i-1} or after it would never receive more than $p^{i-1} < p^i$ of processing.

We are now ready to perform the normalization of a schedule.

1) No job released before time t_k remains in $Q_B(S,t)$; it is easy to see that all of them terminate in the execution of B by time t_k (an easy contradiction follows otherwise). We can delete all these jobs from the schedule and "start

history" at time t_k; the subsequent actions of B remain identical. On the other hand, the only change for A after time t_k will be a possibly smaller number of jobs to execute, this can only decrease the remaining times in $Q_A(S, t)$.

2) Let δ be some very small number. We can decrease the size of each job $j \in Q_B(S, t_0)$ to $p_j(1 + \delta)$. This transformation does not affect the snapshot and will only decrease the residual time for A (and thus $\Delta\Delta(S, t)$).

3) We can release all jobs that were released at time t' in the time interval $[t_i, t_{i-1}]$, at time t_i; this will not affect the snapshot of B; the jobs released during this time interval with size at most p^i were and will be completed; other jobs in both cases receive p^i of processing time. On the other hand, after this change A has more flexibility to schedule the jobs. We can show that if this extra flexibility changes the behavior of A then not only the cost of A decreases, but also the final value of $\Delta_A(S, t)$ (essentially, this is the very reason that algorithm A, greedy in its nature, is optimal — there are no conflicts between the short term goals and the possibility of the new jobs in the future).

4) We can delete every job completed by both algorithms and compress the duration of the intervals (t_i, t_{i-1}) during which B was working on these jobs. By reasoning similar to that of step 3 this can only help A because deleting a job of size s from the schedule and s/v time units from A's and B's respective histories leaves A with an extra $s(1 - 1/v)$ time units to decrease the remaining times (r_j's) of its other jobs.

5) At time t_k, now the "beginning of history", A chooses the smallest of the jobs released at this time, say j, to work upon. Suppose that $s_j \leq p^k$; because B completed j and we performed step 4 of the normalization we know that A did not finish j by time t_{k-1}; and hence did not touch the jobs that B worked on but did not finish. These jobs, say c of them, of size $p^k + \delta$ will forever after (until time t_0) loose the size competition with any new jobs, hence they belong to $Q_A(S, t)$ with residual values $p^k + \delta$. As a result, jobs released at time t_k contribute at least cs to $\Delta_A(S, t)$ and exactly cs/v to $\Delta_B(S, t)$. Hence the deletion of the interval $[t_k, t_{k-1})$ from the schedule (and all the jobs released during this time) decreases $\Delta\Delta(S, t, s)$ by cs, therefore in this case we delete the time interval and decrease the value of k.

6) Now all the jobs which start at time t_k have identical size $p^k + \delta$. We simultaneously decrease their size and the duration of the interval $[t_k, t_{k-1})$ in proportion p^{k-1}/p^k. The outcome of this change for B is neutral; during this interval B will give each of these jobs p^{k-1} time and during the interval (t_{k-1}, t_{k-2}) these jobs will not be competitive with jobs released at t_{k-1}; afterwards neither job released at time t_k nor at t_{k-1} will be scheduled (before $t_0 = t$). The outcome for A is neutral or a gain; the number of jobs completed for A will be the same, the number of those that were not touched and thus never will be will be the same (as each contributed exactly s to $\Delta_A(S, t)$, and the remaining time at time t_{k-1} of the jobs started but not finished by A will be smaller).

7) Because the job released at times t_k and t_{k-1} are identical, we can release all of them at time t_k for the reasons that we argued in step 3, this decreases the value of k.

By applying steps 5,6,7 as long as possible, we convert S into a "batch" schedule: c identical jobs are released at time t_1. From Lemma 4 we have $\Delta_A(S, t_1 + \varepsilon) = \Delta_B(S, t_1 + \varepsilon)$. Furthermore by Lemma 5 we know that for this batch schedule $2\Delta_B(S, t)$ will decrease slower then $v\Delta_A(S, t)$. Thus, the lemma holds.

Our final lemma will show that ϵ-weak competitive ratio implied by Theorem 1 is tight for $v \geq 2$.

In terms of the worst case analysis, Theorem 1 is tights for every $v \geq 2$. To show it, we first prove the following lemma.

Lemma 7. *Let s_i is the size of the ith job in S, then*

$$2C_A(S) - vC_B(S) = \sum_{i=1}^{n} s_i$$

Proof. Let $S = ((0, s_n)(0, s_{n-1}), \cdots, (0, s_1))$ be a batch of jobs released at time 0. Let the jobs be numbered such that $s_{i+1} \leq s_i$.

A's strategy is to process s_n first, since it is the smallest job. While A is processing job n there are $n - 1$ jobs waiting. When job n completes A then starts work on job $n - 1$. A continues working on consecutive jobs in S until they are all completed. The total response time $C_A(S) = \sum_{i=1}^{n} is_i$.

As soon as B start processing it will begin working on all the jobs in parallel. When the smallest job finishes each of the n jobs will experience a delay of (ns_n/v) thus, the increase in the total response time for all the jobs is $(n^2 s_n)/v$. This slow down for the completion time of the first job is compensated for by noting that job $n - 1$ has only $s_{n-1} - s_n$ units of work left. When the second to smallest job finishes, the total response time increases by $(n-1)^2(s_{n-1} - s_n)/v$. Explicitly writing down the total response time of B is illuminating.

$$C_B(S) = (n^2 s_n)/v + (n-1)^2(s_{n-1} - s_n)/v \cdots + 1(s_1 - s_2))/v$$

This sum can be rewritten by gathering all the s_i terms together, multiplying both sides by v yields:

$$vC_B(S) = \sum_{i=1}^{n} s_i(2i - 1)$$

Now

$$2C_A(S) - vC_B(S) = \sum_{i=1}^{n} s_i$$

Lemma 8 can be applied to the following schedule: W jobs of size 1 are released at time 0. Then $C_A(S) = 1 + 2 + \ldots N = (N^2 - N)/2$. By Lemma 8, $vC_B(S) = N^2 - 2N$. Thus $(v - 1)$ competitive ratio (in our paper $1 + \epsilon = v$) is at least $v/2 - O(1/N)$.

3 Conclusions

There are indications that the ϵ-weak competitive ratio will be able to distinguish between various scheduling algorithms. Consider a schedule S consists of $2v$ jobs of size $4\log(N)$ released at time 0 and at each integer time from 1 to N a job with size 1 is released. Consider the Round Robin (RR) algorithm running schedule S on a machine with speed v. One can show that the ϵ-weak competitive ratio of RR is $\Omega(N/v)$. Thus, given any speed v for RR's computer, a schedule can be devised in which the optimal off-line adversary has a lower cost that then RR.

However faster the computer of RR is, there exists a schedule for which the optimal off-line adversary with the slow computer has a lower cost.

References

1. S. Albers, *Better bounds for online scheduling*, STOC (1997), 130-139.
2. S. Irani and A. Karlin, *Online Computation*, in D. Hochbaum (ed.), *Approximation Algorithms for NP Hard Problems*, PWS Publishing, Boston MA, 1997, 521-563.
3. B. Kalyanasundaram and K. Pruhs *Speed is as powerful as clairvoyance*, FOCS (1995), 214-221.
4. T. Matsumoto, *Competitive analysis of the round robin algorithm*, International Symposium on Algorithms and Computation (1992), 71-77.
5. R. Motwani, S. Phillips and E. Torng, *Non-clairvoyant scheduling*, SODA (1993), 422-431.
6. C. Phillips, C. Stein, E. Torng and J. Wein, *Optimal time-critical scheduling via resource augmentation*, STOC (1997), 140-149.
7. W. Smith, *Various optimizers for single-stage production*, Naval Research Logistics Quarterly (1956), 59-66.
8. A. Tannenbaum and A. Woodhull, *Operating Systems*, Prentice Hall, New Jersey, 1997.

Randomized Online Multi-threaded Paging

Steven S. Seiden*

Technische Universität Graz
Institut für Mathematik B
Steyrergasse 30/ 2 Stock
A-8010 Graz, Austria

Abstract. We present the first randomized upper and lower bounds for online multi-threaded paging as introduced by Feuerstein and Strejilevich de Loma [5]. Our main result is a $O(w \log k)$-competitive algorithm for unfair infinite multi-threaded paging, which is optimal to within a constant factor. We also present algorithms and lower bounds for three other sub-models of multi-threaded paging.

1 Introduction

In computer systems, it is often necessary to solve problems with incomplete information. The input evolves with time, and incremental computational decisions must be made based on only part of the input. A typical situation is where a sequence of tasks must be performed. How tasks are performed affects the cost of future tasks. Examples include managing a two level store of memory, performing a sequence of operations on a dynamic data structure, and maintaining data in a multiprocessing environment [8, 10, 16, 19]. An algorithm that decides how to perform a task based only on past requests with no knowledge of the future is said to be an *online algorithm*. In contrast, we refer to an algorithm which has complete information about the tasks to be performed before it makes any decisions as an *offline algorithm*.

In the *online paging problem* we are faced with the following situation: We have a two level hierarchal memory. There is a fast memory, called the *cache*, which can hold up to k pages, and a slow memory which holds $K \gg k$ pages. There is a sequence σ of requests to pages. If a page is in the cache, serving a request to it costs 0. If it is not in the cache, and is requested, it must be brought in at a cost of 1. In order to bring a new page into the cache, some page currently in the cache must be evicted. The problem is online, in that the algorithm serves the ith request without knowledge of requests $i + 1, i + 2, \ldots$. Because of its importance for computer performance, paging is one of the most well studied online problems [1, 3, 6, 7, 10, 11, 15, 21].

Online multi-threaded paging is a generalization of the online paging problem which attempts to model the situation in which page requests are coming from

---

* Email: sseiden@acm.org. This research has been supported by the START program Y43-MAT of the Austrian Ministry of Science.

separate independent threads or processes. This model was developed by Feuerstein and Strejilevich de Loma [5]. The model differs from other online paging models, such as that of Sleator and Tarjan [15], in that the input consists of several (as opposed to one) sequences of page requests. The paging algorithm must decide not only which pages to keep in its cache, but also determine the order in which it serves requests from the sequences. Since the problem is online, the algorithm serves the requests of any given sequence in order, without knowing which pages are requested further on in that given sequence.

We study four different models. The models differ in how 'fairly' the algorithm serves the sequences, and in whether the sequences are finite or infinite. In the most general model, the algorithm is allowed to serve any number of requests from a given sequence before serving another sequence. Further, the algorithm is not restricted to serving the sequences in any fixed order. For instance, the algorithm might serve ten requests of the first sequence, serve five requests of the third sequence, one request of the second sequence seven requests of the third sequence, etc. ... . We evaluate an online algorithm for multi-threaded paging by comparing its performance to that of the optimal offline algorithm. In order to make this comparison equitable, we force the online and optimal offline algorithms to serve the same number n of requests.

Let w be the number of threads and k be the cache size. We call our sequences $\sigma_0, \ldots, \sigma_{w-1}$. The four models are:

UIMTP (Unfair Infinite Multi-Threaded Paging) This is the model just described. Each sequence is infinite in length, and the only restriction is that the online and offline algorithms serve the same number of requests.

UFMTP (Unfair Finite Multi-Threaded Paging) Each sequence σ_i has a length n_i. Both algorithms serve all requests of all sequences.

FIMTP (Fair Infinite Multi-Threaded Paging) Each sequence is infinite in length. However, both algorithms are restricted in that, for a given fixed $t \geq w - 1$, between any two requests from a given sequence they may serve at most t requests from other sequences. In particular, when $t = w - 1$ the algorithms are restricted to serving the sequences in a round-robin fashion.

FFMTP (Fair Finite Multi-Threaded Paging) Each sequence σ_i has a length n_i. The algorithms serve all sequences to completion, and do so fairly as in FIMTP.

For a tuple of sequences $\Sigma = \langle \sigma_0, \ldots, \sigma_{w-1} \rangle$, let $\text{cost}_A(\Sigma)$ and $\text{cost}_{\text{opt}}(\Sigma)$ be the cost incurred by an online algorithm A and an optimal offline algorithm, respectively, who both serve n requests of Σ. A is said to be *c-competitive* if there is a constant d such that for all Σ and n,

$$\text{cost}_A(\Sigma) \leq c \cdot \text{cost}_{\text{opt}}(\Sigma) + d.$$

The constant d may depend on k and w, but not on n. If the algorithm A is randomized then we replace $\text{cost}_A(\Sigma)$ by $\text{E}[\text{cost}_A(\Sigma)]$ in the above definition. The *competitive ratio* of A is the infimum over all c such that A is c-competitive. This approach to analyzing online problems, called *competitive analysis*, was initiated by Sleator and Tarjan, who used it to analyze the List Update problem [15]. The term competitive analysis originated in [9].

A problem related to multi-threaded paging is *application-controlled paging*. In this problem, p applications share a cache. Each application has a sequence of requests to pages. The algorithm must serve an interleaved request sequence. There are three major differences with the model used here: First, the sequences are interleaved is way that is not under the algorithm's control. Second, when a page must be evicted, the algorithm asks the application to decide which of its pages to evict. Third, the application is assumed to complete knowledge of its individual request sequence, and thus can made good decisions about which page to evict. This problem was introduced by Cao, Felten and Li [13], and further studied by Barve, Grove and Vitter [14]. The third difference listed above strongly favors the algorithm. In fact, Barve *et al.* show that the deterministic competitive ratio for this problem is $\Theta(p)$ and that the randomized competitive ratio is $\Theta(\log p)$

Feuerstein and Strejilevich de Loma [5] define multi-threaded paging, and show the following results, all for the deterministic setting: For UIMTP they show a wk-competitive algorithm, and show that it is optimal by proving a matching lower bound. For UFMTP they show a wk-competitive algorithm, and prove a deterministic lower bound of $k + 1 - 1/w$. For FIMTP they show that no deterministic algorithm is competitive for $t \geq w \geq 2$. For the case that $t = w - 1$, they show a $(k+w)$-competitive algorithm, and show a deterministic lower bound of k. For FFMTP they show that a deterministic competitive algorithm exists only for $w = 2$ and $t = 1$. They further show a $(k + w)$-competitive algorithm, and a deterministic lower bound of k for this case. Strejilevich de Loma [17] shows the following results: a lower bound of $w + 1$ for FIMTP with $k = 1$ and w even, a lower bound of w for FIMTP with $k = 1$ and w odd, a lower bound of $w + 1$ for FFMTP with $k = 1$ and $w = 2$, a $(k + 1)$-competitive algorithm for FIMTP with $t = w - 1$ and $w \leq k$, a $(k + 1)$-competitive algorithm for FFMTP with $t = w - 1$ and $w = 2 \leq k$.

This paper is organized as follows: In Section 2, we present material ancillary to the analysis in the succeeding sections. In Section 3, we present a randomized algorithm for UIMTP which is $w(2H_k + \epsilon)$-competitive for any $\epsilon > 0$, and show that it is optimal to within a constant factor. This algorithm achieves the same competitive ratio for UFMTP. In Section 4, we show a randomized lower bound for UFMTP. In Section 5, we show that in general, it is not possible to have a competitive algorithm for FIMTP. We also present an randomized algorithm and a lower bound for the case where competitiveness is possible. In Section 6, we show results analogous to those of the previous section for FFMTP. In Section 7, we present conclusions and open problems.

2   Background

The algorithms we present are derived from the single-thread MARK algorithm of Fiat, Karp, Luby, McGeoch, Sleator, and Young [6]. This randomized algorithm is $2H_k$-competitive against an oblivious adversary, where

$$H_k = \frac{1}{1} + \frac{1}{2} + \cdots + \frac{1}{k} \approx \ln k$$

is the kth harmonic number. MARK is displayed in Figure 1 for the reader's convenience. In our description of the algorithm, the set C corresponds to the

1. Let C be the set of pages initially in the cache.
2. $M \leftarrow \emptyset$.
3. While there are requests:
   (a) Let p be the next requested page.
   (b) $M \leftarrow M \cup \{p\}$.
   (c) If $|M| = k + 1$ then $M \leftarrow \{p\}$.
   (d) If $p \notin C$ then:
      i. Pick a page p' in $C \backslash M$ uniformly at random.
      ii. Bring in p in place of p'.
      iii. $C \leftarrow C \backslash \{p'\} \cup \{p\}$.

Fig. 1. The MARK algorithm.

set of pages currently in the cache. The pages in the set M are called *marked* pages. The algorithm breaks the sequence of requests into phases. The first phase begins at the start of the sequence. When $k + 1$ pages become marked, all pages are un-marked except for the current one and a new phase begins.

We review the relevant parts of the analysis of MARK. Please see Fiat et al. [6] for complete details. We recall the following definition from [6]: A page which has not been requested in this phase is *clean* if it was not requested in the previous phase. For a given phase, let ℓ be the number of requests to clean pages. Fiat et al. show that the cost incurred by MARK is at most $\ell(H_k - H_\ell + 1)$ and that the amortized optimal offline cost is at least $\ell/2$. Thus MARK is $2H_k$-competitive.

We also use the single-thread paging randomized lower bound construction of Motwani and Raghavan in our proofs [12]. This proof differs from the original proof of Fiat et al. [6], and is inspired by a proof of Borodin, Linial and Saks [4]. We present the relevant facts here. For complete details, the reader should see [12]. The construction makes use of Yao's corollary to the von Neumann minimax principle [18, 20]. This principle states that, for a given problem, one can show a lower bound cost of c for any randomized algorithm by showing a distribution over inputs such that the expected cost to any deterministic algorithm is c. For the purposes of this paper, we show that the competitive ratio of any randomized online algorithm is at least c by showing a distribution over inputs X such that

$$\mathrm{E}_X[\mathrm{cost}_\mathcal{A}(X)] \geq c \cdot \mathrm{E}_X[\mathrm{cost}_{\mathrm{opt}}(X)]$$

for all deterministic online algorithms $\mathcal{A}$. To show a lower bound for the single-threaded paging problem, we consider a sequence $\sigma = p_1, p_2, \ldots, p_n$ of pages drawn from a set P of size $k+1$. Page p_i is chosen uniformly at random from $P\setminus\{p_{i-1}\}$. This sequence is broken into phases. The first phase begins at the beginning of the sequence. A phase ends, and a new one begins, just before all $k+1$ pages have been requested. The expected length of a phase is kH_k. The adversary pays one for each phase. If we have a such a sequence σ with $n = |\sigma|$, then we also have

$$\mathrm{E}[\mathrm{cost}_{\mathrm{opt}}(\sigma)] \le \frac{n}{kH_k}.$$

(Equality is approached as $n \to \infty$.) Any deterministic online algorithm incurs an expected cost of $1/k$ for each request for a total of H_k. Combining these facts, we have shown a distribution where any deterministic algorithm incurs an expected competitive ratio of H_k, thus no randomized algorithm is less than H_k competitive.

3   Results for UIMTP

We present the first randomized algorithm for multi-threaded paging. This algorithm, which we call ALTERNATE MARK, is derived from the MARK algorithm of Fiat et $al.$ [6]. It is similar in spirit to the ALTERNATE FLUSH WHEN FULL algorithm of Feuerstein and Strejilevich de Loma [5]. The algorithm has an integer parameter m. It appears in Figure 2. The algorithm has the nice property that

1. $i \leftarrow 0$.
2. While there are requests:
    (a) Restore the state from the previous time that σ_i was served (if this is not the first time).
    (b) Serve σ_i using MARK until the amortized optimal offline cost has increased by m.
    (c) Save the state of MARK.
    (d) $i \leftarrow i + 1 \bmod w$.

Fig. 2. The ALTERNATE MARK algorithm.

it does what a real operating system might do: Save and restore state between threads.

Theorem 1. ALTERNATE MARK is $w(2H_k+k/m)$-$competitive$ $against$ an $oblivious$ $adversary$ for $UIMTP.$

Proof. Note that, ignoring state changes, the algorithm behaves exactly like MARK on each sequence. Ignoring the cost of any state restores, for each sequence, the algorithm incurs an amortized cost of at most $x2H_k$ if the optimal offline cost of the sequence is x. Each state restore costs the algorithm at most k, since at most k pages are brought into the cache. So each time the algorithm executes Steps 2a and 2b, it pays at most $m2H_k + k$, amortized. No other steps cost the algorithm. Suppose that Step 2 has completed at least nw times, but fewer than $(n + 1)w$ times. The algorithm pays at most $(n + 1)w(m2H_k + k)$. Now, as in the proof of Theorem 3.1.2 of Feuerstein and Strejilevich de Loma [5], there is one sequence where the adversary has served as many requests as the algorithm. Therefore, the adversary's cost is at least mn. We have

$$\text{cost}_A(\Sigma) \leq (n + 1)w(m2H_k + k)$$
$$= w(2H_k + k/m)(mn) + w(m2H_k + k)$$
$$\leq w(2H_k + k/m)\text{cost}_{\text{opt}}(\Sigma) + w(m2H_k + k),$$

and the theorem follows. □

Therefore, ALTERNATE MARK is $(2wH_k + \epsilon)$-competitive as long as $m \geq wk/\epsilon$.

We give a randomized lower bound which shows that ALTERNATE MARK is optimal up to a constant factor:

Theorem 2. *No algorithm for UIMTP is better than $(w + 1)H_k/2$-competitive.*

Proof. Recall the lower bound of Motwani and Raghavan for single-threaded paging. That proof implies that we can generate random sequences such that if the length of the sequence is x, the expected cost incurred by any deterministic algorithm is x/k, while the expected optimal cost is at most $x/(kH_k)$. We generate w such sequences independently, where each thread has its own set of $k + 1$ pages which it exclusively requests. Let X be the number of requests to be served. We pick one sequence uniformly at random and modify it so that after X/w requests it becomes *easy* in that it repeatedly requests the same page. Any algorithm serving this sequence incurs zero cost after the first $X/w + 1$ requests.

The adversary knows which of the sequences becomes easy after X/w requests, so he serves just this one sequence. His expected cost is at most

$$(X/w)/(kH_k) = X/(wkH_k).$$

We assume that the online algorithm incurs zero cost once it discovers the easy sequence. Since each sequence requests pages from a different set, the expected cost to any online algorithm for serving any hard request is at least $1/k$. Since this is the case, we can lower bound the algorithm's cost by considering the following simplified problem: Requests are no longer to pages, but simply are hard or easy. Each hard request costs the algorithm exactly $1/k$. Each easy request costs the algorithm nothing. Each sequence begins with X/w hard requests. One sequence, chosen uniformly at random, becomes easy on request $X/w + 1$, while the others remain hard.

We lower bound the cost to any deterministic algorithm for this modified problem. Note that a deterministic algorithm is essentially a search strategy. We can assume that the algorithm decides its entire search strategy in advance. Until the moment the algorithm discovers the easy sequence, there is no reason for it to change strategies. Let ℓ be the number of times that the algorithm would serve X/w requests of some sequence, in his fixed strategy. (The algorithm may in reality search fewer sequences, if it discovers the easy sequence). The first such time, there is a probability of $1/w$ that the algorithm discovers the easy sequence. Given that the algorithm fails in this first attempt, it succeeds on the second attempt with probability $1/(w-1)$. The probability that it fails the first attempt is $(1-1/w)$. So the probability that it succeeds on the second attempt is $(1-1/w)/(w-1)$. In general, if it fails on the first $i-1$ attempts, it succeeds on the ith attempt with probability $1/(w-i+1)$. The probability that it fails the first $i-1$ attempts is $(1-1/w)\cdots(1-1/(w-i+2))$. For each attempt, the algorithm incurs a cost of at least $X/(wk)$. If it fails in all attempts, it pays X/k. Putting all these facts together, the expected cost incurred by the algorithm is

$$\sum_{i=1}^{\ell}\frac{1}{w-i+1}\prod_{j=0}^{i-2}\left(1-\frac{1}{w-j}\right)\frac{iX}{wk}+\prod_{j=0}^{\ell-1}\left(1-\frac{1}{w-j}\right)\frac{X}{k}$$

$$=\sum_{i=1}^{\ell}\frac{1}{w-i+1}\prod_{j=0}^{i-2}\frac{w-j-1}{w-j}\frac{iX}{wk}+\prod_{j=0}^{\ell-1}\frac{w-j-1}{w-j}\frac{X}{k}$$

$$=\sum_{i=1}^{\ell}\frac{iX}{w^2k}+\frac{w-\ell}{w}\frac{X}{k}$$

Clearly, this is minimized when the algorithm makes as many 'attempts' as possible. I.e. when $\ell=w$. We therefore have

$$\sum_{i=1}^{\ell}\frac{iX}{w^2k}+\frac{w-\ell}{w}\frac{X}{k}\geq\sum_{i=1}^{w}\frac{iX}{w^2k}=\frac{X(w+1)}{2wk}.$$

The competitive ratio is $(X(w+1)/(2wk))/(X/(wkH_k))=(w+1)H_k/2$. $\qquad\square$

4 Results for UFMTP

It is not hard to see that ALTERNATE MARK can also be used for UFMTP, and that it remains $w(H_k+k/m)$-competitive. However, we are unable to show as good a lower bound for UFMTP. We do have the following result:

Theorem 3. *For any fixed w, the randomized competitive ratio for UFMTP is at least H_k.*

Proof. The sequence σ_0 is generated using the Motwani and Raghavan construction. The other sequences, $\sigma_1,\ldots,\sigma_{w-1}$, are exact copies of σ_0. Let $n=wm$ be the total number of requests to be served. The adversary serves all the first requests of each sequence, all the second requests, etc.... The expected length of a phase in the adversary's overall sequence of requests served is wkH_k. Therefore, his total expected cost is $n/(wkH_k)=m/(kH_k)$. The online algorithm pays an expected $1/k$ each time it serves the jth request of some sequence for the first time. This occurs at least m times. So the competitive ratio is

$$\frac{m/k}{m/(kH_k)} = H_k.$$

□

5 Results for FIMTP

Our results for this model, like those of Feuerstein and Strejilevich de Loma [5], are mostly negative. We are, however, able to show a competitive algorithm for the cases where this is possible.

Theorem 4. *There is no competitive randomized algorithm for FIMTP with $t \geq w \geq 2$.*

Proof. We again use Yao-Von Neumann. We have pages $p_1, \ldots, p_{k+1}$. Each sequence requests page p_1 for the first $m - 1$ requests. One sequence, picked uniformly at random, continues to request p_1 ad infinitum. The others begin, starting with request m, to request $p_1, p_2, \ldots, p_{k+1}$ repeatedly. The adversary sets the number of requests to be served to be $(w + 1)m$. Since $t > w - 1$, the adversary, who knows which sequence remains easy, serves all requests at a cost of at most one: He serves the easy sequence twice, and then each other sequence, repeating this process m times. Consider the online algorithm. Let x be the number of requests served by the online algorithm when it first reaches the $(m + 1)$st request of some sequence, say σ_i. Such an $x \leq mw$ must exist. With probability $1 - 1/w$, this sequence is not the easy one. Since the algorithm serves at least $(w + 1)m - x \geq m$ more requests, it serves at least $\lfloor m/(t+1) \rfloor$ requests of σ_i. If this sequence is not the easy one, then its remainder cycles through $k + 1$ pages repeatedly, and any algorithm pays at least one for each block of $k + 1$ requests. Therefore the algorithm's expected cost is at least

$$\left(1 - \frac{1}{w}\right) \Omega\left(\frac{m}{tk}\right).$$

Since m can be made arbitrarily large, the ratio of the algorithm's cost to the adversary's cost is unbounded. □

The preceding theorem implies that the only case where a competitive algorithm is possible is the completely fair one, where $t = w - 1$. In this case, the algorithms must commit themselves, at the beginning of processing, to some permutation for serving the sequences and then use this ordering thereafter. However, the situation differs from single-threaded paging in that the algorithms might choose different permutations. At a given point in time, it is not true that the algorithms have served the same requests. However, this is true on every ith request where $i = wj$ for some integer $j \geq 1$. In this case, both algorithms have served the first

j requests of each sequence. After Feuerstein and Strejilevich de Loma [5], we call the tuple of jth requests the jth *row*.

For the case that $t = w-1$, we present an $(2H_k+w+1)$-competitive algorithm, called ROUND ROBIN MARK. It appears in Figure 3.

---

1. Pick an arbitrary permutation in which to serve each row.
2. While there are requests:
   (a) Serve requests using MARK until a single phase is completed.
   (b) Pick an arbitrary cache slot. Let p be the page in that slot.
   (c) Evict p to serve the next request.
   (d) Complete serving requests to the current row, using that single slot.
   (e) Restore p.

Fig. 3. The ROUND ROBIN MARK algorithm.

Theorem 5. ROUND ROBIN MARK *is* $(2H_k + w + 1)$-*competitive for FIMTP with* $t = w - 1$.

Proof. We call each complete execution of Step 2 a *super-phase*. Each super-phase consists of one phase of MARK, plus one additional request, plus a number of requests to complete the current row. We first analyze the optimal offline cost for a super-phase. Note a phase of MARK ends immediately before $k + 1$ pages are requested. This $(k + 1)$st page is the one served in Step 2c. So at least $k + 1$ pages are requested in the super-phase. Further, since in Step 2d we complete the current row, ROUND ROBIN MARK and the adversary serve the same set of requests during the super-phase. Therefore the adversary's cost is at least one.

Let A and A' be the sets of pages which the adversary has in its cache at the beginning and end of the super-phase, respectively. Similarly define M and M' for MARK. Let $O = M \cap A$ and $O' = M' \cap A'$. We define $\phi = |O|/2$ to be our potential value at the beginning of the super-phase. The potential at the end is $\phi' = |O'|/2$.

We define a *clean* page to one which was not requested during the previous phase of MARK. We let ℓ be the number of requests to clean pages during the phase of MARK of this super-phase We show that in an amortized sense, the adversary incurs a cost of at least $\ell/2$. We use the following scheme for charging the adversary: We charge him $\frac{1}{2}$ to bring a page into the cache, and $\frac{1}{2}$ to evict a page from the cache.

There are no clean pages in M. So the adversary has at most $|A \backslash M| = k - 2\phi$ of the clean pages to be requested in his cache. The adversary brings these pages into his cache at some point during the super-phase. Therefore, he pays at least $(\ell - (k - 2\phi))/2 = \ell/2 - k/2 + \phi$ for loading pages.

Further, each page in M' is served by the adversary at some point in the super-phase. The adversary no longer has the pages in $M' \backslash O'$ in his cache. Therefore he pays at least $|M' \backslash O'| = (k - 2\phi')/2 = k/2 - \phi'$ for evictions.

The total cost is $\ell/2+\phi-\phi'$. Summing this over all super-phases, the potential values telescope, and so the amortized cost of a super-phase is $\ell/2$.

The cost for the phase of MARK is at most ℓH_k. The cost of Step 2c is one. The cost of Step 2d is at most $w - 1$. The cost of Step 2e is one. So the competitive ratio is

$$\frac{\ell H_k + w + 1}{\max\{1, \ell/2\}} \leq 2H_k + w + 1.$$

$\square$

We are able to show the following lower bounds on FIMTP with $t = w - 1$. Our bounds show that the H_k term in the competitive ratio of ROUND ROBIN MARK is necessary, and also that there must be some term dependent on w.

Theorem 6. *The randomized competitive ratio for FIMTP with $t = w-1$ is at least H_k.*

Proof. The sequence σ_0 is generated using the Motwani and Raghavan construction. The other sequences, $\sigma_1, \ldots, \sigma_{w-1}$, are exact copies of σ_0. Now the situation is exactly like that in single-threaded paging. $\square$

Theorem 7. *The randomized competitive ratio for FIMTP with $t = w - 1$ is $\Omega(w/(k \log k))$.*

Proof. We use the Yao-Von Neumann principle. We have pages $p_1 \ldots p_{k+1}$. The first two rows consists entirely of requests to p_1. Then each sequence picks a page uniformly and independently at random, and requests it ad infinitum. The algorithm chooses a permutation in the first w requests that it serves and serve each row after the first using this permutation. Order the requests in a single row by the algorithm's permutation. No matter what this ordering is, the expected number of requests before $k + 1$ different pages are seen is $(k + 1)H_{k+1}$. The algorithm pays at least one to serve a block of requests to $k + 1$ distinct pages. The algorithm pays $\Omega(w/((k+1)H_{k+1})) = \Omega(w/(k \log k))$ for each row. The adversary picks a permutation that sorts the requests of each row. All the requests for p_1 come first, followed by all requests for p_2, ... Therefore, the adversary pays at most 2 for each row. $\square$

Corollary 1. *The randomized competitive ratio for FIMTP with $t = w - 1$ is $\Omega(\log k + w/(k \log k))$.*

6 Results for FFMTP

Our results for this model are even more negative than those for FIMTP. The following theorem is analogous to Theorem 4. Its proof is similar, and so it is omitted for brevity.

Theorem 8. *There is no competitive randomized algorithm for FFMTP with $t \geq w \geq 2$.*

Corollary 2. *There is no competitive randomized algorithm for FFMTP with* $t = w - 1$ *and* $w > 2$.

Proof. If $w > 2$ then the adversary can simply have one sequence with a only a single request. In that case, we are in exactly the situation of the theorem after the first row. $\square$

Consider the following algorithm for $w = 2$, $t = 1$. The algorithm behaves as ROUND ROBIN MARK until the first sequence ends, and then behaves as MARK. It is not hard to see that this algorithm is $(2H_k + 3)$-competitive. Further, this is optimal to within a constant factor, since Theorem 3 is applicable in this model.

7 Conclusions

We have shown the first randomized lower and upper bounds for multi-threaded paging. In the cases of UIMTP and FFMTP, we shown upper and lower bounds which match to within a constant factor. It would be interesting to find the answers to the following open questions:

1. How can the gaps between the upper and lower bounds be closed for UFMTP and FIMTP? For UFMTP there is a factor of w between the upper and lower bounds. For FIMTP it is unclear how the competitive ratio should depend on w. There is factor of $\Omega(k \log k)$ between the lower bound and the competitive ratio of our algorithm.
2. We use the MARK algorithm of Fiat *et al.* [6] in our algorithms. Could one instead use the H_k-competitive algorithm of McGeoch and Sleator [11]?
3. How can one devise better algorithms for FIMTP with $t = w - 1$ and $w \leq k$? Strejilevich de Loma [17] shows that this case admits an improved competitive ratio in the deterministic setting.

Acknowledgement

The author would like to thank Esteban Feuerstein and Alejandro Strejilevich de Loma for their many comments on this work. He would also like to thank John Noga and Gerhard Woeginger for pointing out errors in the nearly final draft.

References

1. ALBERS, S. On the influence of lookahead in competitive paging algorithms. *Algorithmica 18*, 3 (Jul 1997), 283–305.
2. BEN-DAVID, S., BORODIN, A., KARP, R., TARDOS, G., AND WIGDERSON, A. On the power of randomization in on-line algorithms. *Algorithmica 11*, 1 (Jan 1994), 2–14.

3. BORODIN, A., IRANI, S., RAGHAVAN, P., AND SCHIEBER, B. Competitive paging with locality of reference. *Journal of Computer and System Sciences 50*, 2 (Apr 1995), 244–258.

4. BORODIN, A., LINIAL, N., AND SAKS, M. An optimal online algorithm for metrical task system. *Journal of the Association for Computing Machinery 39*, 4 (Oct 1992), 745–763.

5. FEUERSTEIN, E., AND STREJILEVICH DE LOMA, A. On multi-threaded paging. In *Proceedings of the 7th International Symposium on Algorithms and Computation* (Dec 1996), pp. 417–426.

6. FIAT, A., KARP, R., LUBY, M., MCGEOCH, L., SLEATOR, D., AND YOUNG, N. Competitive paging algorithms. *Journal of Algorithms 12*, 4 (Dec 1991), 685–699.

7. IRANI, S., KARLIN, A., AND PHILLIPS, S. Strongly competitive algorithms for paging with locality of reference. *SIAM Journal on Computing 25*, 3 (Jun 1996), 477–497.

8. KARLIN, A., MANASSE, M., MCGEOCH, L., AND OWICKI, S. Competitive randomized algorithms for nonuniform problems. *Algorithmica 11*, 6 (Jun 1994), 542–571.

9. KARLIN, A., MANASSE, M., RUDOLPH, L., AND SLEATOR, D. Competitive snoopy caching. *Algorithmica 3*, 1 (1988), 79–119.

10. MANASSE, M., MCGEOCH, L., AND SLEATOR, D. Competitive algorithms for server problems. *Journal of Algorithms 11*, 2 (Jun 1990), 208–230.

11. MCGEOCH, L., AND SLEATOR, D. A strongly competitive randomized paging algorithm. *Algorithmica 6*, 6 (1991), 816–825.

12. MOTWANI, R., AND RAGHAVAN, P. *Randomized Algorithms*. Cambridge University Press, 1997.

13. P. CAO, E.W. FELTEN, K. L. Application-controlled file caching policies. In *Proc. of the Summer USENIX Conference* (1994).

14. R. D. BARVE, E. F. GROVE, J. S. V. Application-controlled paging for a shared cache. In *Proc. 36th Symp. on Foundations of Computer Science* (1995), pp. 204–213.

15. SLEATOR, D., AND TARJAN, R. Amortized efficiency of list update and paging rules. *Communications of the ACM 28*, 2 (Feb 1985), 202–208.

16. SLEATOR, D., AND TARJAN, R. Self adjusting binary search trees. *Journal of the ACM 32*, 3 (Jul 1985), 652–686.

17. STREJILEVICH DE LOMA, A. New results on fair multi-threaded paging. In *Proceedings of the 1st Argentine Workshop on Theoretical Informatics* (1997), pp. 111–122.

18. VON NEUMANN, J., AND MORGENSTERN, O. *Theory of games and economic behavior*, 1st ed. Princeton University Press, 1944.

19. WESTBROOK, J. Randomized algorithms for multiprocessor page migration. *SIAM Journal on Computing 23*, 5 (Oct 1994), 951–965.

20. YAO, A. C. C. Probabilistic computations: Toward a unified measure of complexity. In *Proc. 18th IEEE Symposium on Foundations of Computer Science (FOCS)* (1977), pp. 222–227.

21. YOUNG, N. The k-server dual and loose competitiveness for paging. *Algorithmica 11*, 6 (Jun 1994), 525–41.

Determinant: Old Algorithms, New Insights (Extended Abstract)

Meena Mahajan[1] and V Vinay[2]

[1] Institute of Mathematical Sciences, Chennai 600 113, INDIA.
meena@imsc.ernet.in
[2] Department of Computer Science and Automation,
Indian Institute of Science, Bangalore 560 012, INDIA.
vinay@csa.iisc.ernet.in

Abstract. In this paper we approach the problem of computing the characteristic polynomial of a matrix from the combinatorial viewpoint. We present several combinatorial characterizations of the coefficients of the characteristic polynomial, in terms of walks and closed walks of different kinds in the underlying graph. We develop algorithms based on these characterizations, and show that they tally with well-known algorithms arrived at independently from considerations in linear algebra.

1 Introduction

Computing the determinant, or the characteristic polynomial, of a matrix is a problem which has been studied for several years from the numerical analysis viewpoint. In the mid 40's, a series of algorithms which employed sequential iterative methods to compute the polynomial were proposed, the most prominent being due to Samuelson [Sam42], Krylov, Leverier; see, for instance, the presentation in [FF63]. Then, in the 80's, a series of parallel algorithms for the determinant were proposed, due to Csanky, Chistov, Berkowitz [Csa76,Chi85,Ber84]. This culminated in the result, shown independently by several complexity theorists including Vinay, Damm, Toda, Valiant [Vin91,Dam91,Tod91,Val92], that computing the determinant of an integer matrix is complete for the complexity class GapL, and hence computationally equivalent in a precise complexity-theoretic sense to iterated matrix multiplication or matrix powering.

In an attempt to unravel the ideas that went into designing efficient parallel algorithms for the determinant, Valiant studied Samuelson's algorithm and interpreted the computation combinatorially [Val92]. He presented a combinatorial theorem concerning closed walks in graphs, the correctness of which followed from that of Samuelson's algorithm. This was the first attempt to view determinant computations as graph-theoretic rather than linear algebraic manipulations. Inspired by this, and by Straubing's [Str83] purely combinatorial and extremely elegant proof of the Cayley-Hamilton Theorem, Mahajan and Vinay [MV97] described a combinatorial algorithm for computing the characteristic polynomial. The proof of correctness of this algorithm is also purely combinatorial and does not rely on any linear algebra or polynomial arithmetic.

In this paper, we follow up on the work presented in [Val92,Str83,MV97] and present a unifying combinatorial framework in which to interpret and analyse a host of algorithms for computing the determinant and the characteristic polynomial. We first describe what the coefficients of the characteristic polynomial of a matrix M represent as combinatorial entities with respect to the graph G_M whose adjacency matrix is M. We then consider various algorithms for evaluating the coefficients, and in each case we relate the intermediate steps of the computation to manipulation of similar combinatorial entities, giving combinatorial proofs of correctness of these algorithms.

In particular, in the graph-theoretic setting, computing the determinant amounts to evaluating the signed weighted sum of cycle covers. This sum involves far too many terms to allow evaluation of each, and we show how the algorithms of [Sam42,Chi85,Csa76] essentially expand this sum to include more terms i.e., generalizations of cycle covers, which eventually cancel out but which allow easy evaluation. The algorithm in [MV97] uses clow sequences explicitly; Samuelson's method [Sam42] implicitly uses prefix clow sequences; Chistov's method [Chi85] implicitly uses tables of tour sequences; and Csanky's algorithm [Csa76] hinges around Leverier's Lemma (see, for instance, [FF63]), which can be interpreted using loops and partial cycle covers. In each of these cases, we explicitly demonstrate the underlying combinatorial structures, and give proofs of correctness which are entirely combinatorial in nature. In this paper, we omit the combinatiorial proof of Csanky's algorithm for want of space; the full paper carries the proof.

In a sense, this paper parallels the work done by a host of combinatorialists in proving the correctness of matrix identities using the graph-theoretic setting. Foata [Foa65] used tours and cycle covers in graphs to prove the MacMohan master theorem; Straubing [Str83] reproved the Cayley-Hamilton theorem using counting over walks and cycle covers; Garsia [GE], Orlin [Orl78] and Tempereley [Tem81] independently found combinatorial proofs of the matrix-tree theorem and Chaiken [Cha82] generalized the proof to the all-minor matrix-tree theorem; Foata [Foa80] and then Zeilberger [Zei85] gave new combinatorial proofs of the Jacobi identity; Gessel [Ges79] used transitive tournaments in graphs to prove Vandermonde's determinant identity. More recently, Minoux [Min97] showed an extension of the matrix-tree theorem to semirings, again using counting arguments over arborescences in graphs. For beautiful surveys of some of these results, see Zeilberger's paper [Zei85] and chapter 4 of Stanton and White's book on Constructive Combinatorics [SW86]. Zeilberger ends with a host of "exercises" in proving many more matrix identities combinatorially.

Thus, using combinatorial interpretations and arguments to prove matrix identities has been around for a while. To our knowledge, however, a similar application of combinatorial ideas to interpret, or prove correctness of, or even develop new *algorithms* computing matrix functions, has been attempted only twice before: by Valiant [Val92] in 1992, and by the present authors in our earlier paper in 1997 [MV97]. We build on our earlier work and pursue a new thread of ideas here.

This paper is thus a collection of new interpretations and proofs of known results.

2 Matrices, Determinants and Graphs

Let A be a square matrix of dimension n. For convenience, we state our results for matrices over integers, but they apply to matrices over any commutative ring.

We associate matrices of dimension n with complete directed graphs on n vertices, with weights on the edges. Let G_A denote the complete directed graph associated with the matrix A. If the vertices of G_A are numbered $\{1, 2, \ldots, n\}$, then the weight of the edge $\langle i, j \rangle$ is a_{ij}. We use the notation $[n]$ to denote the set $\{1, 2, \ldots, n\}$.

The determinant of the matrix A, $det(A)$, is defined as the signed sum of all weighted permutations of S_n as follows:

$$det(A) = \sum_{\sigma \in S_n} sgn(\sigma) \prod_i a_{i\sigma(i)}$$

where $sgn(\sigma)$ is $+1$ if the number of inversions in σ is even and -1 otherwise. The number of inversions is the cardinality of the set $\{\langle i, j \rangle \mid i < j, \sigma(i) > \sigma(j)\}$.

Each $\sigma \in S_n$ has a cycle decomposition, and it corresponds to a set of cycles in G_A. Such cycles of G_A have an important property: they are all simple (non-intersecting), disjoint cycles; when put together, they touch each vertex exactly once. Such sets of cycles are called cycle covers. Note that cycle covers of G_A and permutations of S_n are in bijection with each other.

We define weights of cycle covers to correspond to weights of permutations. The weight of a cycle is the product of the weights of all edges in the cycle. The weight of a cycle cover is the product of the weights of all the cycles in it. Thus, viewing the cycle cover C as a set of edges, $w(C) = \prod_{e \in C} w(e)$. Since the weights of the edges are dictated by the matrix A, we can write $w(C) = \prod_{\langle i,j \rangle \in C} a_{ij}$.

We can also define the sign of a cycle cover to be consistent with the sign of the corresponding permutation. A cycle cover is even (resp. odd) if it contains an even number (resp. odd) of even length cycles. Equivalently, the cycle cover is even (resp. odd) if the number of cycles plus the number of edges is even (resp. odd). Define sign of a cycle cover C to be $+1$ if C is even, and -1 if C is odd. Cauchy showed that with this definition, the sign of a permutation (based on inversions) and the sign of the associated cycle cover is the same. For our use, this definition of sign based on cycle covers will be more convenient.

Let $C(G_A)$ denote the set of all cycle covers in the graph G_A. Then we have

$$det(A) = \sum_{C \in C(G_A)} sgn(C)w(C) = \sum_{C \in C(G_A)} sgn(C) \prod_{\langle i,j \rangle \in C} a_{ij}$$

Consider the characteristic polynomial of A,

$$\chi_A(\lambda) = det(\lambda I_n - A) = c_0 \lambda^n + c_1 \lambda^{n-1} + \ldots + c_{n-1}\lambda + c_n$$

To interpret these coefficients, consider the graph $G_A(\lambda)$ whose edges are labeled according to the matrix $\lambda I_n - A$. The coefficient c_l collects part of the contribution to $\det(\lambda I_n - A)$ from cycle covers having at least $(n - l)$ self loops. (A self loop at vertex k now carries weight $\lambda - a_{kk}$.) This is because a cycle cover with i self loops has weight which is a polynomial of degree i in λ. For instance, with $n = 4$, consider the cycle cover $\langle 1, 4 \rangle, \langle 2, 2 \rangle, \langle 3, 3 \rangle, \langle 4, 1 \rangle$ in $G_A(\lambda)$. This has weight $(-a_{14})(\lambda - a_{22})(\lambda - a_{33})(-a_{41})$, contributing $a_{14}a_{22}a_{33}a_{41}$ to c_4, $-a_{14}a_{41}(a_{22} + a_{33})$ to c_3, $a_{14}a_{41}$ to c_2, and 0 to c_1.

Following Straubing's notation, we consider partial permutations, corresponding to partial cycle covers. A partial permutation σ is a permutation on a subset $S \subseteq [n]$. The set S is called the domain of σ, denoted $\operatorname{dom}(\sigma)$. The completion of σ, denoted $\hat{\sigma}$, is the permutation in S_n obtained by letting all elements outside $\operatorname{dom}(\sigma)$ be fixed points. This permutation $\hat{\sigma}$ corresponds to a cycle cover C in G_A, and σ corresponds to a subset of the cycles in C. We call such a subset a partial cycle cover $\mathcal{PC}$, and we call C the completion of $\mathcal{PC}$. A partial cycle cover is defined to have the same parity and sign as its completion. It is easy to see that the completion need not be explicitly accounted for in the parity; a partial cycle cover $\mathcal{PC}$ is even (resp. odd) iff the number of cycles in it, plus the number of edges in it, is even (resp. odd).

Getting back to the characteristic polynomial, observe that to collect the contributions to c_l, we must look at all partial cycle covers with l edges. The $n - l$ vertices left uncovered by such a partial cycle cover $\mathcal{PC}$ are the self-loops, from whose weight the λ term has been picked up. Of the l vertices covered, self-loops, if any, contribute the $-a_{kk}$ term from their weight, not the λ term. And other edges, say $\langle i, j \rangle$ for $i \neq j$, contribute weights $-a_{ij}$. Thus the weights for $\mathcal{PC}$ evidently come from the graph G_{-A}. If we interpret weights over the graph G_A, a factor of $(-1)^l$ must be accounted for independently.

Formally,

Definition 1. *A cycle is an ordered sequence of m edges $C = \langle e_1, e_2, \ldots, e_m \rangle$ where $e_i = \langle u_i, u_{i+1} \rangle$ for $i \in [m - 1]$ and $e_m = \langle u_m, u_1 \rangle$ and $u_1 \leq u_i$ for $i \in [m]$ and all the u_i's are distinct. u_1 is called the head of the cycle, denoted $h(C)$. The length of the cycle is $|C| = m$, and the weight of the cycle is $w(C) = \prod_{i=1}^{m} w(e_i)$. The vertex set of the cycle is $V(C) = \{u_1, \ldots, u_m\}$.*

An l-cycle cover C is an ordered sequence of cycles $C = \langle C_1, \ldots, C_k \rangle$ such that $V(C_i) \cap V(C_j) = \phi$ for $i \neq j$, $h(C_1) < \ldots < h(C_k)$ and $|C_1| + \ldots + |C_k| = l$.

The weight of the l-cycle cover is $wt(C) = \prod_{j=1}^{k} w(C_j)$, and the sign is $sgn(C) = (-1)^{l+k}$.

As a matter of convention, we call n-cycle covers simply cycle covers.

Proposition 1. *The coefficients of $\chi_A(\lambda)$ are given by*

$$c_l = (-1)^l \sum_{\substack{C \text{ is an } l\text{-cycle cover in } G_A}} sgn(C)wt(C)$$

3 Summing over permutations efficiently

As noted in Proposition 1, evaluating the determinant (or for that matter, any coefficient of the characteristic polynomial) amounts to evaluating the signed weighted sum over cycle covers (partial cycle covers of appropriate length). We consider four efficient algorithms for computing this sum. Each expands this sum to include more terms which mutually cancel out. The differences between the algorithms is essentially in the extent to which the sum is expanded.

3.1 From Cycle Covers to Clow Sequences

Genaralize the notion of a cycle and a cycle cover as follows:

A *clow* (for closed-walk) is a cycle in G_A (not necessarily simple) with the property that the minimum vertex in the cycle – called the *head* – is visited only once. An l-clow sequence is a sequence of clows where the heads of the clows are in strictly increasing order and the total number of edges (counting each edge as many times as it is used) is l. Formally,

Definition 2. *A clow is an ordered sequence of edges $C = \langle e_1, e_2, \ldots, e_m \rangle$ such that $e_i = \langle u_i, u_{i+1} \rangle$ for $i \in [m-1]$ and $e_m = \langle u_m, u_1 \rangle$ and $u_1 \neq u_j$ for $j \in \{2, \ldots, m\}$ and $u_1 = min\{u_1, \ldots, u_m\}$. The vertex u_1 is called the head of the clow and denoted $h(C)$. The length of the clow is $|C| = m$, and the weight of the clow is $w(C) = \prod_{i=1}^{m} w(e_i)$.*

An l-clow sequence C is an ordered sequence of clows $\mathcal{C} = \langle C_1, \ldots, C_k \rangle$ such that $h(C_1) < \ldots < h(C_k)$ and $|C_1| + \ldots + |C_k| = l$.

The weight of the l-clow sequence is $wt(\mathcal{C}) = \prod_{j=1}^{k} w(C_j)$, and the sign is $sgn(\mathcal{C}) = (-1)^{l+k}$.

Note that the set of l-clow sequences properly includes the set of l-cycle covers on a graph. And the sign and weight of a cycle cover are consistent with its sign and weight when viewed as a clow sequence.

The theorem below establishes a connection between the coefficients of a characteristic polynomial and clow sequences.

Theorem 1 (Theorem 2.1 in [MV97]).

$$c_l = (-1)^l \sum_{C \text{ is an } l\text{-clow sequence}} sgn(\mathcal{C}) wt(\mathcal{C})$$

3.2 Clow Sequences with the Prefix Property: Getting to Samuelson's Method

The generalization from cycle covers to clow sequences has a certain extravagance. The reason for going to clow sequences is that evaluating their wieghted sum is easy, and this sum equals the sum over cycle covers. However, there are several clow sequences which we can drop from consideration without sacrificing ease of computation. One such set arises from the following consideration:

In a cycle cover, all vertices are covered exactly once. Suppose we enumerate the vertices in the order in which they are visited in the cycle cover (following the order imposed by the cycle heads). If vertex h becomes the head of a cycle, then all vertices in this and subsequent cycles are larger than h. So all the lower numbered vertices must have been already visited. So at least $h - 1$ vertices, and hence $h - 1$ edges, must have been covered.

We can require our clow sequences also to satisfy this property. We formalize the prefix property: a clow sequence $C = \langle C_1, \ldots, C_k \rangle$ has the prefix property if for $1 \leq r \leq k$, the total lengths of the clows $C_1, \ldots, C_{r-1}$ is at least $h(C_r) - 1$. A similar prefix property can be formalized for partial cycle covers. Formally,

Definition 3. *An l-clow sequence $C = \langle C_1, \ldots, C_k \rangle$ is said to have the prefix property if it satisfies the following condition:*

$$\forall r \in [k], \quad \sum_{t=1}^{r-1} |C_t| \geq h(C_r) - 1 - (n - l)$$

The interesting fact is that the involution constructed in the previous subsection for clow sequences works even over this restricted set!

Theorem 2 (Theorem 2 in [Val92]).

$$c_l = (-1)^l \sum_{\substack{C \text{ is an } l\text{-clow sequence} \\ \text{with the prefix property}}} sgn(C) wt(C)$$

In [Val92], Valiant observes that prefix clow sequences are the terms computed by Samuelson's method for evaluating $\chi_\lambda(A)$. Hence the correctness of the theorem follows from the correctness of Samuelson's method. And the correctness of Samuelson's method is traditionally shown using linear algebra. A simple alternative combinatorial proof of this theorem is shown in the appendix.

Algorithm using prefix clow sequences To compute c_l using this characterization, we must sum up the contribution of all l-clow sequences with the prefix property. One way is to modify the dynamic programming approach used in the previous sub-section for clow sequences. This can be easily done. Let us instead do things differently; the reason will become clear later.

Adopt the convention that there can be clows of length 0. Then each l-clow sequence C has exactly one clow C_i with head i, for $i = 1$ to n. So we write $C = \langle C_1, \ldots, C_n \rangle$.

Define the signed weight of a clow C as $sw(C) = -w(C)$ if C has non-zero length, and $sw(C) = 1$ otherwise. And define the signed weight of an l-clow sequence as $sw(C) = \prod_{i=1}^{n} sw(C)$. Then $sgn(C)w(C) = (-1)^l sw(C)$. So from the above theorem,

$$c_l = \sum_{\substack{C \text{ is an } l\text{-clow sequence} \\ \text{with the prefix property}}} sw(C)$$

We say that a sequence of non-negative integers $l_1, \ldots, l_n$ satisfies property prefix(l) if

(1). $\sum_{t=1}^{n} l_t = l$, and

(2). For $r \in [n]$, $\sum_{t=1}^{r-1} l_t \geq r - 1 - (n - l)$. Alternatively $\sum_{t=r}^{l} l_t \leq n - r + 1$.

Such sequences are "allowed" as lengths of clows in the clow sequences we construct; no other sequences are allowed.

We group the clow sequences with prefix property based on the lengths of the individual clows. In a clow sequence with prefix property $\mathcal{C}$, if the length of clow C_i (the possibly empty clow with head i) is l_i, then any clow with head i and length l_i can replace C_i in $\mathcal{C}$ and still give a clow sequence satisfying the prefix property. Thus, if $z(i, p)$ denotes the total signed weight of all clows which have vertex i as head and length p, then

$$c_l = \sum_{l_1, \ldots, l_n : \text{prefix}(l)} \prod_{i=1}^{n} z(i, l_i)$$

To compute c_l efficiently, we place the values $z(i, p)$ appropriately in a series of matrices $B_1, \ldots, B_n$. The matrix B_k has entries $z(k, p)$. Since we only consider sequences satisfying prefix(l), it suffices to consider $z(k, p)$ for $p \leq n - k + 1$. Matrix B_k is of dimension $(n - k + 2) \times (n - k + 1)$ and has $z(k, p)$ on the pth lower diagonal as shown below.

$$B_k = \begin{bmatrix} z(k,0) & 0 & 0 & \cdots & 0 & 0 \\ \vdots & \vdots & \vdots & & \vdots & 0 \\ z(k, n-k+1) & z(k, n-k) & z(k, n-k-1) & \cdots & z(k,2) & z(k,1) \end{bmatrix}$$

Now from the equation for c_l, it is clear that

$$[\, c_0\ c_1\ c_2\ c_3\ \cdots\ \cdots\ c_n\,]^T = \prod_{k=1}^{n} B_k$$

It remains now to compute $z(i, p)$, the entries in the B matrices. We know that $z(i, 0) = 1$ and $z(i, 1) = -a_{ii}$. For $p \geq 2$, a clow of length p with head i must first visit a vertex $u > i$, then perform a walk of length $p - 2$ via vertices greater than i to some vertex $v > i$, and then return to i. To construct the path, we exploit the fact that the (j, k)th entry in a matrix A^p gives the sum of the weights of all paths in G_A of length exactly p from j to k. So we must consider the induced subgraph with vertices $i + 1, \ldots, n$. This has an adjacency matrix A_{i+1} obtained by removing the first i rows and the first i columns of A. So $A_1 = A$. Consider the submatrices of A_i as shown below.

$$A_i = \begin{pmatrix} a_{ii} & R_i \\ S_i & A_{i+1} \end{pmatrix}$$

Then the clows contributing to $z(i, p)$ must use an edge in R_i, perform a walk corresponding to A_{i+1}^{p-2}, and then return to i via an edge in S_i. In other words,

$$z(i, p) = -R_i\, A_{i+1}^{p-2}\, S_i$$

So the matrices B_k look like this:

$$B_k = \begin{bmatrix} 1 & 0 & 0 & \cdots & 0 & 0 \\ -a_{kk} & 1 & & \cdots & 0 & 0 \\ -R_k S_k & -a_{kk} & 1 & \cdots & 0 & 0 \\ \vdots & \vdots & \vdots & & \vdots & 0 \\ \vdots & \vdots & \vdots & & \vdots & 0 \\ -R_k A_{k+1}^{n-k-2} S_k & -R_k A_{k+1}^{n-k-3} S_k & -R_k A_{k+1}^{n-k-4} S_k & \cdots & -a_{kk} & 1 \\ -R_k A_{k+1}^{n-k-1} S_k & -R_k A_{k+1}^{n-k-4} S_k & -R_k A_{k+1}^{n-k-3} S_1 & \cdots & -R_k S_k & -a_{kk} \end{bmatrix}$$

This method of computing $\chi_A(\lambda)$ is precisely Samuelson's method [Sam42] (see [FF63,Ber84,Val92]). Samuelson arrived at this formulation using Laplace's Theorem on the matrix $\lambda I - A$, whereas we have arrived at it via clow sequences with the prefix property. This interpretation of the Samuelson-Berkowitz algorithm is due to Valiant[Val92]; the combinatorial proof of correctness (proof of Theorem 2) is new. (It is mentioned, without details, in [MV97].)

3.3 From Clows to Tour Sequences Tables: Getting to Chistov's Algorithm

We now move in the other direction – generalize further beyond clow sequences. Firstly, we relax the condition that the head of a clow may be visited only once. This gives us more generalized closed walks which we call tours. To fix a canonical representation, we do require the edges of the tour to be listed beginning from an occurrence of the head. Since there could be multiple such occurrences, we get different tours with the same multiset of edges. eg the tour corresponding to the vertex sequence 253246 is different from the tour corresponding to the vertex sequence 246253. Secondly, we deal with not just sequences but ordered lists, or tables, of sequences. Within a sequence the tours are ordered by their heads (and all heads are distinct), but there is no ordering amongst the different sequences. And the parity of a tour sequence table depends on the number of sequences in it, not the number of tours in it. A clow sequence is thus a tour sequence table where each sequence contains a single tour which is a clow and the sequences are ordered by their tour heads. Formally,

Definition 4. *A tour is an ordered sequence of edges $C = \langle e_1, e_2, \ldots, e_p \rangle$ such that $e_i = \langle u_i, u_{i+1} \rangle$ for $i \in [p-1]$ and $e_p = \langle u_p, u_1 \rangle$ and $u_1 = min\{u_1, \ldots, u_m\}$. The vertex u_1 is called the head of the tour and denoted $h(C)$. The length of the tour is $|T| = p$, and the weight of the tour is $wt(T) = \prod_{i=1}^{m} w(e_i)$.*

A j-tour sequence $\mathcal{T}$ is an ordered sequence of tours $\mathcal{T} = \langle T_1, \ldots, T_k \rangle$ such that $h(T_1) < \ldots < h(T_k)$ and $|T_1| + \ldots + |T_k| = j$. The weight of the tour sequence is $wt(\mathcal{T}) = \prod_{j=1}^{k} wt(T_j)$, and the length is $|\mathcal{T}| = j$.

An *l-tour sequence table* TST is an ordered sequence of tour sequences $\mathcal{F} = \langle \mathcal{T}_1, \ldots, \mathcal{T}_r \rangle$ such that $|\mathcal{T}_1| + \cdots + |\mathcal{T}_r| = l$. The *weight* of the TST is $wt(\mathcal{F}) = \prod_{j=1}^{r} wt(\mathcal{T}_j)$, and the *sign* is $(-1)^{l+r}$.

The following theorem shows that TSTs can be used to compute the characteristic polynomial.

Theorem 3.

$$c_l = (-1)^l \sum_{\mathcal{F} \text{ is an } l\text{-TST}} sgn(\mathcal{F})wt(\mathcal{F})$$

Proof. We demonstrate an involution on the set of l-TSTs with all l-clow sequences being fixed points, and all other l-TSTs being mapped to TSTs of the same weight but opposing sign. Since l-clow sequences which are not cycle covers also yield a net contribution of zero (Theorem 1), the sum over all l-TSTs is precisely c_l.

Given an l-TST $\mathcal{F} = \langle \mathcal{T}_1, \ldots, \mathcal{T}_r \rangle$, let H be the set of all vertices which occur as heads of some tour in the table. For $S \subseteq H$, we say that S has the *clow sequence property* if the following holds: There is an $i \leq r$ such that
(1). The tour sequences $\mathcal{T}_{i+1}, \ldots, \mathcal{T}_r$ are all single-tour sequences
    (say tour sequence $\mathcal{T}_j$ is the tour T_j),
(2). No tour in any of the tour sequences $\mathcal{T}_1, \ldots, \mathcal{T}_i$ has a head vertex in S,
(3). Each vertex in S is the head of a tour T_j for some $i + 1 \leq j \leq r$. i.e. $\{h(T_j) \mid j = i+1, \ldots, r\} = S$,
(4). The tour sequence table $\langle \mathcal{T}_{i+1}, \ldots, \mathcal{T}_r \rangle$ actually forms a clow sequence.
    i.e. the tours T_j for $i + 1 \leq j \leq r$ are clows, and $h(T_{i+1}) < \cdots < h(T_r)$.
In other words, all tours in $\mathcal{F}$ whose heads are in S are actually clows which occur in a contiguous block of single-tour sequences, arranged in strictly increasing order of heads, and this block is not followed by any other tour sequences in $\mathcal{F}$.

Now, in H, find the smallest vertex v such that $H_{>v} = \{h \in H \mid h > v\}$ has the clow sequence property but $H_{\geq v} = \{h \in H \mid h \geq v\}$ does not.

If no such v exists, then H satisfies the clow sequence property and hence $\mathcal{F}$ is an l-clow sequence. In this case, map it to itself.

If such a v exists, then locate the first tour sequence $\mathcal{T}_i = \langle T_1, \ldots, T_k \rangle$ where v appears (as a head). Then v is the head of the last tour T_k, because all tours with larger heads occur in a contiguous block of single-tour sequences at the end. The tour T_k can be uniquely decomposed as TC, where T is a tour and C a clow, both with head v.

Case 1: $T \neq \phi$. Map this l-TST to an l-TST where $\mathcal{T}_i$ is replaced, at the same position, by the following two tour sequences: $\langle C \rangle, \langle T_1, \ldots, T_{k-1}, T \rangle$. This preserves weight but inverts the sign. In the modified l-TST, the newly introduced sequence containing only C will be chosen for modification as in Case 3.

Case 2: $T = \phi$, and $k > 1$. Map this l-TST to an l-TST where $\mathcal{T}_i$ is replaced, at the same position, by the following two tour sequences: $\langle C \rangle, \langle T_1, \ldots, T_{k-1} \rangle$. This too preserves weight but inverts the sign. In the modified l-TST, the newly introduced sequence containing only C will be chosen for modification as in Case 3.

Case 3: $T = \phi$ and $k = 1$. Then a tour sequence T_{i+1} must exist, since otherwise $H_{\geq v}$ would satisfy the clow sequence property. Now, if T_{i+1} has a tour with head greater than v, then, since $H_{>v}$ satisfies the clow sequence property, the TST $T_{i+1}, \ldots, T_r$ must be a clow sequence. But recall that T has the first occurrence of v as a head and is itself a clow, so then $T_i, \ldots, T_r$ must also be a clow sequence, and $H_{\geq v}$ also satisfies the clow sequence property, contradicting our choice of v. So T_{i+1} must have all tours with heads at most v. Let $T_{i+1} = \langle P_1, \ldots, P_s \rangle$. Now there are two sub-cases depending on the head of the last tour P_s.

Case 3(a). $h(P_s) = v$. Form the tour $P'_s = P_s C$. Map this l-TST to a new l-TST where the tour sequences T_i and T_{i+1} are replaced, at the same position, by a single tour sequence $\langle P_1, \ldots, P_{s-1}, P'_s \rangle$. The weight is preserved and the sign inverted, and in the modified l-TST, the tour P'_s in this new tour sequence will be chosen for modification as in Case 1.

Case 3(b). $h(P_s) \neq v$. Map this l-TST to a new l-TST where the tour sequences T_i and T_{i+1} are replaced, at the same position, by a single tour sequence $\langle P_1, \ldots, P_s, C \rangle$. The weight is preserved and the sign inverted, and in the modified l-TST, the tour C in this new tour sequence will be chosen for modification as in Case 2.

Thus l-TSTs which are not l-clow sequences yield a net contribution of zero.

Algorithm using tour sequence tables We show how grouping the l-TSTs in a carefully chosen fashion gives a formulation which is easy to compute.

Define $e_l = (-1)^l c_l$, then

$$e_l = \sum_{\mathcal{F} \text{ is an } l\text{-TST}} sgn(\mathcal{F})wt(\mathcal{F})$$

To compute c_l and hence e_l using this characterization, we need to compute the contributions of all l-TSTs. This is more easily achieved if we partition these contributions into l groups depending on how many edges are used up in the first tour sequence of the table. Group j contains l-TSTs of the form $\mathcal{F} = \langle T_1, \ldots, T_r \rangle$ where $|T_1| = j$. Then $\mathcal{F}' = \langle T_2, \ldots, T_r \rangle$ forms an $(l-j)$-TST, and $sgn(\mathcal{F}) = -sgn(\mathcal{F}')$ and $wt(\mathcal{F}) = wt(T_1)wt(\mathcal{F}')$. So the net contribution to e_l from this group, say $e_l(j)$, can be factorized as

$$e_l(j) = \sum_{\substack{T: j\text{-tour sequence} \\ \mathcal{F}': (l-j)\text{-TST}}} -sgn(\mathcal{F}')wt(\mathcal{F}')wt(T)$$

$$= -\left(\sum_{T: j\text{-tour sequence}} wt(T) \right) \left(\sum_{\mathcal{F}': (l-j)\text{-TST}} sgn(\mathcal{F}')wt(\mathcal{F}') \right)$$

$$= -d_j e_{l-j}$$

where d_j is the sum of the weights of all j-tour sequences.

Now we need to compute d_j.

It is easy to see that $A^l[1,1]$ gives the sum of the weights of all tours of length l with head 1. To find a similar sum over tours with head k, we must consider the induced subgraph with vertices $k, k+1, \ldots, n$. This has an adjacency matrix A_k obtained by removing the first $k-1$ rows and the first $k-1$ columns of A.

(We have already exploited these properties in Section 3.2.) Let $y(l,k)$ denote the sum of the weights of all l-tours with head k. Then $y(l,k) = A_k^l[1,1]$.

The weight of a j-tour sequence T can be split into n factors: the kth factor is 1 if T has no tour with head k, and is the weight of this (unique) tour otherwise. So

$$d_j = \sum_{0 \le l_i \le j:\ l_1 + \ldots + l_n = j} \prod_{i=1}^n y(l_i, i)$$
$$= \sum_{0 \le l_i \le j:\ l_1 + \ldots + l_n = j} \prod_{i=1}^n A_i^{l_i}[1,1]$$

Let us define a power series $D(x) = \sum_{j=0}^\infty d_j x^j$. Then, using the above expression for d_j, we can write

$$D(x) = \left(\sum_{l=0}^\infty x^l A_1^l[1,1] \right) \left(\sum_{l=0}^\infty x^l A_2^l[1,1] \right) \cdots \left(\sum_{l=0}^\infty x^l A_n^l[1,1] \right)$$

Since we are interested in d_j only for $j \le n$, we can ignore monomials of degree greater than n. This allows us to evaluate the first $n+1$ coefficients of $D(x)$ using matrix powering and polynomial arithmetic. And now e_l can be computed inductively using the following expression:

$$e_l = \sum_{j=1}^l e_l(j) = \sum_{j=1}^l -d_j e_{l-j}$$

But this closely matches Chistov's algorithm [Chi85]! The only difference is that Chistov started off with various algebraic entities, manipulated them using polynomial arithmetic, and derived the above formulation, whereas we started off with TSTs which are combinatorial entities, grouped them suitably, and arrived at the same formulation. And at the end, Chistov uses polynomial arithmetic to combine the computation of $D(x)$ and e_l. The full paper has a self-contained algebraic exposition of Chistov's algorithm to expose the contrast of the two proofs; the reader may also consult [Koz92] for an algebraic treatment.

Acknowledgment

We thank Ashok Subramanian for showing how our combinatorial interpretations and construction in the clow sequences approach extend to Chistov's algorithm as well.

References

[Ber84] S. J. Berkowitz. On computing the determinant in small parallel time using a small number of processors. *Information Processing Letters*, 18:147–150, 1984.

[Cha82] S. Chaiken. A combinatorial proof of the all minors matrix theorem. *SIAM J. Algebraic Discrete Methods*, 3:319–329, 1982.

[Chi85] A. L. Chistov. Fast parallel calculation of the rank of matrices over a field of arbitrary characteristic. In *Proc Int. Conf. Foundations of Computation Theory, LNCS 199*, pages 63–69. Springer, 1985.

[Csa76] L. Csanky. Fast parallel inversion algorithm. *SIAM J of Computing*, 5:818–823, 1976.

[Dam91] C. Damm. DET=L$^{(\#L)}$. Technical Report Informatik–Preprint 8, Fachbereich Informatik der Humboldt–Universität zu Berlin, 1991.

[FF63] D. Fadeev and V. Fadeeva. *Computational Methods in Linear Algebra*. Freeman, San Francisco, 1963.

[Foa65] D. Foata. Etude algébrique de certains problèmes d'analyse combinatoire et du calcul des probabilités. *Publ. Inst. Statist. Univ. Paris*, 14:81–241, 1965.

[Foa80] D. Foata. A combinatorial proof of Jacobi's identity. *Ann. Discrete Math.*, 6:125–135, 1980.

[GE] A. Garsia and Ö. Eğecioğlu. Combinatorial foundations of computer science. unpublished collection.

[Ges79] I. Gessel. Tournaments and Vandermonde's determinant. *J Graph Theory*, 3:305–307, 1979.

[Koz92] D. Kozen. *The Design and Analysis of Algorithms*. Springer-Verlag, New York, 1992.

[Lei92] T. F. Leighton. *Introduction to Parallel Algorithms and Architectures: Arrays, Trees, Hypercubes*. Morgan Kaufmann Publishers Inc., San Mateo, 1992.

[Min97] M. Minoux. Bideterminants, arborescences and extension of the matrix-tree theorem to semirings. *Discrete Mathematics*, 171:191–200, 1997.

[MV97] M. Mahajan and V Vinay. Determinant: combinatorics, algorithms, complexity. *Chicago Journal of Theoretical Computer Science* http://cs-www.uchicago.edu/publications/cjtcs, 1997:5, 1997. A preliminary version appeared as "A combinatorial algorithm for the determinant" in Proceedings of the *Eighth Annual ACM-SIAM Symposium on Discrete Algorithms* SODA97.

[Orl78] J. B. Orlin. Line-digraphs, arborescences, and theorems of Tutte and Knuth. *J. Combin. Theory Ser. B*, 25:187–198, 1978.

[Sam42] P. A. Samuelson. A method of determining explicitly the coefficients of the characteristic polynomial. *Ann. Math. Stat.*, 13:424–429, 1942.

[Str73] V. Strassen. Vermeidung von divisionen. *Journal of Reine U. Angew Math*, 264:182–202, 1973.

[Str83] H. Straubing. A combinatorial proof of the Cayley-Hamilton theorem. *Discrete Maths.*, 43:273–279, 1983.

[SW86] D. Stanton and D. White. *Constructive Combinatorics*. Springer-Verlag, 1986.

[Tem81] H. N. V. Tempereley. *Graph Theory and Applications*. Ellis Horwood, Chichester, 1981.

[Tod91] S. Toda. Counting problems computationally equivalent to the determinant. manuscript, 1991.

[Val92] L. G. Valiant. Why is boolean complexity theory difficult? In M. S. Paterson, editor, *Boolean Function Complexity*. Cambridge University Press, 1992. London Mathematical Society Lecture Notes Series 169.

[Vin91] V Vinay. Counting auxiliary pushdown automata and semi-unbounded arithmetic circuits. In *Proc. 6th Structure in Complexity Theory Conference*, pages 270–284, 1991.

[Zei85] D. Zeilberger. A combinatorial approach to matrix algebra. *Discrete Mathematics*, 56:61–72, 1985.

Solving Fundamental Problems on Sparse-Meshes

Jop F. Sibeyn*

Abstract. A sparse-mesh, which has PUs on the diagonal of a two-dimensional grid only, is a cost effective distributed memory machine. Variants of this machine have been considered before, but none of them is so simple and pure as a sparse-mesh. Various fundamental problems (routing, sorting, list ranking) are analyzed, proving that sparse-meshes have a great potential. The results are extended for higher dimensional sparse-meshes.

1 Introduction

On ordinary two-dimensional meshes we must accept that, due to their small bisection width, for most problems the maximum achievable speed-up with n^2 processing units (*PUs*) is only $\Theta(n)$. On the other hand, networks such as hypercubes impose increasing conditions on the interconnection modules with increasing network sizes. Cube-connected-cycles do not have this problem, but are harder to program due to their irregularity. Anyway, because of a basic theorem from VLSI lay-out [19], all planar architectures have an area that is quadratic in their bisection-width. But this in turn means that, except for several special problems with much locality (the Warshall algorithm most notably), we must accept that the hardware cost is quadratic in the speed-up that we may hope to obtain (later we will also deal with three-dimensional lay-outs). Once we have accepted this, we should go for the simplest and cheapest architecture achieving this, and a good candidate is the sparse-mesh considered in this paper.

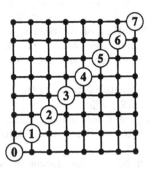

Fig. 1. A sparse-mesh with 8 PUs and an 8 × 8 network of buses. The large circles indicate the PUs with their indices, the smaller circles the connections between the buses.

Network. A *sparse-mesh* with n PUs consists of a two-dimensional $n \times n$ grid of buses, with PUs connected to them at the diagonal. The horizontal buses are called *row-buses*,

* Max-Planck-Institut für Informatik, Im Stadtwald, Saarbrücken, Germany, jopsi@mpi-sb.mpg.de, http://www.mpi-sb.mpg.de/ ~ jopsi/.

the vertical buses are called *column-buses*. PU_i, $0 \leq i < n$, can send data along the i-th row-bus and receive data from the i-th column-bus. In one *step*, PU_i can send one packet of standard length to an arbitrary PU_j. The packet first travels along the i-th row-bus to position (i, j) and then turns into the j-th column-bus. Bus conflicts are forbidden: the algorithm must guarantee that in every step a bus is used for the transfer of only one packet. See Figure 1 for an illustration.

The *standard length* of the packets should be chosen such that the *start-up time*, the time to establish a connection, does *not* dominate the *transfer time*, the time to actually transfer the bytes over the connections. For large systems, this will imply that the standard length increases with n, because of a distance dependent delay. Currently, common startup times are $10\mu s$ or more. Even if this would be reduced by a factor 1000, which would mean a tremendous progress, then still a signal can travel $10 \cdot 10^{-9} \cdot 3 \cdot 10^8 = 3$ meters in one start-up time. So, in comparison to the start-up time, the size of a computer is not very important yet, and for some time to come the standard length may even decrease.

In comparison to a mesh, we have reduced the number of PUs from n^2 to n, thus saving substantially on the hardware cost. In comparison with other networks, we have a very simple interconnection network that can be produced easily and is scalable without problem. The sparse-mesh is very similar to the Parallel Alternating Direction Machine, *PADAM*, considered in [2, 3] and the coated-mesh considered in [10]. Though similar in inspiration, the sparse-mesh is simpler. In Section 6 the network is generalized for higher dimensions.

Problems. Parallel computation is possible only provided that the PUs can exchange data. Among the many possible communication patterns, patterns in which each PU sends and receives at most h packets, *h-relations*, have attracted most attention. *h*-relations are of outstanding importance by their role as a basic communication operation in PRAM-simulation of nicely distributed problems and in the BSP-model and the variants thereof. An *h*-relation is called *balanced*, if every PU sends h/n packets to each PU.

Lemma 1 *On a sparse-mesh with n PUs, balanced h-relations can be performed in h steps.*

Unfortunately, not all *h*-relations are balanced, and thus there is a need for algorithms that route arbitrary *h*-relations efficiently. Also for cases that the PUs have to route less than n packets, the above algorithm does not work. Offline, all *h*-relations can be routed in h steps. The algorithm proceeds as follows:

Algorithm OFFLINE_ROUTE

1. Construct a bipartite graph with n vertices on both sides. Add an edge from Node i on the left to Node j on the right for each packet going from PU_i to PU_j.
2. Color this *h*-regular bipartite graph with h colors.
3. Perform h routing steps. Route all packets whose edges got Color t, $0 \leq t < h$, in Step t.

This coloring idea is standard since [1]. Its feasibility is guaranteed by Hall's theorem. As clearly at least h steps are required for routing an *h*-relation, the offline algorithm is optimal. This shows that, in principle, the *h*-relation routing problem is nothing more than the problem of constructing a conflict-free bus-allocation schedule.

In [3] routing h-relations is considered for the PADAM. This network is so similar to the sparse-mesh, that most results carry over. It was shown that for $h \geq n$, h-relations can be routed in $\mathcal{O}(h)$ time. Further it was shown that 1-optimality can be achieved for $h = \omega(n \cdot \log n \cdot \log \log n)$. Here and in the remainder we call a routing algorithm c-*optimal*, if it routes all h-relations in at most $c \cdot (h + o(h))$ time.

Other fundamental problems that should be solved explicitly on any network, are sorting and list ranking. Sorting has many functions, in parallel computation it can also be utilized for all kinds of rearrangements. List ranking is a key-subroutine in many important problems on graphs and trees. With h keys/nodes per PU, their communication complexity is comparable to that of an h-relation. In [3], the list-ranking problem was solved asymptotically optimally for $h = \omega(n \cdot \log n \cdot \log \log n)$.

New Results. In this paper we address the problems of routing, sorting and list ranking. Applying randomization, it is rather easy to route n^ϵ-relations, $0 < \epsilon \leq 1$, $(2/\epsilon)$-optimally. An intricate deterministic algorithm is even $(1/\epsilon)$-optimal. This might be the best achievable. We present a deterministic sampling technique, by which all routing algorithms can be turned into sorting algorithms with essentially the same time consumption. In addition to this, we give an algorithm for ranking randomized lists, that runs in $6/\epsilon \cdot (h + o(h))$ steps for lists with $h = n^\epsilon$ nodes per PU. This appears to be a strong result. Our generalization of the sparse-mesh to higher-dimensions is based on the generalized diagonals that were introduced in [6]. At least for three dimensions this makes practical sense: n^2 PUs are interconnected by a cubic amount of hardware, which means an asymptotically better ratio than for two-dimensional sparse-meshes. On higher dimensional sparse-meshes everything becomes much harder, because it is no longer true that all one-relations can be routed in a single step. We consider the presented deterministic routing algorithm for higher dimensional sparse-meshes to be the most interesting of the paper.

All results constitute considerable improvements over those in [3]. The results of this paper demonstrate that networks like the sparse-mesh are versatile, not only in theory, but even with a great practical potential: we show that for realistic sizes of the network, problems with limited parallel slackness can be solved efficiently. All proofs are omitted due to a lack of space.

2 Randomized Routing

Using the idea from [20] it is easy to obtain a randomized 2-optimal algorithm for large h. The algorithm consists of two *randomizations*, routings in which the destinations are randomly distributed. In Round 1, all packets are routed to randomly chosen intermediate destinations; in Round 2, all packets are routed to their actual destinations.

Lemma 2 *On a sparse-mesh with n PUs, h-relations can be routed in $2 \cdot h + o(h)$ steps, for all $h = \omega(n \cdot \log n)$, with high probability.*

We show how to route randomizations for $h = \omega(\sqrt{n} \cdot \log n)$. First the PUs are divided into $\sqrt{n}$ subsets of PUs, $S_0, \ldots, S_{\sqrt{n}-1}$, then we do the following:

Algorithm RANDOM_ROUTE
1. Perform $\sqrt{n} - 1$ supersteps. In Superstep t, $1 \leq t < \sqrt{n}$, PU_i in S_j, $0 \leq i, j < \sqrt{n}$, sends all its packets with destination in $S_{(j+t) \bmod \sqrt{n}}$ to PU_i in $S_{(j+t) \bmod \sqrt{n}}$.
2. Perform $\sqrt{n} - 1$ supersteps. In Superstep t, $1 \leq t < \sqrt{n}$, PU_i in S_j, $0 \leq i, j < \sqrt{n}$, sends all its packets with destination in $PU_{(i+t) \bmod \sqrt{n}}$ in S_j to their destinations.

This approach can easily be generalized. For $h = \omega(n^\epsilon \cdot \log n)$, the algorithm consists of $1/\epsilon$ routing rounds. In Round r, $1 \leq r \leq 1/\epsilon$, packets are routed to the subsets, consisting of $n^{1-\epsilon \cdot r}$ PUs, in which their destinations lie.

Theorem 1 *On a sparse-mesh with n PUs,* RANDOM_ROUTE *routes randomizations with $h = \omega(n^\epsilon \cdot \log n)$ in $(h + o(h))/\epsilon$ steps, with high probability. Arbitrary h-relations can be routed in twice as many steps.*

3 List Ranking and Edge Coloring

List ranking is the problem of determining the rank for every node of a set of linked lists. [5] devotes a chapter to this problem. By *rank* we mean the distance to the final node of its list. h denotes the number of nodes per PU.

The edges of a regular bipartite graph of degree m can be colored by splitting the graph in two subgraphs $\log m$ times, each with half the previous degree [4]. Lev, Pippenger and Valiant [11] have shown that each halving step can be performed by solving two problems that are very similar to list ranking (determining the distance to the node with minimal index on a circle). Thus, if list-ranking is solved in time T, then bipartite graphs of degree m can be colored in time $\mathcal{O}(\log m \cdot T)$. In [3] list ranking on PADAMs is performed by simulating a work-optimal PRAM algorithm. For coloring the eges of a bipartite regular graph, the above idea was used:

Lemma 3 *[3] On a sparse-mesh with n PUs, list ranking can be solved in $\mathcal{O}(h)$ steps, for all $h \geq n \cdot \log \log n$. A bipartite regular graph with $h \cdot n$ edges, can be colored in $\mathcal{O}(\log(h \cdot n) \cdot h)$ steps.*

The algorithms are asymptotically optimal, but the hidden constants are quite bad, and the range of applicability is limited to large h. In the following we present a really good and versatile randomized list-ranking algorithm. It is based on the repeated-halving algorithm from [14]. This algorithm has the unique property, that not only the number of participating nodes is reduced in every round, but that the size of the processor network is reduced as well. Generally, the value of this property is limited, but on the sparse-mesh, where we need a certain minimal h for efficient routing, this is very nice.

3.1 Repeated-Halving

It is assumed that the nodes of the lists are randomly distributed over the PUs. If this is not the case, they should be randomized first. The algorithm consists of $\log n$ reduction steps. In each step, first the set of PUs is divided in two halves, S_0 and S_1. The nodes in S_0 with current successor in S_1 and vice-versa, are called *masters*, the other nodes are *non-masters*. The current successor of a node is stored in its *crs* field. For the time being, we assume that each PU holds $h = \omega(n \cdot \log n)$ nodes.

Algorithm REDUCE

1. Each non-master p follows the links until a master or a final node is reached and sets $crs(p)$ to this node.
2. Each master p asks $p' = crs(p)$ for $crs(p')$.
3. Each master p asks $p' = crs(p)$ for $crs(p')$.

In a full algorithm one must also keep track of the distances. After Step 2, each master in S_i, $i = 0$ or 1, points to the subsequent master in $S_{(i+1) \bmod 2}$. Thus, after Step 3, each master in S_i, points to the subsequent master in S_i itself. Now recursion can be applied on the masters. This does not involve communication between S_0 and S_1 anymore. Details are provided in [14]. The expected number of masters in each subset equals $h \cdot n/4$. Thus, after $\log n$ rounds, we have n subproblems of expected size h/n, that can be solved internally. Hereafter, the reduction must be reversed:

Algorithm EXPAND

1. Each non-master p asks $p' = mst(p)$ for $crs(p')$.

For Step 1 of REDUCE we repeat pointer-jumping rounds as long as necessary. In every such round, each node p who has not yet reached a master or a final node, asks $p' = crs(p)$ for $crs(p')$ and the distance thereto. Normally pointer-jumping is inefficient, but in this case, because the expected length of the lists is extremely short, the number of nodes that participates decreases rapidly with each performed round [16].

Lemma 4 *On a sparse-mesh with n PUs, Step 1 of REDUCE can be performed in $2.5 \cdot (h + o(h))$, for all $h = \omega(n \cdot \log n)$, with high probability.*

Lemma 5 *On a sparse-mesh with n PUs, Step 2 and Step 3 of REDUCE and EXPAND can be performed in $3/2 \cdot h + o(h)$ steps for all $h = \omega(n \cdot \log n)$, with high probability.*

Theorem 2 *On a sparse-mesh with n PUs, list ranking can be solved in $14 \cdot h + o(h)$ steps, for all $h = \omega(n \cdot \log n)$, with high probability.*

For smaller h, the average number of participating nodes per PU decreases to less than 1. This is not really a problem: for such a case, the maximum number of participating nodes in any PU can easily be estimated on $\mathcal{O}(\log n)$, and the routing operations can be performed in $\mathcal{O}(\log^2 n)$. In comparison to the the total routing time, the cost of these later reduction rounds is negligible. Using Theorem 1, this gives

Corollary 1 *On a sparse-mesh with n PUs, list ranking can be solved in $14/\epsilon \cdot (h+o(h))$ steps, for all $h = \omega(n^\epsilon \cdot \log n)$, with high probability.*

3.2 Sparse-Ruling-Sets

The performance of the previous algorithm can be boosted by first applying the highly efficient sparse-ruling-sets algorithm from [13] to reduce the number of nodes by a factor of $\omega(1)$. Practically, this does not always make sense (see [17] for an analysis of list-ranking on the Intel Paragon), but for sufficiently large n, it gives a great improvement of the performance. We summarize the main ideas of the algorithm.

Algorithm SPARSE_RULING_SETS

1. In each PU randomly select h' nodes as rulers.
2. The rulers initiate waves that run along their list until they reach the next ruler or a final element. If a node p is reached by a wave from a ruler p', then p sets $mst(p) = p'$.
3. The rulers send their index to the ruler whose wave reached them.

By a *wave*, we mean a series of sending operations in which a packet that is destined for a node p is forwarded to $crs(p)$. Hereafter, we can apply the previous list-ranking algorithm for ranking the rulers. Finally each non-ruler p asks $mst(p)$ for the index of the last element of its list and the distance thereof. Details can be found in [13], particularly the initial nodes must be handled carefully.

Theorem 3 *On a sparse-mesh with n PUs, list ranking can be solved in $6/\epsilon \cdot (h + o(h))$ steps, for all $h = \omega(n^\epsilon \cdot \log n)$, with high probability. Under the same conditions, a bipartite regular graph with $h \cdot n$ edges, can be colored in $\mathcal{O}(\log(h \cdot n) \cdot \epsilon \cdot (h + o(h)))$ steps.*

4 Faster Routing

Our goal is to construct a deterministic $\mathcal{O}(1/\epsilon)$-optimal algorithm for routing n^ϵ-relations. As in [3], we apply the idea from [18] for turning offline routing algorithms into online algorithms by solving Step 2 of OFFLINE_ROUTE after sufficient reduction of the graph online. However, here we obtain much more interesting results.

Let $h = \omega(n^\epsilon \cdot \log n)$, and define f by $f(n) = (h/n^\epsilon \cdot \log n)^{1/2}$. As in RANDOM_ROUTE, we are going to perform $1/\epsilon$ rounds. In Round r, $1 \leq r \leq 1/\epsilon$, all packets are routed to the subsets, consisting of $n^{1-\epsilon \cdot r}$ PUs each, in which their destinations lie. We describe Round 1. First the PUs are divided in n^ϵ subsets, $S_0, \ldots, S_{n^\epsilon-1}$, each consisting of $n^{1-\epsilon}$ PUs. Then, we perform the following steps:

Algorithm DETERMINE_DESTINATION

1. Each PU_i, $0 \leq i < n$, internally sorts its packets on the indices of their destination subsets. The packets going to the same S_j, $0 \leq j < n^\epsilon$, are filled into superpackets of size $f \cdot \log n$, leaving one partially filled or empty superpacket for each j. The superpackets p, going to the same S_j, are numbered consecutively and starting from 0 with numbers a_p. α_{ij} denotes the total number of superpackets going from S_i to S_j.

2. For each j, $0 \leq j < n^\epsilon$, the PUs perform a parallel prefix on the α_{ij}, to make the numbers $A_{ij} = \sum_{i' < i} \alpha_{ij}$ available in PU_i, $0 \leq i < n$.

3. For each superpacket p in PU_i, $0 \leq i < n$, with destination in S_j, $0 \leq j < n^\epsilon$, set $dest_p = (A_{ij} + a_p) \bmod n^{1-\epsilon}$.

For each superpacket p with destination in S_j, $dest_p$ gives the index of the PU in S_j, to which p is going to be routed first. Because the PUs in S_j are the destination of $h \cdot n^{1-\epsilon}/(f \cdot \log n) + n$ superpackets, the bipartite graph with n nodes on both sides and one edge for every superpacket has degree $h/(f \cdot \log n) + n^\epsilon$. Its edges can be colored in $o(h)$ time. If the edge corresponding to a superpacket p gets Color t, $1 \leq t < h/(f \cdot \log n) + n^\epsilon$, then p is going to be routed in superstep t.

Lemma 6 *On a sparse-mesh with n PUs, and $h = \omega(n^\epsilon \cdot \log n)$ all packets can be routed to their destination subsets of size $n^{1-\epsilon}$ in $h + o(h)$ steps.*

Repeating the above steps $1/\epsilon$ times, the packets eventually reach their destinations:

Theorem 4 *On a sparse-mesh with n PUs, h-relations with $h = \omega(n^\epsilon \cdot \log n)$ can be routed in $(h + o(h))/\epsilon$ steps.*

Our algorithm is not entirely deterministic: the underlying list-ranking algorithm is randomized. It is likely that deterministic list-ranking for a problem with $h = n^\epsilon$ nodes per PU can be performed in $\mathcal{O}(h/\epsilon)$ time. But, for our main claim, that n^ϵ-relations can be routed $(1/\epsilon)$-optimally, we do not need this: if $h = \omega(n^\epsilon \cdot \log^2 n)$ the size of the superpackets can be taken $f \cdot \log^2 n$, and we can simply apply pointer-jumping for the list-ranking.

5 Faster Sorting

On meshes the best deterministic routing algorithm is more or less a sorting algorithm [9, 8]. For sparse-meshes the situation is different: the routing algorithm presented in Section 4 is in no way a sorting algorithm. However, it can be enhanced to sort in essentially the same time.

We first consider $h = w(n \cdot \log n)$. This case also gives the final round in the sorting algorithm for smaller h hereafter. Define $f(n) = (h/(n \cdot \log n))^{1/2}$, and let $m = h/f = n \cdot f \cdot \log(n)$. In [15] it is shown how to deterministically select a high-quality sample. The basic steps are:

Algorithm REFINED_SAMPLING

1. Each PU internally sorts all its keys and selects those with ranks $j \cdot h/m$, $0 \leq j < m$, to its sample.
2. Perform $\log n$ merge and reduce rounds: in Round r, $0 \leq r < \log n$, the samples, each of size m, in two subsets of 2^r PUs are merged, and only the keys with even ranks are retained.
3. Broadcast the selected sample $\mathcal{M}$ of size m to all PUs.

Lemma 7 *On a sparse-mesh with n PUs, REFINED_SAMPLING selects a sample of size m in $o(h)$ steps. The global rank, $Rank_p$, of an element p, with rank $rank_p$ among the elements of $\mathcal{M}$ satisfies*

$$h \cdot n/m \cdot rank_p - \mathcal{O}(h \cdot n/m \cdot \log n) \leq Rank_p \leq h \cdot n/m \cdot rank_p + \mathcal{O}(h \cdot n/m \cdot \log n).$$

Thus, by merging the sample and its packets, a PU can estimate for each of its packets p its global rank $Rank_p$ with an error bounded by $\mathcal{O}(h \cdot n/m \cdot \log n) = \mathcal{O}(h/f) = o(h)$. This implies that if we set

$$dest_p = \lfloor n \cdot rank_p/m \rfloor,$$

for each packet p, where $rank_p$ is defined as above, that then at most $h + o(h)$ packets have the same *dest*-value. Thus all packets can be routed to the PU given by their *dest*-values in $h + o(h)$ steps. Furthermore, after each PU has sorted all the packets it received, the packets can be rearranged in $o(h)$ steps so that each PU holds exactly h packets.

Lemma 8 *On a sparse-mesh with n PUs, the sorting algorithm based on REFINED_SAMPLING sorts an h-relation in $h + o(h)$ steps, for all $h = w(n \cdot \log n)$.*

For $h = w(n^\epsilon \cdot \log n)$, we stick close to the pattern of the routing algorithm from Section 4: we perform $1/\epsilon$ rounds of bucket-sort with buckets of decreasing sizes. Define $f(n) = (h/(n^\epsilon \cdot \log n))^{1/2}$. For each round of the bucket sorting, we run REFINED_SAMPLING with $m = h/f = n^\epsilon \cdot f \cdot \log(n)$. This sample selection takes $\mathcal{O}(m) = o(h)$ steps. In Round r, $1 \leq r \leq 1/\epsilon$, the sample is used to guess in which subset of size $n^{1-r/\epsilon}$ each packet belongs. Then they are routed to these subsets with the algorithm of Section 4.

Theorem 5 *On a sparse-mesh with n PUs, the sorting algorithm based on REFINED_SAMPLING sorts an h-relation in $(h + o(h))/\epsilon$ steps, for all $h = w(n^\epsilon \cdot \log n)$.*

The remark at the end of Section 4 can be taken over here: $h = w(n^\epsilon \cdot \log^2 n)$ is required for making the algorithm deterministic, but this has no impact on the claim that $(1/\epsilon)$-optimality can be achieved for $h = n^\epsilon$.

6 Higher Dimensional Sparse-Meshes

One of the shortcomings of the PADAM [2, 3] is that it has no natural generalization for higher-dimensions. The sparse-mesh can be generalized easily.

Definition 1 *A subset of a d-dimensional $n \times \cdots \times n$ grid is a diagonal, if the projection of all its points on any of the $(d-1)$-dimensional coordinate planes is a bijection.*

For this we need the following definition (a simplification of the definition in [6]):

$$Diag_d = \{(x_0, x_1, \ldots, x_{d-1}) | 0 \leq x_0, x_1, \ldots, x_{d-1} < n,$$
$$\text{and } (x_0 + x_1 + \cdots + x_{d-1}) \bmod n = 0\}.$$

Lemma 9 [6] *In a d-dimensional $n \times \cdots \times n$ grid, $Diag_d$ is a diagonal.*

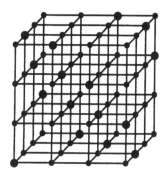

Fig. 2. A three-dimensional $4 \times 4 \times 4$ sparse-mesh: 16 PUs, such that if they were occupied by towers in a three-dimensional chess game, none of them could capture another one.

The d-dimensional sparse-mesh, consists of the n^{d-1} PUs on $Diag_d$, interconnected by $d \cdot n^{d-1}$ buses of length n. The PU at position $(x_0, \ldots, x_{d-1})$ is denoted $PU_{x_0, \ldots, x_{d-1}}$. The routing is *dimension-ordered*. Thus, after t substeps, $0 \leq t < d$, a packet traveling from $PU_{x_0, \ldots, x_{t-1}, x_t, \ldots, x_{d-1}}$ to $PU_{x'_0, \ldots, x'_{t-1}, x'_t, \ldots, x'_{d-1}}$ has reached position $(x'_0, \ldots, x'_{t-1}, x_t, \ldots, x_{d-1})$. The scheduling should exclude bus conflicts. $Diag_3$ is illustrated in Figure 2.

6.1 Basic Results

Now it is even harder to find a conflict-free allocation of the buses. Different from before, it is no longer automatically true, that every one-relation can be routed offline in one step (but see Lemma 12!). For example, in a four-dimensional sparse-mesh, under the permutation $(x_0, x_1, x_2, x_3) \mapsto (x_2, x_3, x_0, x_1)$, all packets at the positions $(0, 0, x_2, n - x_2)$, $0 \leq x_2 < n$, have to pass through position $(0, 0, 0, 0)$. However, for the most common routing pattern, balanced h-relations, this is no problem, because it can be written as the composition of n^{d-1} shifts. A *shift* is a routing pattern under which, for all $x_0, \ldots, x_{d-1}$, the packets in $PU_{x_0, \ldots, x_{d-1}}$ have to be routed to $PU_{(x_0+s_0) \bmod n, \ldots, (x_{d-1}+s_{d-1}) \bmod n}$, for some given $s_0, \ldots, s_{d-1}$.

Lemma 10 *A shift in which each PU contributes one packets can be routed in one step.*

Lemma 10 implies that the randomized routing algorithm from Section 2 carries on with minor modifications. One should first solve the case of large $h = \mathcal{O}(n^{d-1} \cdot \log n)$ and then extend the algorithms to smaller h:

Theorem 6 *On a d-dimensional sparse-mesh with n^{d-1} PUs, the generalization of* RANDOM_ROUTE *routes randomizations with $h = \omega(n^{\epsilon} \cdot \log n)$ in $(d-1) \cdot (h + o(h))/\epsilon$ steps, with high probability. Arbitrary h-relations can be routed in twice as many steps.*

The algorithm for ranking randomized lists of Section 3 has the complexity of a few routing operations, and this remains true. The same holds for our technique of Section 5 for enhancing a routing algorithm into a sorting algorithm with the same complexity. The only algorithm that does *not* generalize is the deterministic algorithms of Section 4, because it is not based on balanced h-relations.

6.2 Deterministic Routing

In this section we analyze the possibilities of deterministic routing on higher-dimensional sparse-meshes. Actually, for the sake of a simple notation, we describe our algorithms for two-dimensional sparse-meshes, but we will take care that they consist of a composition of shifts.

For $h = \omega(n \cdot \log n)$, we can match the randomized result. The algorithm is a deterministic version of the randomized one: it consists of two phases. In the first phase the packets are smoothed-out, in the second they are routed to their destinations. *Smoothing-out* means rearranging the packets so that every PU holds approximately the same number of packets with destinations in each of the PUs. Let $f = (h/(n \cdot \log n))^{1/2}$.

Algorithm DETERMINISTIC_ROUTE

1. Each PU_i, $0 \le i < n$, internally sorts its packets on the indices of their destination PU. The packets going to the same PU_j, $0 \le j < n$, are filled into superpackets of size $f \cdot \log n$, leaving one partially filled or empty superpacket for each j.
2. Construct a bipartite graph with n nodes on both sides, and an edge from Node i on the left to Node j on the right, for every superpacket going from PU_i to PU_j. Color this graph with $(f+1) \cdot n$ colors. A superpacket p with Color j, $0 \le j < (f+1) \cdot n$ has $dest_p = j \bmod n$.
3. Perform $n-1$ shifts. In Shift t, $1 \le t < n$, each PU_i, $0 \le i < n$, routes the $f+1$ superpackets p with $dest_p = (i+t) \bmod n$ to $PU_{(i+t) \bmod n}$.
4. Perform $n-1$ shifts. In Shift t, $1 \le t < n$, each PU_i, $0 \le i < n$, routes the $f+1$ superpackets with destination in $PU_{(i+t) \bmod n}$ to this PU.

The algorithm works, because the coloring assures that for each source and destination PU there are exactly $f+1$ packets with the same *dest*-value. This implies that both routings are balanced.

Theorem 7 *On a d-dimensional sparse-mesh with n^{d-1} PUs* DETERMINISTIC_ROUTE *routes h-relations with $h = \omega(n^{d-1} \cdot \log n)$ in $2 \cdot (h + o(h))$ steps.*

For smaller h, ideas from DETERMINISTIC_ROUTE are combined with ideas from RANDOM_ROUTE: the algorithm consists of rounds in which the packets are smoothed-out and then routed to the subsets in which their destinations lie. The fact that this second phase must be balanced, implies that the smoothing must be almost perfect. Another essential observation is that if packets are redistributed among subsets of PUs, and if this routing has to be balanced, that then the number of subsets should not exceed h. Let $h = w(n^\epsilon \cdot \log n)$, and set $f = (h/n)^{1/2}$. The PUs are divided in n^ϵ subsets S_i of $n^{1-\epsilon}$ PUs each. Then all packets are routed to the subset in which their destinations lie, by performing

Algorithm DETERMINISTIC_ROUTE

1. Each PU_i, $0 \le i < n$, internally sorts its packets on the indices of their destination PU. The packets going to the same S_j, $0 \le j < n^\epsilon$, are filled into superpackets of size $f \cdot \log n$, leaving one partially filled or empty superpacket for each j.

2. Construct a bipartite graph with n^ϵ nodes on both sides, and an edge from Node i on the left to Node j on the right, for every superpacket going from S_i to S_j. Color this graph with $(f+1) \cdot n$ colors. For a superpacket p with Color j, $0 \le j < (f+1) \cdot n$, set $dest_p = j \bmod n^\epsilon$.

3. In each S_i, $0 \le i < n^\epsilon$, determine for all j, $0 \le j < n^\epsilon$, the rank, $rank_p$, of each packet p with $dest_p = j$ among the packets p' with $dest_{p'} = j$.

4. Rearrange the packets within the S_i, $0 \le i < n^\epsilon$, such that thereafter a packet p with $rank_p = j$, $0 \le j < (f+1) \cdot n^{1-\epsilon}$ stands in $PU_{j \bmod n^{1-\epsilon}}$ of S_i.

5. Perform $n^\epsilon - 1$ shifts. In Shift t, $1 \le t < n^\epsilon$, each PU_i, $0 \le i < n^{1-\epsilon}$, in S_j, $0 \le j < n^\epsilon$, routes the $f + 1$ superpackets p with $dest_p = (j + t) \bmod n^\epsilon$ to PU_i in $S_{(j+t) \bmod n^\epsilon}$.

6. In each S_i, $0 \le i < n^\epsilon$, determine for all j, $0 \le j < n^\epsilon$, the rank, $rank_p$, of each packet p with destination in S_j among the packets with destination in this subset.

7. Rearrange the packets within the S_i, $0 \le i < n^\epsilon$, such that thereafter a packet p with $rank_p = j$, $0 \le j < (f+1) \cdot n^{1-\epsilon}$ stands in $PU_{j \bmod n^{1-\epsilon}}$ of S_i.

8. Perform $n^\epsilon - 1$ shifts. In Shift t, $1 \le t < n^\epsilon$, each PU_i, $0 \le i < n^{1-\epsilon}$, in S_j, $0 \le j < n^\epsilon$, routes the $f + 1$ superpackets with destination in $S_{(j+t) \bmod n^\epsilon}$ to PU_i in this subset.

The ranks can be computed as in Section 4 in $\mathcal{O}(n^\epsilon \cdot \log n) = o(h)$ steps. The coloring guarantees that

Lemma 11 *After Step 4, for each j, $0 \le j < n^\epsilon$, every PU holds exactly $f + 1$ super-packets p with $dest_p = j$. After Step 7, for each j, $0 \le j < n^\epsilon$, every PU holds exactly $f + 1$ superpackets p with destination in S_j.*

The rearrangements within the subsets are routings with a large h as described by Theorem 7 (the superpackets do not need to be filled into supersuperpackets!), and thus they can be performed without further recursion.

Theorem 8 *On a d-dimensional sparse-mesh with n^{d-1} PUs DETERMINISTIC_ROUTE routes h-relations with $h = w(n^\epsilon \cdot \log n)$ in $(6 \cdot (d-1)/\epsilon - 4) \cdot (h + o(h))$ steps.*

¿From an algorithmic point of view DETERMINISTIC_ROUTE is the climax of the paper: all developed techniques are combined in a non-trivial way, to obtain a fairly strong result. As in Section 4, $h = w(n^\epsilon \cdot \log^2 n)$ is required for also making the coloring deterministic.

6.3 Three-Dimensional Sparse-Meshes

Three-dimensional sparse-meshes are practically the most interesting generalization. It is a lucky circumstance that for them even the results from Section 4 carry on, because of the following

Lemma 12 *On a three-dimensional sparse-mesh, each one-relation can be routed in a single step.*

Lemma 12 leads to the following analogue of Theorem 4:

Theorem 9 *On a three-dimensional $n \times n \times n$ sparse-mesh, h-relations with $h = \omega(n^\epsilon \cdot \log n)$ can be routed in $(2 \cdot h + o(h))/\epsilon$ steps.*

The coated-mesh [10] can be trivially generalized for higher dimensions. However, whereas routing on a two-dimensional coated-mesh is almost as easy as on a sparse-mesh, there is no analogue of Lemma 12 for three-dimensional coated-meshes.

7 Conclusion

Our analysis reveals that the sparse-mesh has a very different character than the mesh. Whereas on a mesh several approaches essentially give the same result, we see that on a sparse-mesh, there is a great performance difference. For $h = n^\epsilon$, column-sort performs poorly, because the operations in subnetworks are costly, while on a mesh they are for free. On a mesh also the randomized algorithm inspired by [20] performs optimal [12, 7], because there the worst-case time consumption is determined by the time the packets need to cross the bisection. For the sparse-mesh this argument does not apply, and our deterministic routing algorithm is twice as fast.

References

1. Baumslag, M., F. Annexstein, 'A Unified Framework for Off-Line Permutation Routing in Parallel Networks,' *Mathematical Systems Theory*, 24(4), pp. 233–251, 1991.
2. Chlebus, B.S., A. Czumaj, L. Gąsieniec, M. Kowaluk, W. Plandowski, 'Parallel Alternating-Direction Access Machine,' *Proc. 21st Mathematical Foundations of Computer Science*, LNCS. 1113, pp. 267–278, Springer-Verlag, 1996.
3. Chlebus, B.S., A. Czumaj, J.F. Sibeyn, 'Routing on the PADAM: Degrees of Optimality,' *Proc. 3rd Euro-Par Conference*, LNCS. 1300, pp. 272–279, Springer-Verlag, 1997.
4. Cole, R., J. Hopcroft, 'On Edge Coloring Bipartite Graphs,' *SIAM Journal on Computing*, 11, pp. 540–546, 1982.
5. JáJá, J., *An Introduction to Parallel Algorithms*, Addison-Wesley Publishing Company, Inc., 1992.
6. Juurlink, B., P.S. Rao, J.F. Sibeyn, 'Worm-Hole Gossiping on Meshes and Tori,' *Proc. 2nd Euro-Par Conference*, LNCS 1123, pp. 361–369, Springer-Verlag, 1996.
7. Kaufmann, M., J.F. Sibeyn, 'Randomized Multipacket Routing and Sorting on Meshes,' *Algorithmica*, 17, pp. 224-244, 1997.
8. Kaufmann, M., J.F. Sibeyn, T. Suel, 'Derandomizing Algorithms for Routing and Sorting on Meshes,' *Proc. 5th Symposium on Discrete Algorithms*, pp. 669–679, ACM-SIAM, 1994.
9. Kunde, M., 'Block Gossiping on Grids and Tori: Deterministic Sorting and Routing Match the Bisection Bound,' *Proc. European Symposium on Algorithms*, LNCS 726, pp. 272–283, Springer-Verlag, 1993.
10. Leppänen V., M. Penttonen, 'Work-Optimal Simulation of PRAM Models on Meshes,' *Nordic Journal of Computing*, 2, pp. 51–69, 1995.

11. Lev, G., N. Pippenger, L.G. Valiant, 'A Fast Parallel Algorithm for Routing in Permutation Networks,' *IEEE Transactions on Computers*, 30(2), pp. 93–100, 1981.
12. Rajasekaran, S., '*k-k* Routing, *k-k* Sorting, and Cut-Through Routing on the Mesh,' *Journal of Algorithms*, 19(3), pp. 361–382, 1995.
13. Sibeyn, J.F., 'List Ranking on Interconnection Networks,' *Proc. 2nd Euro-Par Conference*, LNCS 1123, pp. 799–808, Springer-Verlag, 1996.
14. Sibeyn, J.F., 'Better Trade-offs for Parallel List Ranking,' *Proc. 9th Symposium on Parallel Algorithms and Architectures*, pp. 221–230, ACM, 1997.
15. Sibeyn, J.F., 'Sample Sort on Meshes,' *Proc. 3rd Euro-Par Conference*, LNCS 1300, pp. 389–398, Springer-Verlag, 1997.
16. Sibeyn, J.F., 'From Parallel to External List Ranking,' *Techn. Rep. MPI-I-97-1021*, Max-Planck Institut für Informatik, Saarbrücken, Germany, 1997.
17. Sibeyn, J.F., F. Guillaume, T. Seidel, 'Practical Parallel List Ranking,' *Proc. 4th Symposium on Solving Irregularly Structured Problems in Parallel*, LNCS 1253, pp. 25–36, Springer-Verlag, 1997.
18. T. Suel, Permutation Routing and Sorting on Meshes with Row and Column Buses, *Parallel Processing Letters*, 5, pp. 63–80, 1995.
19. Thompson, C.D., 'Area-Time Complexity for VLSI,' *Proc. 11th Symposium on Theory of Computing*, pp. 81–88, ACM, 1979.
20. Valiant, L.G., G.J. Brebner, 'Universal Schemes for Parallel Communication,' *Proc. 13th Symposium on Theory of Computing*, pp. 263–277, ACM, 1981.

Output-Sensitive Cell Enumeration in Hyperplane Arrangements

Nora Sleumer

Institut für Theoretische Informatik
ETH Zentrum, 8092 Zürich, Switzerland
sleumer@inf.ethz.ch

Abstract. We present a simple and practical algorithm for enumerating the set of cells C of an arrangement of m hyperplanes. For fixed dimension its time complexity is $O(m \cdot |C|)$. This is an improvement by a factor of m over the reverse search algorithm by Avis and Fukuda. The algorithm needs little space, is output-sensitive, straightforward to parallelize and the implementation is simple for all dimensions.

1 Introduction

A finite set of m distinct hyperplanes partitions the d-dimensional Euclidean space $\mathbb{R}^d$ into faces of all dimensions. These convex polyhedral regions collectively define an *arrangement*. Listing the faces of such an arrangement is a fundamental problem of computational geometry. Enumerating the d-faces, called *cells*, is the key building block for the construction of arrangements. The number of cells is $O(m^d)$, though it can be considerably smaller for certain classes of arrangements. The cell information can be used to enumerate the acyclic orientations of a graph [BVS+93]. It has applications such as hidden surface removal [McK87] and robot motion planning. By duality, solutions to higher-dimensional problems formulated for configurations of point sets can be found by solving the corresponding problem for arrangements [Ede87].

Until recently two enumeration algorithms were known that run in time proportional to the maximum number of cells in an arrangement. Topological sweeps [EG86] of two-dimensional arrangements can be used to enumerate all the cells in $O(m^d)$ time. The incremental algorithm [EOS86] additionally determines the incidences between all the faces of an arrangement but needs space proportional to the size of the output, which is $O(m^d)$. Both use symbolic perturbation to handle degenerate cases, an approach that creates empty cells, thus increasing the number of cells that are handled to the maximum possible in an arrangement.

Avis and Fukuda [AF96] developed the first output-sensitive algorithm for the problem. We improved it by a factor of m to obtain a time complexity of $O(m \cdot |C|)$, for fixed dimension d. Our algorithm has low memory requirements, is easy to implement for all dimensions if a linear programming solver is available, and it is well-suited for running in parallel. It runs in time proportional to the size of the input and output and the output is produced at a steady rate as the algorithm proceeds. It is also simple to adapt the algorithm to constrain the area in which the cells are to be enumerated.

2 Background

Avis and Fukuda [AF96] developed a *reverse search* method that runs in time $O(m \cdot d \cdot \text{lp}(m, d) \cdot |C|)$, where $\text{lp}(m, d)$ is the time needed to solve an $m \times d$ linear program. Their algorithm has a time complexity of $O(m^2 \cdot |C|)$ when the dimension d is fixed. We have increased the efficiency of its two main procedures, thereby improving the overall time complexity by a factor of m.

Consider the state space of all the cells that could possibly exist in an arrangement. We can identify a cell by the following method. Each hyperplane $h_i, 1 \le i \le m$, is given by a linear equality $h_i := \{x : a_i x = b_i\}$ where a_i is the i'th row of an $m \times d$ matrix A, and b_i the i'th element of a column vector b of size m. We do not make any assumptions about the hyperplanes except that they are distinct. For each point in $\mathbb{R}^d$ we define a *sign vector* of length m consisting of $+$, 0 and $-$ signs. Position i in the sign vector of a point x is a $+$ if x lies on the positive side of hyperplane i, $(a_i x > b_i)$, a 0 if it lies on the hyperplane and a $-$ if it lies on the negative side, $(a_i x < b_i)$. We identify a cell with the sign vector of its points, which consists only of $+$ and $-$ signs, when the meaning is clear. The set of indices of the hyperplanes on whose positive (negative) side a cell c lies is called the positive (negative) support of the cell and is denoted by c^+ (c^-). We pick one cell to be the *root cell* and reverse the orientations of some of the hyperplanes, if necessary, so its sign vector consists only of $+$'s.

Two cells are called *neighbors* if only one hyperplane separates them, in other words, if their sign vectors differ in exactly one element. A cell can have at most m neighbors. A *parent* cell of c is a unique neighbor which has one more $+$ in its sign vector and a *child* cell of c is any neighbor of which it is the parent. The output of the algorithm is the set of all sign vectors representing the cells.

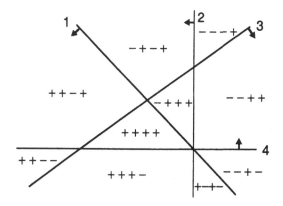

Fig. 1. An arrangement of four lines with sign vectors.

The state space consists of 2^m possible sign vectors of which at most $O(m^d)$ can exist in an arrangement. The state space can be searched efficiently by assigning to each cell (except for the root cell) a unique parent cell, thereby obtaining a directed tree structure for the cells. The reverse search algorithm traverses this tree backwards, enumerating the cells exactly once. One procedure, called the *parent search* (ParentSearch), is needed to identify the unique parent of a cell, and another procedure, the *adjacency oracle* (AllAdj), is necessary to find all the neighbors of a cell. The traversal visits each cell once so during this visit we can output its sign vector to a file. At each level lower in the search tree one element in the sign vector must change from a + to a −. This property bounds the height of the search tree by m.

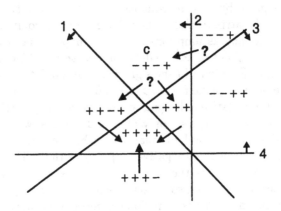

Fig. 2. Is $++-+$ or $-+++$ the parent of c ? Is c the parent of $---+$?

For example, consider the cell $c = -+-+$ in Figure 2. Its positive and negative supports are $c^+ = \{2, 4\}$ and $c^- = \{1, 3\}$, respectively, and it is bounded by hyperplanes 1, 2 and 3. In order to obtain a useful structure on the set of cells we need to answer two questions: Who are the neighbors of c? Which cell is the parent of c? The first question is answered by the adjacency oracle and the second by the parent search function.

3 The Algorithm

When Avis and Fukuda first applied the reverse search method to the cell enumeration problem they gave implementations of the two main procedures, the adjacency oracle and the parent search function, both of which we improved to obtain a better overall time complexity. Where they used a lexicographic approach, we applied a geometric method, thereby decreasing the time required.

The procedure ParentSearch returns the sign vector of the parent cell. The procedure AllAdj returns a list of the indices of the non-redundant hyperplanes that bound a cell, which is enough to identify the neighbors. These hyperplanes are called *tight*. When the procedure CellEnum is called with the root cell, it generates all the cells of an arrangement. This reverse search algorithm differs from the first one in that we generate all the neighbors at once instead of one at a time.

Procedure CellEnum(c);
Input: a cell c represented as a sign vector
             the hyperplanes given by the matrix A, b
Output: all cells in the subtree rooted at c
begin
    output c;
(a) *adjlist* := AllAdj(c);   (* find tight hyperplanes *)
    for each hyperplane h in *adjlist* **do**   (* for each neighbor *)
      if $(h \in c^+)$ **then**   (* if c is on the positive side of h *)
        $e := c$; $e[h] := -$;   (* create a new sign vector *)
(b)   **if** (ParentSearch(e) $= c$) **then** (* if it is a child *)
(c)   CellEnum(e);   (* recurse on it *)
        endif
      endif
    endfor
end.

The overall time complexity depends on the time needed for the adjacency oracle and the parent search. We propose a new parent search function exploiting interior points explicitly. When this is combined with an efficient implementation of the adjacency oracle, the running time of the whole algorithm is improved by a factor of m.

3.1   An Interior Point

Throughout the algorithm it is assumed that there is a function IntPt(c) which returns a unique interior point of a cell c. Such a point can be found by solving the following linear program:

$$\begin{aligned}
\text{maximize } & z \\
\text{subject to } \quad a_s \cdot x + z &\leq b_s \quad && \forall s \in c^+ \\
-a_t \cdot x + z &\leq -b_t \quad && \forall t \in c^- \\
z &\leq 1 \quad && (\text{* or any other constant *})
\end{aligned}$$

The hyperplanes are oriented so that c is on the positive side and the interior point is the value of x for which z is maximum. The linear program is of size $m \times d$ so the time complexity is $O(\mathrm{lp}(m, d))$.

3.2   The Parent Search

The purpose of the parent search is to identify a unique parent cell for each cell, except for the root cell. In Avis and Fukuda's algorithm this required m linear programs of size $m \times d$ to be solved in the worst case. We use a geometric construction for which only one linear program needs to be solved, namely, the linear program for determining a unique point inside each cell. The interior point r of the root cell is a global variable which is calculated once at the beginning.

The parent of a cell c is found as follows: Connect the interior point p of cell c to the interior point r of the root cell. The first cell that is hit when traversing the line segment pr is defined to be the parent cell of c. (See Figure 3.) In case of a tie, symbolic perturbation of p can be used.

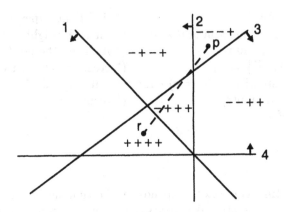

Fig. 3. The cell $-+-+$ is the parent of $---+$.

Procedure *ParentSearch(c)*;
Input: a cell c represented as a sign vector
          the hyperplanes given by the matrix A, b
Output: the sign vector f representing the unique parent cell of c
begin
    $p := \text{IntPt}(c)$;   (* calculate the interior point of c *)
    find the hyperplane h_k which intersects pr closest to p;
    $f^+ := c^+ \cup \{k\}$; $f^- := c^- \setminus \{k\}$; (* set the sign vector of the parent cell *)
    return f;
end.

It is guaranteed that $k \in c^-$ since there must be a hyperplane separating the interior point of the root cell r from p, so there is an intersection of a hyperplane with the line pr between r and p. Now r must lie on the positive side of all hyperplanes separating the root cell from c so p must lie on the negative side of these hyperplanes; in particular, p must lie on the negative side of the hyperplane h_k that intersects line pr at the point closest to p.

The parent search finds an interior point and calculates m intersections. The time needed for this is $O(\text{lp}(m, d) + m \cdot d)$. This is a reduction by a factor of m over the previous parent search. The space needed for the parent search function is that needed for the linear program.

3.3 The Adjacency Oracle

In the reverse search algorithm of Avis and Fukuda the adjacency oracle finds the neighbors one at a time, requiring m linear programs of size $m \times d$ to be solved. By trading some space for time, the neighbors can be found more efficiently all at once. We use a method similar to the one described in [OSS95] which finds the l non-redundant bounding hyperplanes of a polyhedron in time $O(m \operatorname{lp}(l, d))$. These tight hyperplanes separate a cell from its neighbors so once we find this set of hyperplanes, we know the sign vectors of the adjacent cells.

We first find the interior point p of a cell and an initial set V of d hyperplanes that determine a vertex of c, by using the raindrop algorithm as presented in [BFM97], for example. For each of the other hyperplanes, we determine whether it is tight w.r.t. the hyperplanes in V whose orientation is determined by c, denoted by $c|V$, by solving a linear program of size $|V| \times d$. If it is, then we have a witness point that shows that the hyperplane is tight. The witness point q is obtained by moving the hyperplane away from the interior point and optimizing in that direction. We shoot a ray from the interior point p to the point q, find the intersection closest to p and the first hyperplane hit must be tight w.r.t. the whole cell. It is added to V. If there are several hyperplanes that cross the line pq at the same point closest to p then p can be perturbed symbolically to identify one which is tight w.r.t. the whole cell. If the initial hyperplane tested is not the first one hit then it needs to be tested for tightness again. At the end V contains all the indices of the hyperplanes which are tight w.r.t. the polyhedron represented by the cell, and on the other side of each of these hyperplanes lies a neighboring cell.

Consider the example in Figure 4. If lines 1 and 2 determine the vertex and line 3 is tested then the first line hit will be line 4, which bounds the cell.

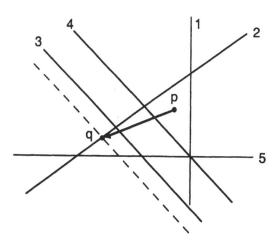

Fig. 4. Determining the bounding hyperplanes of a cell.

The pseudocode is as follows:

Procedure AllAdj(c);
Input: a cell c represented as a sign vector
the hyperplanes given by the matrix A, b
Output: the set V of indices which identifies the neighboring cells of c
begin
    $p := \text{IntPt}(c)$;   (* calculate the interior point of c *)
    $V := \text{Raindrop}(c, p)$; (* an initial set of vertex-defining hyperplanes *)
    $U := \{1, \ldots, m\}\backslash V$;
    while $U \neq \emptyset$
        select a $j \in U$;   (* for each hyperplane not yet considered *)
        if (h_j is tight w.r.t. $c|V$) **then** (the witness point is q)
            find the hyperplane $h_k, k \in U$, which intersects pq closest to p;
            $U := U\backslash\{k\}$; $V := V\cup\{k\}$;
        else
            $U := U\backslash\{j\}$;
        endif
    endwhile
    return V;
end.

The time complexity of AllAdj consists of the time needed for finding the interior point, $O(\text{lp}(m, d))$, the time needed for the raindrop algorithm, $O(m \cdot d^2)$ and the time required to test each hyperplane for tightness w.r.t. the bounding hyperplanes that have been found already. The latter is bounded by $O(\text{lp}(l, d))$ where l is the maximum number of neighbors of a cell. Thus, the total time complexity is $O(\text{lp}(m, d) + m \cdot d^2 + m \cdot \text{lp}(l, d))$, and the space requirement is that needed for solving an $m \times d$ linear program.

3.4 Complexity of the Cell Enumeration Algorithm

The output is produced at a steady rate as the algorithm progresses, and we will prove that it is a *polynomial-rate* algorithm, that is, the first k cells are produced in time k times a polynomial in the size of the input.

Theorem 1. *The total time complexity of our algorithm for enumerating the set C of cells of an arrangement of m hyperplanes in $\mathbb{R}^d$ is*

$$O((d \cdot \text{lp}(m, d) + m \cdot d^2 + m \cdot \text{lp}(l, d)) \cdot |C|),$$

where l is the maximum number of neighbors of a cell and $\text{lp}(m, d)$ indicates the time (or space) complexity needed for solving an $m \times d$ linear program. The worst-case space complexity is

$$O(m^2 + m \cdot d + \text{lp}(m, d)).$$

Proof. The procedure CellEnum is called once for every cell. Its complexity is dominated by the time needed for lines (a) and (b). Let l be the maximum number of neighbors a cell has in an arrangement. The AllAdj procedure, whose time complexity is $O(\mathrm{lp}(m, d) + m \cdot d^2 + m \cdot \mathrm{lp}(l, d))$, is called exactly once for each cell. The ParentSearch procedure, which has a time complexity of $O(\mathrm{lp}(m, d) + m \cdot d)$, is called once for each $(d-1)$-face in the arrangement. There are $O(d)$ times as many $(d-1)$-faces as cells, so ParentSearch is called $O((\mathrm{lp}(m, d) + m \cdot d) \cdot d \cdot |C|)$ times. Adding these two we get the resulting complexity of $O((d \cdot \mathrm{lp}(m, d) + m \cdot d^2 + m \cdot \mathrm{lp}(l, d)) \cdot |C|)$.

The algorithm can recurse at most m times, since at every level in the recursion (c), a $+$ in the sign vector is turned into a $-$. Thus the maximum amount of space needed to store the neighbors of each cell in the path of the search tree that is being investigated is at most $O(m^2)$. In addition to this $O(m \cdot d)$ space is needed for the input plus the space requirements of the linear program, giving the total space complexity.

Theorem 2. *If the dimension d is fixed, then the algorithm has a time complexity of*

$$O(m \cdot |C|).$$

It is a polynomial-rate algorithm with a complexity of

$$O(m \cdot l \cdot k),$$

for the k'th cell, $k \leq |C|$, and it needs space $O(m^2)$.

Proof. If we fix the dimension we can apply Megiddo's linear programming algorithm [Meg84] whose time complexity is linear in the number of constraints so $\mathrm{lp}(m, d)$ becomes $O(m)$ in our analysis. Since the time needed to test the tightness of each hyperplane is now linear in the number of neighbors, we can sum it over all the neighbors of the cells. The total number of neighbor relations in the whole arrangement is two times the number of $(d-1)$-faces, which is $O(d)$ times the number of cells. So the work needed to find all the neighbors is $O(m \cdot |C|)$. The total time complexity then reduces to $O(m \cdot |C|)$.

To show that the algorithm is polynomial-rate we need to consider how many calls can be made to the AllAdj and ParentSearch procedures before we reach the k'th cell. AllAdj is called once for each of the $k-1$ previous cells that were calculated before. ParentSearch can be called $l-1$ times with negative results in line (b), though it is not very likely. This brings us to a time complexity of $O((m \cdot d^2 + m \cdot \mathrm{lp}(l, d) + l \cdot \mathrm{lp}(m, d)) \cdot k)$, for the k'th cell. If we keep d fixed and use a linear-time algorithm such as Megiddo's for the linear programming then this reduces to $O(m \cdot l \cdot k)$. The space requirement for Meggido's algorithm is linear in the number of constraints (hyperplanes) but we still need $O(m^2)$ space for remembering the neighbors in the recursion.

4 Implementation and Applications

We have implemented the algorithm in C using the dual simplex method of the CDD package [Fuk95] for solving the linear programs. Even though we cannot assert that the simplex algorithm runs in time linear in the number of constraints, as experience has shown, the worst-case instances rarely arise, and this is also true in our case. The algorithm has also been implemented in parallel using ZRAM, a library of parallel algorithms [BMFN96]. The following table shows the times required for one processor and the speedup obtained with random instances on 10 to 150 processors.

m	d	# cells	1 proc (s)	10 speedup	50 speedup	100 speedup	150 speedup
10	4	386	9	6	12	12	12
15	4	1,941	69	8	24	25	33
20	4	6,196	299	9	33	50	60
25	4	15,276	924	9	40	62	84
30	4	31,931	2350	9	42	73	107
35	4	59,536	5140	10	45	82	122

For instances with less than 2000 cells it is not efficient to use more than 50 processors; for instances with more than 6000 cells even using 150 processors is efficient.

The duality between a configuration of points and an arrangement of hyperplanes means that problems of either type can be transformed into problems of the other type. Indeed, many problems that are formulated for point configurations are easier to approach as problems about arrangements. This is especially true for problems in higher dimensions so this area will provide interesting applications [Ede87].

An orientation of the edges of a graph is said to be acyclic if the resulting directed graph contains no directed cycle. By a well-known relation [BVS+93], it is possible to enumerate the acyclic orientations of graphs by transforming the problem into a cell enumeration problem. Each hyperplane represents an edge of the graph, the dimension is equal to the number of vertices, and it can be shown that the sign vectors of the cells give us directly the acyclic orientations of the graph.

5   Conclusions and Future Work

A tool based on a simple, efficient and robust algorithm for computing arrangements can be used to increase our understanding of arrangements and many other applications. Our algorithm is a big step towards an arrangement engine capable of computing all the faces of large-scale arrangements.

We would like to compare our time and space complexities with implementations of the incremental algorithm and the topological sweep algorithm, using degenerate and non-degenerate instances of arrangements, but we have not been able to find any implementations, neither for two nor for higher dimensions. Indeed, O'Rourke mentions in the Computational Geometry Column [O'R96] "Holes in [the collection of available software] include code for constructing arrangements."

Aside from the experiments mentioned above, we would like to extend the algorithm to enumerate the faces of all dimensions, use exact arithmetic and visualize two- and three-dimensional arrangements to investigate the properties of arrangements.

Acknowledgments

I am deeply indebted to Komei Fukuda for introducing me to the subject and for the many fruitful discussions. The ideas for finding the interior point and for efficiently finding all neighbors at once are due to him, as well as the linear programming code of the CDD package and a lot of useful tips.

Thanks also go to Lutz Kettner and Frank Wagner for their constructive and motivating criticism.

References

[AF96]   D. Avis and K. Fukuda. Reverse search for enumeration. *Discrete Applied Mathematics*, 65:21–46, 1996.

[BFM97]   D. Bremner, K. Fukuda, and A. Marzetta. Primal-dual methods for vertex and facet enumeration. In *Proceedings of the 13th Annual ACM Symposium on Computational Geometry (SoCG)*, pages 49–56, 1997.

[BMFN96]   A. Brüngger, A. Marzetta, K. Fukuda, and J. Nievergelt. *The Parallel Search Bench ZRAM and its Applications.* To appear in: *Annals of Operations Research.* PS file available from ftp://ftp.ifor.ethz.ch/pub/fukuda/reports, 1996.

[BVS+93]   A. Bjorner, M. Las Vergnas, B. Sturmfels, N. White, and G. Ziegler. *Oriented Matroids*, volume 46. Cambridge University Press, 1993.

[Ede87]   H. Edelsbrunner. *Algorithms in Combinatorial Geometry.* Springer-Verlag, Heidelberg, 1987.

[EG86]   H. Edelsbrunner and L. Guibas. Topologically sweeping an arrangement. In *Proceedings of the 18th ACM Symposium on the Theory Computing (STOC)*, pages 389–403, 1986.

[EOS86]   H. Edelsbrunner, J. O'Rourke, and R. Seidel. Constructing arrangements of lines and hyperplanes with applications. *SIAM Journal on Computing*, 15(2):341–363, 1986.

[Fuk95]   K. Fukuda. *cdd+ Reference Manual.* PS file available from ftp://ftp.ifor.ethz.ch/pub/fukuda/cdd, 1995.

[McK87]   M. McKenna. Worst-case optimal hidden-surface removal. *ACM Transactions on Graphics*, 6:19–28, 1987.

[Meg84]   N. Megiddo. Linear programming in linear time when dimension is fixed. *Journal of the ACM*, 31:114–127, 1984.

[O'R96]   J. O'Rourke. Computational geometry column 28. *International Journal of Computational Geometry and Applications*, 6(2):243–244, 1996.

[OSS95]   T. Ottmann, S. Schuierer, and S. Soundaralakshmi. Enumerating extreme points in higher dimensions. In *Proceedings of the 12th Symposium on the Theoretical Aspects of Computer Science (STACS)*, volume 900 of *LNCS*, pages 562–570. Springer-Verlag, 1995.

Fast and Efficient Computation of Additively Weighted Voronoi Cells for Applications in Molecular Biology

Hans-Martin Will

Institut für Theoretische Informatik, ETH Zürich,
CH-8092 Zürich, Switzerland
will@inf.ethz.ch

Abstract. This paper is concerned with the efficient computation of additively weighted Voronoi cells for applications in molecular biology. We propose a projection map for the representation of these cells leading to a surprising insight into their geometry. We present a randomized algorithm computing one such cell amidst n other spheres in expected time $O(n^2 \log n)$. Since the best known upper bound on the complexity such a cell is $O(n^2)$, this is optimal up to a logarithmic factor. However, the experimentally observed behavior of the complexity of these cells is linear in n. In this case our algorithm performs the task in expected time $O(n \log^2 n)$. A variant of this algorithm was implemented and performs well on problem instances from molecular biology.

1 Introduction

Consider a set of spheres $S = \{\sigma_i, 1 \leq i \leq n\}$, each sphere $\sigma_i = (c_i, r_i)$ defined by its center c_i and radius r_i. If we denote by $d(\cdot, \cdot)$ the Euclidean metric, then we can introduce the distance function $d(x, \sigma) = d(x, c_i) - r_i$. For a point x outside σ this function measures the distance to the surface of the sphere. Finally, we can assign to each of the spheres $\sigma_i \in S$ the set of all points "nearer" to σ_i than to all other spheres by defining

$$V_i = V(\sigma_i) = \left\{ x \in R^3 : \ d(x, \sigma_i) \leq d(x, \sigma_j) \forall 1 \leq j \leq n \right\}.$$

The set V_i is called the additively weighted Voronoi cell of the sphere σ_i, the collection of all the cells $\mathbf{V} = \{V_i, 1 \leq i \leq n\}$ the additively weighted Voronoi tesselation induced by S. We will also refer to V_i as the cell defined by σ_i. This paper aims for the efficient computation of these cells.

1.1 Applications in molecular biology

Additively weighted Voronoi cells have been proposed as a local measure of the packing density of large macro-molecules such as proteins and nucleic acids. This density, in turn, is used to derive entropy-related energy terms or as a screening method for the verification of experimentally gained crystal structures. For an extensive discussion of the advantages of additively weighted Voronoi diagrams compared to other spatial tesselations see the recent paper [10]. For further discussions see [8, 22].

1.2 Previous work

All previous attempts to actually compute spatial additively weighted Voronoi cells for applications in molecular biology discretized the problem [9, 11, 15, 26], or they tried to adjust the vertices of the Voronoi cells of the centers of the spheres [27].

Aurenhammer observed that a cell of the additively weighted Voronoi diagram in R^d can be represented as the projection of the intersection of a cell of a suitably defined power diagram in R^{d+1} with a $d+1$-dimensional cone [2]. If we use an optimal algorithm for computing convex hulls [6], we obtain an $O(n^2)$ algorithm for computing the implicit representation of one or all n additively weighted Voronoi cells in R^3. In practice, however, it turns out to be quite complicated to actually extract an explicit description of the cells from this implicit representation, and we are not aware of any complete implementation of this algorithm in R^3. Therefore, we sought for an approach suitable for an actual implementation.

Finally, there are general methods for computing lower envelopes of algebraic surfaces. In fact, our algorithm is based on ideas taken from [5, 4, 20]. These algorithms have running times — depending on the type of the geometric objects involved — between $O(n^2 \log n)$ and $O(n^{2+\epsilon})$, for arbitrarily small positive ϵ. All these algorithms make use of a vertical decomposition scheme. Since it was reported that using vertical decompositions of spherical planar maps in the context of molecular modelling leads to numerical problems [13], we decided to search for a data structure which does not distinguish a specific direction. Again, we are not aware of an implementation of any of these algorithms for non-linear input objects.

1.3 Outline of paper

We begin our presentation by recalling some characteristic properties of additively weighted Voronoi cells and propose a projection map for their representation. With the new insights we gain from this representation, we are prepared to formulate an incremental algorithm for the computation of these cells. Besides theoretical considerations, we will outline the setup in which this algorithm is used for the intended applications. We conclude with first experiences obtained from an implementation.

2 Geometric properties

The geometric properties of additively weighted Voronoi cells can be found in [3, 18, 21]: A non-empty cell is a star-shaped region, and the center of its defining sphere is contained in the (visibility) kernel. The faces are hyperbolic surface patches and the edges are elliptic or hyperbolic arcs, which may degenerate to parabolic arcs or line segments.

Since a non-empty cell is a star-shaped region, it is a natural approach to represent its surface using a spherical parameterization. We chose to parameterize a cell V_i by a unit sphere around the center c_i of its defining sphere σ_i. Observe, that the diagram does not change if we add a constant Δr to the radii of all spheres in S. To simplify the following presentation, we consider a cell V defined by a sphere $\sigma \in S$ of radius 0 around the origin.

Let $\pi : p \to \frac{p}{\|p\|}$ denote the map projecting a point $p \in f$, f a face of V, onto the parameter range S^2. The collection $P := \{\pi(f) : f \text{ is face of } V\}$ is a subdivision of S^2. For any element $x \in P$ let $\phi(x)$ denote its lifting back to the current boundary of the cell. As it is common parlance, we identify a map with its image.

Lemma 1. *The projection of an edge in $V(\sigma) \cap V(\sigma_i) \cap V(\sigma_j)$ of an additively weighted Voronoi cell $V(\sigma)$ onto a unit sphere around the center c of σ is a circular arc.*

Proof: Let f be a face of V that is defined by a neighboring sphere $\sigma_i = (c_i, r_i)$. Then the lifting map ϕ_i such that $\phi_i(p) = x$ is given by

$$\phi_i : p \mapsto \frac{c_i^2 - r_i^2}{2\left(r_i + \langle p, c_i \rangle\right)} \cdot p.$$

We obtain this mapping by plugging the equations $x = d \cdot p$ and $\|p\| = 1$ into the equation of the graph of the distance function $(d + r_i)^2 = (x - c_i)^2$, where $d = d(x, \sigma_i)$. Equating the distance equations

$$d = \frac{c_i^2 - r_i^2}{2\left(r_i + \langle p, c_i \rangle\right)} = \frac{c_j^2 - r_j^2}{2\left(r_j + \langle p, c_j \rangle\right)}$$

for two neighboring spheres $\sigma_i = (c_i, r_i)$ and $\sigma_j = (c_j, r_j)$, $1 \leq i, j \leq n$, $i \neq j$ yields the equation describing the projection of $\phi_i \cap \phi_j$:

$$\langle a_{i,j}, p \rangle = b_{i,j}, \quad \text{where}$$
$$a_{i,j} = c_i \cdot \left(c_j^2 - r_j^2\right) - c_j \cdot \left(c_i^2 - r_i^2\right)$$
$$b_{i,j} = r_i \cdot \left(r_j^2 - c_j^2\right) - r_j \cdot \left(r_i^2 - c_i^2\right)$$

$\square$

We denote the halfspaces defined in this way by $h_{i,j} = \{x \in R^3 : \langle a_{i,j}, x \rangle \geq b_{i,j}\}$. In the same manner, the boundary of the projection of a bisecting hyperboloid $V(\sigma) \cap V(\sigma_i)$ is a circle on S^2, namely the circle defined by the vanishing points of the denominator of the lifting map ϕ_i satisfying $\langle p, c_i \rangle = -r_i$. We set $h_{i,0} = \{x \in R^3 : \langle c_i, x \rangle \leq -r_i\}$, $h_{0,i} = \{x \in R^3 : \langle c_i, x \rangle \geq -r_i\}$. To simplify notation, we define a symbolic lifting map $\phi_0 : p \to \infty$. For an illustration of this representation see figure 1.

3   An incremental algorithm

Let $\tau(n)$ denote a function bounding the expected complexity of the additively weighted Voronoi cell $V(\sigma)$ amidst n other spheres. The best known upper bound on the worst case complexity of such a cell is $O(n^2)$ [2]. On the other hand, the interior of a protein possesses a packing density comparable to crystal structures [16]. Probablistic models of crystals and quasi-crystals based on additively weighted Voronoi cells suggest a constant bound on the expected complexity of such a cell, regardless of n [18, 19].

In this section we describe an algorithm computing $V(\sigma)$ with a running time dependent on $\tau(r) = \tau_f(r) + \tau_e(r) + \tau_v(r)$, $\tau_f(r)$, $\tau_e(r)$ and $\tau_v(r)$ bounding the expected number of faces, edges and vertices of an additively weighted Voronoi cell amidst r other spheres, respectively. Again, w.l.o.g. we assume that the defining sphere σ is centered at the origin.

The algorithm works incrementally by adding the spheres σ_i from the input set $S = \{\sigma_i, 1 \le i \le n\}$ in their given order. At each step i, the algorithm maintains a subdivision P_i of S^2 that describes the lower envelope $\min_{1 \le i \le n} d_i$ of the functions

$$d_i(p) = \frac{c_i^2 - r_i^2}{2(r_i + \langle p, c_i \rangle)}$$

defined by $\sigma_1, \ldots, \sigma_i$. At certain steps, namely at steps $i = 2^k$, $1 \le k \le \lfloor \log n \rfloor$, the algorithm performs a cleanup operation on P_i to guarantee a running time close to $\tau(n)$.

Additionally, it maintains a set of conflicts C_i. C_i is a relation between the combinatorial elements of P_i, i.e. the vertices, edges, and faces, and all sites from the set $\bar{S}_i = S \setminus S^{(i)}$, $S^{(i)} = \{\sigma_1, \ldots, \sigma_i\}$. It is used to tell the algorithm, where P_{i-1} has to be changed when adding σ_i. For the analysis, we will turn this algorithm into a randomized one by randomly permuting S in the beginning.

The subdivision. We describe our algorithm restricted to computing $V(\sigma)$ within the first octant. Eight similar copies will compute $V(\sigma)$.

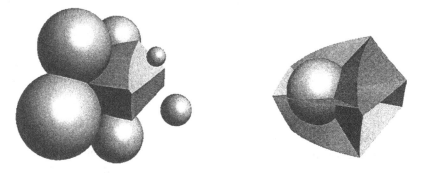

Fig. 1. The edges of an additively weighted Voronoi cell project down as circular arcs.

For $1 \leq i \leq n$ the subdivision P_i is represented by a collection of polytopes $H_j(i)$, $1 \leq j \leq i$, such that $\phi_j(H_j(i) \cap S^2) = \{x \in \partial V : d(x, \sigma) = d(x, \sigma_j)\}$. $H_0(i)$ represents the unbounded portion of the cell. Set $\Delta_0 = \{(x, y, z) \in R^3 : x \geq 0, y \geq 0, z \geq 0, x + y + z \leq 3\}$. For $1 \leq j \leq i \leq n$ let $I_j(i)$ be the minimum subset of $\{0, \ldots, i\}$ identifying non-redundant halfspaces, i.e.

$$S^2 \cap \Delta_0 \cap \bigcap_{0 \leq k \leq i, k \neq j} h_{j,k} = S^2 \cap \Delta_0 \cap \bigcap_{k \in I_j(i)} h_{j,k}.$$

Define $H_i(i) = \Delta_0 \cap \bigcap_{k \in I_i(i)} h_{i,k}$. Recursively we define for $0 \leq j < i$

$$H_j(i) = \begin{cases} \Delta_0 \cap \bigcap_{k \in I_j(i)} h_{j,k} & \text{if } i \text{ is a power of 2} \\ H_j(i-1) \cap h_{j,i} & \text{if } S^2 \cap H_j(i-1) \cap h_{j,i} \neq S^2 \cap H_j(i-1) \\ H_j(i-1) & \text{otherwise} \end{cases}$$

A halfspace $h_{j,i}$ is redundant in step i if it does not define an edge of P_i. The cleanup removes redundant halfspaces.

At each step i, the algorithm maintains a canonical triangulation of all $H_j(i)$, $0 \leq j \leq i$, for which $H_j(i) \cap S^2 \neq \emptyset$:

Definition 1. *Let $H \subset R^3$ be a convex polytope and $a \in R^3$ an arbitrary but fixed direction. For each facet $f \subset H$, choose the minimum vertex $v_{\min}(f)$ with respect to direction a and triangulate f towards $v_{\min}(f)$. The canonical triangulation[1] is obtained as the collection of all 3-dimensional simplices being the convex hull of one of these triangles and $v_{\min}(H)$, the minimum vertex of H with respect to a.*

Let $\Delta_j(i)$ denote the collection of all those simplices of $H_j(i)$. We store adjacency information between all simplices $\Delta \in \Delta_j(i)$, but not between simplices from different sets $\Delta_j(i)$ and $\Delta_k(i)$, $j \neq k$. The subdivision P_i is obtained as the collection of all intersections of these simplices with S^2. Let $\Delta \in \Delta_j(i)$ be a simplex obtained by lifting a triangle t of the triangulation of a facet f of $H_j(i)$ towards the minimal vertex $v_{\min}(H_j(i))$. Then we set $v_{\min(f)}(\Delta) = v_{\min}(f)$ and $v_{\min}(\Delta) = v_{\min}(H_j(i))$.

Search structures. On each facet $f \in H_j(i)$, $0 \leq j \leq i$, we maintain a binary tree search structure for point location within f. Using red-black trees [12], this structure can be constructed in time linear in the number of vertices $|f|$, queries can be answered in time $O(\log |f|)$, and once the location of insertion is known, the structure can be updated in amortized constant time.

At certain steps, namely if $v_{\min}(H_j(i)) \neq v_{\min}(H_j(i-1))$, we establish a *static* point location structure for the complete polytope $H_j(i)$ using the following result:

---

[1] One can argue that the present algorithm also introduces an additional prescribed direction. However, contrary to the situation when constructing tangent planes to circular arcs as it occurs in vertical decomposition schemes, ties with respect to a can be broken very easily using, say, a lexicographical ordering of the coordinates, and we do not have to deal with algebraic numbers (or their approximations).

Theorem 1 ([14]). *For any n vertex planar graph, one can build a point location structure of $O(n)$ size in $O(n \log n)$ time, guaranteeing $O(\log n)$ query time.*

So, given a polytope H with n facets and a set M of m points, we can determine for each point $p \in M$ the simplex Δ of the canonical triangulation of H such that $p \in \Delta$, or verify that no such Δ exists in time $O((n + m) \log n)$.

Conflict information. The algorithm maintains three kinds of conflicts that are associated with the vertices, edge framents and face fragments of all $\Delta \in \bigcup_{j=0}^{i} \Delta_j(i)$:

1. Vertex conflicts: A vertex $v \in P_i$ conflicts with $\sigma_j \in \bar{S}_i$ if ϕ_j will cut the vertex $\phi(v)$ off the cell.
2. Edge conflicts: An edge $e \in P_i$ conflicts with $\sigma_j \in \bar{S}_i$ if ϕ_j intersects $\phi(e)$. We maintain a distinct conflict for each point of intersection. For a single edge e the set of all conflicts $\{(e, \cdot)\}$ is linearly ordered along e. This allows us to split an edge e in constant time regardless of the number of conflicts allocated to it.
3. Face conflicts: A face $f \in P_i, \phi(f) \subset \phi_k$ for some $0 \le k \le i$ conflicts with $\sigma_j \in \bar{S}_i$ if $\phi_k \cap \phi_j \ne \emptyset$ and $\phi_k \cap \phi_j \subset \operatorname{int}(\phi(f))$. For each Δ each face conflict is represented by a point contained in the conflicting region $\Delta \cap S^2$.

We make the following assumptions concerning general position: Any edge conflict is defined as the projection of the intersection of exactly three surfaces $\phi_i, \phi_j, \phi_k, 0 \le i < j < k \le n$, and all points of intersection are not points of tangency.

Initialization. The algorithm begins by constructing an unbounded cell represented by $H_0(0) = \Delta_0$, $\Delta_0(0) = \{\Delta_0\}$ and setting $P_0 = S^2 \cap \Delta_0$. For each $1 \le i \le n$ all conflicts of σ_i with respect to P_0 are calculated. All this can be done in time $O(n \log n)$ and space $O(n)$.

Update step. The i-th update step when adding sphere σ_i to P_{i-1} resulting in P_i is as follows: Let

$$C_{i-1,j}(\sigma_i) = \{\Delta : \Delta \in \Delta_j(i-1), \sigma_i \text{ conflicts } \Delta\}$$

$$C_{i-1}(\sigma_i) = \bigcup_{j=1}^{i-1} C_{i-1,j}(\sigma_i)$$

$$A_{i-1}(\sigma_i) = \{j : 1 \le j < i, \exists \Delta \in \Delta_j(i-1), \sigma_i \text{ conflicts } \Delta\}$$

For each $j \in A_{i-1}(\sigma_i)$ we compute the polytope $H_j(i)$ from $H_j(i-1)$ using the conflict information $C_{i-1,j}(\sigma_i)$. We retriangulate the updated part of $H_j(i)$ and update the search structures on the facets. If no minimal vertex $v_{\min}(f)$ of a facet $f \subset H_j(i-1)$ or even the total minimal vertex $v_{\min}(H_j(i-1))$ is deleted then this update can be performed within $O\left(|C_{i-1,j}(\sigma_i)| \log(n)\right)$ time.

For each conflict c associated with a deleted simplex in $\Delta_j(i-1)$ we can determine in constant time whether it remains a conflict for a simplex $\Delta \in \Delta_j(i)$, and, if so, reinsert it as conflict in the updated structure in time $O(\log n)$. Conflicts which are located on a new face that is created for site σ_i are collected into a set C_{new}.

If we delete a minimum vertex $v_{min}(f)$ of a facet $f \subset H_j(i-1)$, then we retriangulate f and reallocate all affected conflicts to their new locations. This can be done in time $O(|f| + m \log n)$, m being the number of conflicts to be reallocated, and requires no additional space.

Similarly, if the vertex $v_{min}(H_j(i-1))$ happens to be deleted, then we triangulate $H_j(i)$ from scratch. We compute the point location structure described above and use it to assign all conflicts associated with any $\Delta \in \Delta_j(i-1)$ to their new locations. This amounts to $O(|H_j(i)| \log n + m \log n)$ time and temporarily requires space $O(|H_j(i)|)$.

If $C_{i-1}(\sigma_i) \neq \emptyset$, we compute the polytope $H_i(i)$, its canonical triangulation, the subdivision P_i, and the point location structures. Note, that exactly sites of simplices in $C_{i-1}(\sigma_i)$ have facet defining halfspaces $h_{i,j}$ for $H_i(i)$. This amounts to $O(|C_{i-1}(\sigma_i)| \log n)$ time and $O(|C_{i-1}(\sigma_i)|)$ space. Then for each conflict $c \in C_{new}$, we find $\Delta \in \Delta_i(i)$ such that $c \in \Delta$ in $O(n)$ time, and check if its corresponding site σ_k, $k > i$, associated with c still conflicts Δ. If so, we allocate this conflict to Δ and recursively visit all neighboring simplices, as long as they have not been visited yet and provided they also conflict with σ_k. This traversal can be charged onto the number of newly created conflicts times a factor of $O(\log n)$ for the sorted insertion.

Finally, if $i \in \{2^k, 1 \leq k \leq \lfloor \log n \rfloor\}$, we peform a cleanup operation. For each polytope $H_j(i)$, such that $H_j(i) \cap S^2 \neq \emptyset$, we determine the set $I_j(i)$. We compute the $H_j(i)$, $0 \leq j \leq i$, their triangulations, the point location structures, and we reallocate all conflicts to their new locations.

Probabilistic analysis. We want to analyze the expected work performed by the algorithm if the σ_i are inserted in random order. We do this, as it is common practice, in the manner described in [7] and [25].

For $R = \{\sigma_{i_1}, \ldots, \sigma_{i_r}\} \subset S$ let $f(R) = \sum_{i=1}^{r} |\Delta_{i_j}(r)|$ be the total complexity of the representation of the cell defined by R. Let

$$f_r = \frac{1}{\binom{n}{r}} \sum_{R \subset S, |R| = r} f(R)$$

be the expectation of this value. In a first step, we will analyze the behavior of our algorithm in terms of n and f_r. Then we will bound f_r in terms of $\tau(r)$.

Proposition 1. *Let $S = \{\sigma_i, 1 \leq i \leq n\}$. Then the expected total number of conflicts created or reallocated by the above algorithm when adding the elements from S in random order is bounded by*

$$11(n-1)f_1 + \frac{11}{n+1}f_{n+1} + 110 n \sum_{r=1}^{n} \frac{f_{r+1}}{r(r+1)} - 121 \sum_{r=1}^{n} \frac{f_{r+1}}{r+1}.$$

Proof: Observe, that every simplex of our subdivision is defined by at most 11 spheres. Setting the parameter $d = 11$ in the analysis as given in [25], we obtain the claimed bound. □

Proposition 2. *Let $S = \{\sigma_i, 1 \leq i \leq n\}$. Then the expected total number of simplices created due to retriangulation operations by the above algorithm when adding the elements from S in random order is bounded by*

$$6 \sum_{r=1}^{n} \frac{f_r}{r}.$$

Proof: As in the proof of the previous proposition we apply backwards analysis [25]. Let Δ be a simplex that is created during step r of the algorithm because either one facet or a complete polytope requires retriangulation. Running the algorithm backwards, this is equivalent to the situation that removing σ_r from the cell at step r would delete at least one of the two minimal vertices $v_{\min(f)}(\Delta)$ or $v_{\min}(\Delta)$. Each of these vertices is defined by at most 3 sites from S. Therefore, the expected value T_r of the number of simplices created due to retriangulation in step r is bounded by

$$T_r \leq \frac{1}{\binom{n}{r}} \sum_{R \subset S, |R|=r} \frac{6}{r} f(R) = \frac{6}{r} f_r.$$

Summing up for $r = 1 \ldots n$ we obtain the claimed bound. □

Because the functions $\tau_f(r)$ and $\tau_e(r)$ might not be monotone increasing in r, we set $\hat{\tau}_X(r) = \max_{1 \leq i \leq r} \tau_X(i)$ for $X \in \{e, f, v, \cdot\}$.

Proposition 3. *Let $S = \{\sigma_i, 1 \leq i \leq n\}$. Then at each step r of the algorithm the following bound holds:*

$$f_r \leq 2\,\hat{\tau}_f(r) + 8\,\hat{\tau}_e(r)$$

Proof: Let f be the number of faces and e the number of edges of an additively weighted Voronoi cell V. Then, by Euler's relation, the number of simplices needed to represent V is bounded by $f + 4e$. Therefore, if r is a power of 2, the cleanup operation guarantees $f_r \leq \tau_f(r) + 4\tau_e(r)$. Otherwise, let $r^* = \lfloor \log r \rfloor$ be the index of the latest cleanup. Any creation of a new polytope is caused by the creation of at least one new face of the cell, and any addition of a halfspace to a polytope is caused by the creation of at least one new edge of the cell. Hence

$$f_r \leq \tau_f(r^*) + 4\tau_e(r^*) + \sum_{i=r^*+1}^{r} \frac{\tau_f(i) + 4\tau_e(i)}{i}$$

$$\leq \tau_f(r^*) + 4\tau_e(r^*) + \sum_{i=r^*+1}^{r} \frac{\tau_f(r) + 4\tau_e(r)}{r^* + 1}$$

$$= \tau_f(r^*) + 4\tau_e(r^*) + \frac{r - r^*}{r^* + 1}\left(\hat{\tau}_f(r) + 4\hat{\tau}_e(r)\right)$$

$$\leq 2\,\hat{\tau}_f(r) + 8\,\hat{\tau}_e(r)$$

□

Theorem 2. *Let $S = \{\sigma_i, 1 \leq i \leq n\}$. Then the described algorithm computes the additively weighted Voronoi cell of a sphere σ in expected time*

$$O\left(\sum_{r=1}^{n} \frac{n+r}{r^2}\hat{\tau}(r)\log n\right).$$

Observe, that the time requirements are simply the space requirements times a factor of $\log n$.

Proof: According to propositions 1 and 3 the expected total number of conflicts created is $O\left(n\sum_{r=1}^{n}\frac{\hat{\tau}(r)}{r^2}\right)$. Similarly, as shown in proposition 2, the expected total number of simplices created due to retriangulation is bounded by $O\left(\sum_{i=1}^{r}\frac{\hat{\tau}(r)}{r}\right)$. Hence, the expected total computational effort required by these steps is bounded by

$$O\left(\sum_{r=1}^{n} \frac{n+r}{r^2}\hat{\tau}(r)\log n\right).$$

A cleanup operation at step i, i a power of 2, obviously involves a subset of all operations performed by the algorithm up to step i. Therefore, the expected total computational effort required for cleanup is bounded by the sum $O\left(\sum_{k=1}^{\lfloor\log n\rfloor}\sum_{r=1}^{2^k}\frac{n+r}{r^2}\hat{\tau}(r)\log n\right)$. $\square$

Corollary 1. *The described algorithm computes the additively weighted Voronoi cell of a sphere σ amidst n other spheres in expected time $O(n^2\log n)$. If $\tau(n) = O(n)$, the algorithm accomplishes the computation in expected time $O(n\log^2 n)$.*

4 Practical considerations

We implemented a version of the algorithm which does not maintain the triangulation of the polytopes $H_j(i)$, but rather uses a planar map on the sphere as underlying data structure. This implies that finding an object $p \in P_i$ a conflict c has to be allocated to needs a search time linear in the size of the face. For this reason, we get an additional $\frac{\tau(n)}{n}$ factor into our running time estimates. The geometric primitives used by the implemented algorithm are formulated in terms of oriented circles, much in the same way as recently proposed in [1].

4.1 Further processing

Both visualization and volume computation require a triangulation of the cell. We do this as follows, see also figure 2:

1. First, we split all edges into line segments. The resulting simple polygons are triangulated.
2. Using the standard Lawson-flip ([17] this triangulation is transformed into a constrained Delaunay-triangulation on the sphere.
3. Finally, circumcenters of large or skinny triangles are added to the triangulation (similar to [24]).

4.2 Preprocessing

Biological macromolecules, such as proteins or enzymatic RNA, usually have thousands of atoms. On the other hand, a typical additively weighted Voronoi cell of one of its atoms has on average about ten neighbors, since these molecules tend to be tightly packed. Therefore, it is important to quickly identify a small set of possible neighbors of the atomic sphere whose cell is to be computed. We achieve this identification using the lifting map from [2]. Typically, we can identify a set of less than 15 atoms if no hydrogens atoms are present, and less than 30 if they are.

4.3 The Implementation

We implemented the algorithm using the C++ programming language. The algorithm including triangulation requires about 9000 lines of code. Currently, the algorithm uses double precision floating point arithmetic to perform its calculations. Moreover, we incorporated our algorithm with all pre- and post-processing steps included into an interactive program for the examination of data sets from the Brookhaven Protein Database (PDB). See figure 3 for a screen-shot.

For a typical input of up to 30 spheres the algorithm runs well below 100ms, about half of the time being spent on the triangulation. Computing the initial neighbor sets for all atoms of a molecule having about 5000 atoms requires less than 10 seconds. Thus, in less than 10 minutes the volumes of all additively weighted Voronoi cells of all atoms can be calculated. This timing data was obtained on a current mid-range PC offering a 266MHz Pentium II processor and 64MB of RAM operating Windows NT workstation 4.0. Profiling tools (MS-profile, Intel VTune) reveal that about 60% of the overall time of both the triangulation and the computation of the cell is spent on the evaluation of the numerical primitives.

5 Conclusions

In this paper, we provided new insights into the geometry of additively weighted Voronoi cells. We used these insights to derive a new and relatively simple algorithm for their efficient computation. The algorithm was implemented and

Fig. 2. The subdivision of the surface is refined to a Delaunay triangulation, which is further refined by introducing additional points.

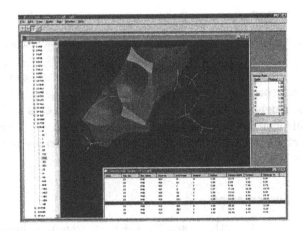

Fig. 3. A screen-shot of our interactive program exemplifying the use of additively weighted Voronoi cells in molecular biology.

performs well on problem instances from molecular biology. Up to now, the design of our algorithm with numerical robustness in mind has proved its worth and could be verified by application to a large collection of data sets taken from the PDB.

Acknowledgments

I am very grateful to the members of the research group of Prof. Frömmel for answering my many questions concerning the applications, especially J. Fauck, A. Goede, R. Preiß-ner, and, of course, Prof. Frömmel himself. My advisor E. Welzl steadily encouraged me to actually implement this algorithm. Finally, I would like to thank an anonymous referee for many helpful comments.

References

1. Andrade, M. V., & Stolfi, J. (1998). Exact Algorithms for Circles on the Sphere. To appear in Proc. 14th Annu. ACM Sympos. Comput. Geom.
2. Aurenhammer, F. (1987). Power diagrams: properties, algorithms and applications. SIAM J. Comput., 16, 78–96.
3. Aurenhammer, F. (1991). Voronoi diagrams: A survey of a fundamental geometric data structure. ACM Comput. Surv., 23, 345–405.
4. de Berg, M., Dobrindt, K., & Schwarzkopf, O. (1995). On lazy randomized incremental construction. Discrete Comput. Geom., 14, 261–286.
5. Boissonnat, J. D., & Dobrindt, K. T. G. (1996). On-line construction of the upper envelope of triangles and surface patches in three dimensions. Comput. Geom. Theory Appl., 5, 303–320.
6. Chazelle, B. (1993). An optimal convex hull algorithm in any fixed dimension. Discrete Comput. Geom., 10, 377–409.

7. Clarkson, K. L., & Shor, P. W. (1989). Applications of random sampling in computational geometry, II, Discrete Comput. Geom., 4, 387–421.
8. Gerstein, M., Tsai, J., & Levitt, M. (1995). The Volume of Atoms on the Protein Surface: Calculated from Simulation, using Voronoi Polyhedra. Journal of Molecular Biology, 249, 955–966.
9. Geysen, H. M., Tainer, J. A., Rodda, S. J., Mason, T. J., Alexander, H., Getzoff, E. D., & Lerner, R. A. (1987). Chemistry of antibody binding to a protein. Science, 235, 1184–1190.
10. Goede, A., Preißner, R., & Frömmel, C. (1997). Voronoi Cell - A new method for the allocation of space among atoms. Journal of Computational Chemistry.
11. Gschwend, D. A. (1995). Dock, version 3.5 . San Francisco: Department of Pharmaceutical Chemistry, University of California.
12. Guibas, L. J., & Sedgewick, R. (1978). A diochromatic framework for balanced trees. Proc. 19th Annu. Sympos. Foundations of Computer Science. (pp. 8–21).
13. Halperin, D., & Shelton, C. (1997). A perturbation scheme for spherical arrangements with application to molecular modeling, Proc. 13th Annu. ACM Sympos. Comput. Geom. (pp. 183–192).
14. Kirkpatrick, D. G. (1983). Optimal search in planar subdivisions. SIAM J. Comput., 12, 28–35.
15. Kleywegt, G. T., & Jones, T. A. (1994). Detection, delineation, measurement and display of cavities in macromolecular structures. Acta Crystallographica, D50, 178–185.
16. Kyte, J. (1995). Structure in Protein Chemistry. Garland Publishing.
17. Lawson, C. L. (1977). Software for C1 surface interpolation. In J. R. Rice (Ed.), Math. Software III (pp. 161–194). New York, NY: Academic Press.
18. Meijering, J. L. (1953). Interface area, edge length, and number of vertices in crystal aggregates with random nucleation : Philips Research Report.
19. Møller, J. (1992). Random Johnson-Mehl tesselations. Adv. Appl. Prob., 24, 814–844.
20. Mulmuley, K. (1994). An Efficient Algorithm for Hidden Surface Removal, II. Journal of Computer and Systems Sciences, 49, 427–453.
21. Okabe, A., Boots, B., & Sugihara, K. (1992). Spatial Tessellations: Concepts and Applications of Voronoi Diagrams. Chichester, UK: John Wiley & Sons.
22. Pontius, J., Richelle, J., & Wodak, S. J. (1996). Deviations from Standard Atomic Volumes as a Quality Measure for Protein Crystal Structures. Journal of Molecular Biology, 264, 121–136.
23. Preparata, F. P., & Hong, S. J. (1977). Convex hulls of finite point sets in two and three dimensions. Comm. ACM 20, (pp. 87–93)
24. Ruppert, J. (1993). A New and Simple Algorithm for Quality 2-Dimensional Mesh Generation, Proc. 4th ACM-SIAM Sympos. Discrete Algorithms (pp. 83–92).
25. Seidel, R. (1993). Backwards analysis of randomized geometric algorithms. In J. Pach (Ed.), New Trends in Discrete and Computational Geometry, (pp. 37–67). Berlin: Springer-Verlag
26. Tilton, R. F., Singh, U. C., Weiner, S. J., Connolly, M. L., Kuntz, I. D., Kollman, P. A., Max, N., & Case, D. A. (1986). Computational Studies of the interaction of myoglobin and xenon. Journal of Molecular Biology, 192, 443–456.
27. Yeates, T. O. (1995). Algorithms for evaluating the long range accessibility of protein surfaces. Journal of Molecular Biology, 249(4), 804–815.

On the Number of Regular Vertices of the Union of Jordan Regions*

Boris Aronov,[1] Alon Efrat,[2] Dan Halperin,[2] and Micha Sharir[2,3]

[1] Department of Computer and Information Science, Polytechnic University,
Brooklyn, NY 11201-3840, USA. aronov@ziggy.poly.edu
[2] School of Mathematical Sciences, Tel Aviv University, Tel-Aviv 69978, Israel.
{alone,halperin,sharir}@math.tau.ac.il
[3] Courant Institute of Mathematical Sciences,
New York University, New York, NY 10012, USA.

Abstract. Let C be a collection of n Jordan regions in the plane in general position, such that each pair of their boundaries intersect in at most s points, where s is a constant. If the boundaries of two sets in C cross exactly twice, then their intersection points are called *regular vertices* of the arrangement $\mathcal{A}(C)$. Let $R(C)$ denote the set of regular vertices on the boundary of the union of C. We present several bounds on $|R(C)|$, determined by the type of the sets of C. (i) If each set of C is convex, then $|R(C)| = O(n^{1.5+\varepsilon})$ for any $\varepsilon > 0$.[4] (ii) If C consists of two collections C_1 and C_2 where C_1 is a collection of n convex *pseudo-disks* in the plane (closed Jordan regions with the property that the boundaries of any two of them intersect at most twice), and C_2 is a collection of polygons with a total of n sides, then $|R(C)| = O(n^{4/3})$, and this bound is tight in the worst case. (iii) If no further assumptions are made on the sets of C, then we show that there is a positive integer t that depends only on s such that $|R(C)| = O(n^{2-1/t})$.

1 Introduction

Let C be a collection of n Jordan regions (the interiors of closed Jordan curves) in the plane, with the property that any two of boundaries intersect in at most

---

* The first and the fourth authors have been supported by a grant from the U.S.-Israeli Binational Science Foundation. Boris Aronov has also been supported by a Sloan Research Fellowship. Micha Sharir has also been supported by NSF Grants CCR-94-24398 and CCR-93-11127, and by a grant from the G.I.F., the German-Israeli Foundation for Scientific Research and Development. Dan Halperin has been supported by an Alon Fellowship, by ESPRIT IV LTR Project No. 21957 (CGAL), by the USA-Israel Binational Science Foundation, and by The Israel Science Foundation founded by the Israel Academy of Sciences and Humanities. Dan Halperin and Micha Sharir have also been supported by the Hermann Minkowski – Minerva Center for Geometry at Tel Aviv University. Part of the work on this paper has been done when Boris Aronov was visiting Tel Aviv University in March 1997.

[4] Throughout this paper, ε stands for an arbitrarily small positive constant; the constants of proportionality in bounds that involve ε also depend on ε and, generally, tend to infinity as ε approaches zero.

some constant number, s, of points. We assume that the sets of C are in general position, so that no point is incident to more than two boundaries, and at each intersection point where the boundaries meet they cross transversally. Let U denote the *union* of C. We consider the *arrangement* $\mathcal{A}(C)$, formed by the boundaries of the sets in C, and define a *vertex* of $\mathcal{A}(C)$ (an intersection point between two boundaries) to be *regular* if the two boundaries cross exactly twice; all other vertices are called *irregular*. The goal is to obtain sharp bounds on the maximal number of regular vertices that appear on ∂U. We denote, as above, the set of regular vertices on ∂U by $R(C)$.

The interest in this problem goes back to the work of Kedem et al. [11], where it was shown that if all vertices of $\mathcal{A}(C)$ are regular (such a collection C is called a family of *pseudo-disks*), then the number of (regular) vertices of ∂U is at most $6n - 12$, for $n \geq 3$, and this bound is tight in the worst case. Recently, Pach and Sharir [15] have shown that if C is an arbitrary finite collection of convex sets (so that any two of their boundaries can intersect in an arbitrary number of points) then one has $|R(C)| \leq 2|I(C)| + 6n - 12$, where $I(C)$ is the set of irregular vertices on ∂U. This result was instrumental in a recent paper by Efrat and Sharir [6], showing that the complexity of the union of n planar "fat" convex sets, each pair of whose boundaries intersect in at most some constant number of points, is nearly linear in n. However, since $I(C)$ can be $\Omega(n^2)$ for a general collection C, even when no pair of boundaries cross at more than four points the bound of [15] only yields the trivial $O(n^2)$ upper bound on $|R(C)|$. As an example of such a construction, consider a collection of n narrow rectangles arranged in an $n/2 \times n/2$ grid.

Pach and Sharir [15] also give a construction where the number of regular vertices on the boundary of the union of n rectangles and n congruent disks is $\Omega(n^{4/3})$. This is the best known lower bound for the general problem stated above, with a constant number of intersections between any pair of boundaries. Without this last constraint, it is easy to obtain examples with $\Omega(n^2)$ regular vertices on the boundary of the union; see, e.g., [12, 15].

In Section 2 we show that the $\Omega(n^{4/3})$ lower bound is tight for the special class (ii) of collections C as in the abstract. Specifically, we show:

Theorem 1. *The maximum possible number of regular vertices on the boundary of the union of a family of n convex pseudo-disks and of polygons with a total of n edges is $\Theta(n^{4/3})$.*

For the other two main results, we first present in Section 3 a technique for transforming the family of regions so that every regular vertex in $R(C)$ becomes essentially a point of tangency between the two regions, and so that the number of intersections between any pair of boundaries does not increase (see Lemma 4).

Next, in Section 4, we consider the case of general convex regions, and show:

Theorem 2. *The maximum possible number of regular vertices on the boundary of the union of a family of n convex Jordan regions in the plane where any two boundaries intersect in at most a constant number of points, is $O(n^{1.5+\varepsilon})$ for any $\varepsilon > 0$.*

Finally, in Section 5, we study the case of general Jordan regions, and show:

Theorem 3. *The maximum possible number of regular vertices on the boundary of the union of a family of n Jordan regions in the plane, where any two boundaries intersect in at most a constant number s of points, is $O(n^{2-1/t})$, where t is a positive integer that depends only on s.*

In other words, we show that in a fairly general setting, the number of regular vertices to ∂U is subquadratic. This is likely to have implications on the analysis of the complexity of the union of geometric objects in two and three dimensions as indicated by our current work [6]. Our experience has been that improved combinatorial bounds on the complexity of the union of geometric objects often entails efficient algorithms for computing such boundaries. This in turn could have useful implications in applications like robot motion planning [9], solid modeling and others.

Remark: Note that using a straightforward perturbation scheme, one can show that the maximum number of regular vertices on the union is achieved when the given regions are in general position, so this assumption involves no loss of generality.

2 Pseudo-disks and Polygons: Proof of Theorem 1

Let $C = C_1 \cup C_2$ where C_1 is a collection of n convex *pseudo-disks* in the plane (closed Jordan regions with the property that the boundaries of any two of them intersect at most twice), and C_2 is a collection of polygons with a total of n sides. We assume henceforth that the polygons in C_2 are convex; if they are not all convex we first decompose them into convex polygons whose total complexity is $O(n)$. For simplicity of notation we assume that the overall number of sides of the resulting convex polygons is n. We form the union U_1 of C_1. By [11], the boundary of U_1 consists of at most $6n$ convex arcs, each lying on the boundary of a single pseudo-disk. Without loss of generality, we will treat only arcs that lie on the top boundary of their respective pseudo-disks; arcs that include leftmost and/or rightmost points of pseudo-disks are split at those points. Let Γ denote the set of these arcs. Clearly, the relative interiors of the arcs in Γ are pairwise disjoint. Let $m \leq 6n$ denote the size of Γ.

Let U_2 denote the union of C_2, and put $U = U_1 \cup U_2$. Clearly, the number of regular vertices on the boundary of U that are incident to two boundaries of sets in C_1 is at most $6n$.

It is also easy to show that the number of regular vertices that lie on the boundary of U and are incident to two boundaries of polygons in C_2 is $O(n)$. Indeed, let v be such a vertex, and let c, c' be the two polygons in C_2 whose boundaries contain v. Let $K = c \cap c'$. Recall that K must be a convex polygon whose boundary contains only two vertices (one of which is v) where ∂c and $\partial c'$ meet. Hence it must also contain at least one vertex w of c or of c', such that vw is an edge of K. We charge v to such a vertex w, and note that any vertex can be charged at most twice in this manner. Hence the total number of vertices of ∂U of this type is $O(n)$. It thus remains to bound the number of *mixed* regular vertices of ∂U, namely, those that are incident to the boundaries of a pseudo-disk in $\mathcal{C}_1$ and of a polygon in $\mathcal{C}_2$.

Let $\gamma \in \Gamma$, let $d \in \mathcal{C}_1$ be the pseudo-disk whose boundary contains γ, and consider its interaction with a polygon c of $\mathcal{C}_2$ that has a regular vertex on γ. Either c contains the extreme left or right point of d, or $c \cap \partial d$ lies completely in the top portion of ∂d; $c \cap \partial d$ cannot be contained in the bottom portion of ∂d as arcs of Γ were chosen to lie on the top boundaries of their respective pseudo-disks. Polygons $c \in \mathcal{C}_2$ that contain the leftmost (rightmost) point of d contribute at most one mixed vertex of ∂U on γ (two if c contains both extreme points of d), for a total of at most $2n$ vertices. Similarly we can eliminate polygons c which contain one of the endpoints of γ, as those produce at most $2m \leq 12n$ mixed vertices on ∂U. From this point on we restrict our attention to polygons c such that $c \cap \partial d \subset \gamma$. Let v, w be the two points of intersection of γ with ∂c_1, say. If v, w lie on different edges of c_1 then the portion of c_1 inside d must contain a vertex of c_1, and we can then charge v and w to such a vertex, in the same manner as in the preceding paragraph. It follows that the number of such vertices v, w is $O(n)$, and we can thus ignore such cases in our analysis. We can thus assume that γ crosses each ∂c_i at two points that lie on the same edge, e_i, of c_i. We refer to these remaining regular vertices as *edge-touching* vertices.

We adapt the analysis technique of [7, 8]. First, we derive a weaker upper bound on the number of mixed edge-touching regular vertices on ∂U. To this end, we construct a bipartite graph H whose nodes are the arcs in Γ and the polygon edges, and each of its edges connects an arc γ to a polygon edge e if γ crosses e twice, at two regular vertices, at least one of which lies on ∂U. We claim that H does not contain a $K_{2,3}$ as a subgraph (composed of two arcs in Γ and

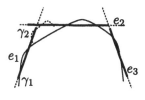

Fig. 1.

of three polygon edges). Indeed, suppose that H did contain such a subgraph, consisting of two arcs $\gamma_1, \gamma_2 \in \Gamma$, and of three polygon edges e_1, e_2, e_3 of polygons c_1, c_2, c_3, respectively. Suppose $\gamma \subset \partial d$, $d \in \mathcal{C}_1$. Since γ_1 intersects each of the three edges twice and each edge lies on the lower boundary of its polygon, the portion of γ_1 outside $c_1 \cup c_2 \cup c_3$ must lie below the lower envelope L of the lines containing e_1, e_2, e_3. Moreover, the regular vertices formed by γ_1 with the three edges must lie on L. Assume, with no loss of generality, that e_1, e_2, e_3 appear in this left-to-right order along L; see Figure 1.

L is partitioned by γ_1 into at most seven sections: the three segments $d \cap e_i \cap L$, $i = 1, 2, 3$, and at most four maximal complement sections of L where the edge-touching vertices of γ_2 and the edges e_i may appear. Since γ_1 and γ_2 are openly disjoint and $\gamma_2 \setminus (c_1 \cup c_2 \cup c_3)$ has to lie below L, all edge-touching vertices induced by γ_2 must appear along one of the latter four complement sections of the envelope. However, none of these sections contain portions of all three edges. Hence γ_2 cannot connect to all of e_1, e_2, and e_3 in the graph.

Since the graph H is bipartite, with at most m and n nodes, respectively, in each class, and since it contains no $K_{2,3}$ as a subgraph, it follows from standard extremal graph-theoretic arguments (see [14]) that the number of its edges, and hence the number of mixed edge-touching regular vertices on ∂U is $O(mn^{1/2}+n)$.

We next choose an integer parameter r, to be fixed below, and construct a $(1/r)$-cutting on the edges of the polygons in C_2 (see [10] for details). This yields a tiling of the plane by $O(r^2)$ pairwise openly disjoint trapezoids, each crossed by at most n/r edges. For each trapezoid τ, consider the set Γ_τ of all the arcs in Γ that cross τ, clipped to within τ; some of these arcs may intersect τ in two connected portions, and we regard each such portion as a separate arc.

We classify the arcs in Γ_τ according to the pairs of sides of τ that they cross. One class consists of *short* arcs that have at least one endpoint inside τ; any other, *long* arc meets $\partial\tau$ exactly twice. There are at most $2m$ short arcs. One class of long arcs is referred to as the class of *bottom* arcs; these are the arcs that have both endpoints on the bottom side of τ.

We claim that, for any class Γ' of arcs, other than those of the short arcs or of the bottom arcs, the number of mixed edge-touching regular vertices on ∂U that are formed within τ by arcs of Γ' is $O(n/r)$. Let $CH(X)$ denote the convex hull of the set X. The proof is similar to that in [7]. It is based on the observation that none of the at most n/r polygon edges that cross τ can form mixed edge-touching regular vertices on ∂U with more than one arc of Γ'. Indeed, if this could happen for one such edge e and two such arcs γ, γ', then $CH(e \cap \gamma)$ and $CH(e \cap \gamma')$ must be disjoint, with, say, $CH(e \cap \gamma)$ lying to the left of $CH(e \cap \gamma')$. (Indeed, the convexity of the pseudo-disks rules out the case that these two intervals are nested within each other; on the other hand, if these intervals overlapped without being nested, then γ and γ' would have to cross, which is impossible.) If we trace γ from the right endpoint of $CH(e \cap \gamma)$ to the right, and trace γ' from the left endpoint of $CH(e \cap \gamma')$ to the left, then these curves must cross within τ—a contradiction. For this not to happen, either one of these curves has to end inside τ (so Γ' is the class of short arcs), or both extensions must cross the bottom side of τ (so Γ' is the class of bottom arcs). Since Γ' is neither of these classes, the claim follows; see [7, 8] for a similar argument. We have thus shown that the number of mixed edge-touching regular vertices on ∂U that are formed within τ by arcs that are neither short nor bottom is $O(n/r)$, for an overall bound of $O(r^2 \cdot (n/r)) = O(nr)$, over all trapezoids τ.

We next claim that the total number of bottom arcs, over all trapezoids τ, is $O(r^2 + m)$. We show this using a graph-planarity argument, similar to that used in [8]. The details are given in the full version.

Let m_τ denote the number of short and bottom arcs in τ. The preceding analysis implies that $\sum_\tau m_\tau = O(r^2 + m)$. The weaker bound obtained above implies that the number of mixed edge-touching regular vertices on ∂U that are formed within a trapezoid τ by its m_τ short and bottom arcs is $O(m_\tau(n/r)^{1/2} + n/r)$, for a total of

$$O\left(\sum_\tau \left[m_\tau \left(\frac{n}{r}\right)^{1/2} + \frac{n}{r}\right]\right) = O\left((m + r^2)\left(\frac{n}{r}\right)^{1/2} + nr\right).$$

As argued above, the overall number of all other mixed edge-touching regular vertices on ∂U is $O(m + nr)$. If we choose $r = \lceil m^{2/3}/n^{1/3} \rceil = \lceil n^{1/3} \rceil$, the overall bound becomes $O(n^{4/3})$. This, combined with the lower bound construction in [15], completes the proof of Theorem 1. $\square$

3 Transforming the Regions

The next lemma is used as a first step in the proofs of Theorems 2 and 3.

Lemma 4. *Let C be a collection of n Jordan regions in the plane, so that each pair of boundaries intersect in a finite number of points. Then we can transform this collection so that if c and c' are a pair of regions in C whose boundaries originally crossed regularly, with at least one of these two crossing points lying on ∂U, then after the transformation c and c' are openly disjoint and touch at a single point that lies on the new union boundary. Moreover, the number of intersections between any two region boundaries does not increase after the transformation. Finally, if all original regions in C are convex, then they remain so after the transformation.*

Proof: The transformation process is iterative: Order the regions in C arbitrarily as $(c_1, \ldots, c_n)$. Let C_i denote the collection after the first i steps of the transformation, with $C_0 = C$. We now describe how to transform C_{i-1} into C_i.

Put $c = c_i$, and apply a homeomorphism τ_c of the plane that maps c onto the closed unit disk D; the existence of such a homeomorphism is a consequence of Schönfliess' Theorem [13]. If all the sets are convex, then we take τ_c to be the identity and for uniformity put $D = c$. We now apply the following steps:

(i) For each $c' \neq c \in C$, we replace each maximal connected arc of $D \cap \partial \tau_c(c')$ by the line segment connecting its endpoints. See Figure 2(ii).

(ii) We shortcut each maximal connected arc γ of $\partial D \cap int(\tau_c(U))$ as follows. Let u and v denote the endpoints of γ, and let $a, b \in C$, $a, b \neq c$, be the sets whose transformed boundaries contain u, v, respectively. If $a \neq b$, we replace γ by the line segment uv; see the left side of Figure 2(iii). If $a = b$, we choose a point v' on ∂D that lies outside $\tau_c(a)$ very close to v (and does not lie inside any other transformed set), and again replace the portion of ∂D between u and v' by the line segment uv'; see the right side of Figure 2(iii).

(iii) We now transform the plane back using τ_c^{-1}.

We iterate these three stages, applying it to each $c_i \in C$ in the above order.

We claim that this transformation does not increase the number of intersections between any pair of boundaries. Indeed, consider the step where a set c_i is processed, and let a, b be two distinct sets in C_{i-1}. If $c_i \neq a, b$ then the portions of ∂a and ∂b outside c do not change, while their portions inside c_i, after applying τ_{c_i} and our "straightening" step (step (ii)), consist of straight segments. If two such segments uv and wz lying on the modified respective boundaries ∂a, ∂b inside D, cross at some point x, then, since u, w, v, z all lie on ∂D and must appear there in this cyclic order, and since the original respective portions of $\partial \tau_{c_i}(a)$ and $\partial \tau_{c_i}(b)$ connecting u to v and w to z lie fully inside D, these original

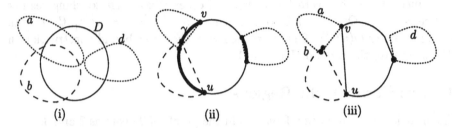

Fig. 2. Demonstration of the transformation rules: (i) The disk D with the images of three other sets, referred to as a, b, and d. (ii) The modified sets a, b and d; the arcs γ of ∂D shortcut in the next stage are highlighted. (iii) D after applying stage (ii) of the transformation.

portions must cross each other at some point x'. We charge x to x', and note that this charging is unique, implying that the number of intersections has not increased.

Suppose then that $c_i = a$, say, and let x be a new intersection between ∂D and $\partial \tau_{c_i}(b)$, after the straightening step. Then x must lie (in the τ_{c_i} image) on one of the new straight segments uv or uv' on ∂a in the transformed plane, and on some new straight portion wz of the transformed ∂b. If x is a common endpoint of these two straight segments, then it must have been an intersection point between ∂a and ∂b in C_{i-1}, so no new intersection arises in this case. Otherwise, as above, it follows that one of w, z, say w, must lie on the portion of ∂D that has been replaced by uv or uv', so we can charge x to w, which was an old intersection point of ∂a and ∂b (in the transformed plane), and this charging is unique, again implying that the number of intersections has not increased.

In particular, every pair of new boundaries intersect (transversally) at most s times, and every pair of boundaries that originally intersected in two (regular) vertices either continue to do so, or just touch each other in a single point, or do not intersect at all after the transformation. Specifically, in the case of regular intersection between two boundaries ∂a, ∂b, if at least one of the two intersections lies on ∂U, then the two transformed boundaries touch at a single point that lies on ∂U; see Figure 2(iii). Indeed, suppose that ∂a and ∂b intersect at exactly two points u, v, one of which lies on ∂U, and suppose that a is processed before b. When a is processed, ∂a and ∂b become touching by construction, and it is easy to verify that this situation does not change after any subsequent transformation step (one needs to verify this only for the step that processes b because no other step affects the neighborhood of this touching point).

All these considerations complete the proof of the lemma. □

Observe that U might be changed by these transformations: The original ∂U is still a portion of the new boundary (not taking into account the slight perturbation introduce in the second case of step (ii)), but U could have gained additional 'holes,' as shown in Figure 2(iii). Note also that the result of the transformation may depend on the order in which the sets are processed.

4 Convex Regions: Proof of Theorem 2

We first apply the transformation described in the previous section to the given family C of convex sets. As already noted, if the sets are convex, then there is no need to apply a homeomorphism to the plane at each step of this process, and the sets remain convex after the transformation. The transformed sets have the following properties: They are convex. Any two boundaries intersect at most s times. Any two sets that intersected regularly become disjoint or touch at a single point. If two sets intersected regularly with at least one point of intersection of their boundaries on ∂U, the transformed sets are openly disjoint and touch on ∂U. If they intersected regularly without creating vertices on ∂U, they are now disjoint. From now on, we assume that C has the properties just noted.

For each $c \in C$, let e_c denote its 'equator', namely the segment connecting the leftmost point and the rightmost point in c. (By an appropriate general position assumption, or by appropriate tilting of the plane, e_c is uniquely defined for each $c \in C$; these segments were called 'sentinels' in [1].)

We first construct a *hereditary* segment tree Q on the x-projections of the equators, as in [3] (consult [3] for more details, and for the terminology that we use below). Each node v of Q stores the standard segment-tree list L_v of so-called 'long' equators, and also a list S_v of 'short' equators, those that are stored in some list L_w, for a proper descendant w of v, and thus have at least one endpoint in the interior of the vertical strip σ_v associated with v. It is easily verified that for any boundary touching between two sets $a, b \in C$, there is a unique node v of Q (on the path to the leaf w whose strip σ_w contains the touching), such that the touching point lies in σ_v and either both a and b belong to L_v or one of them belongs to L_v and the other to S_v. We have $\sum_v (|L_v| + |S_v|) = O(n \log n)$. Also, any such touching occurs between the upper boundary of one set and the lower boundary of the other.

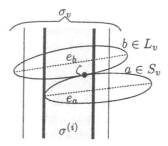

Fig. 3. Illustrating prop.(b)

We now fix a node v and bound the number of boundary touchings within σ_v formed between two sets in $L_v \cup S_v$, at least one of which lies in L_v. We will only describe the case where the other set lies in S_v, because the case where both of them lie in L_v is simpler and can be handled by a similar approach. Moreover, with no loss of generality, it suffices to consider only boundary touchings where the set in L_v lies above the set in S_v.

We apply fairly standard range-searching techniques to obtain a finite collection $\{A_i \times B_i\}_i$ of complete bipartite graphs, such that
(a) For each i, $A_i \subseteq S_v$ and $B_i \subseteq L_v$.
(b) For each i there is a substrip $\sigma^{(i)} \subseteq \sigma_v$, such that for each $a \in A_i$, $b \in B_i$, the x-projection of a contains that of $\sigma^{(i)}$, the equator e_b lies fully above a within σ_v, and the equator e_a lies fully below b within $\sigma^{(i)}$). See Figure 3.
(c) For each pair of sets $a \in S_v$, $b \in L_v$, such that ∂a touches ∂b at a point ζ

within σ_v and a lies below b, there is an index i such that $a \in A_i$, $b \in B_i$, and $\zeta \in \sigma^{(i)}$.

(d) $\sum_i (|A_i| + |B_i|) = O(|S_v||L_v|^{1/2+\varepsilon} + |L_v||S_v|^{1/2+\varepsilon})$, for any $\varepsilon > 0$, where the constant of proportionality depends on ε and on s.

Suppose this has been done. Then fix an index i, and note that any boundary touching ζ that occurs within $\sigma^{(i)}$ between a set a in A_i and a set b in B_i must occur along the upper envelope of the upper boundaries of the sets of A_i, and along the lower envelope of the lower boundaries of the sets of B_i. Indeed, suppose to the contrary that, say, ζ does not lie on this upper envelope. Since ζ lies on the boundary of the union of A_i, the set a' appearing on the envelope at the x-coordinate of ζ must be such that $e_{a'}$ lies above ζ, which contradicts the property that $e_{a'}$ has to lie fully below b within $\sigma^{(i)}$. The argument for the lower envelope is fully symmetric. The number of such touchings is therefore proportional to the combined complexity of these envelopes within $\sigma^{(i)}$, which is at most $\lambda_s(|A_i|) + \lambda_s(|B_i|)$, where $\lambda_s(n)$ denotes the maximum length of (n, s)-Davenport-Schinzel sequences [16]. Summing over all i's, we obtain that the total number of boundary touchings 'associated' with v is proportional to $O(|S_v||L_v|^{1/2+\varepsilon} + |L_v||S_v|^{1/2+\varepsilon})$, for a slightly larger, but still arbitrarily small $\varepsilon > 0$. Summing these bounds over all nodes v of Q, we obtain the overall bound $O(n^{3/2+\varepsilon})$, again for slightly larger, but still arbitrarily small $\varepsilon > 0$, with the constant of proportionality depending on ε and on s. This therefore completes the proof of the theorem.

To obtain the decomposition $\{A_i \times B_i\}$, we use a multi-level range-searching structure, where each node w at each level of the structure will store a complete bipartite graph $A'_w \times B'_w$, such that $A'_w \subseteq S_v$ and $B'_w \subseteq L_v$. Each subsequent level of the structure 'enforces' some more of the desired constraints, and the bipartite graphs within each subsequent level form a refinement of the graphs obtained at the previous level. See [2] for more details concerning multi-level range-searching structures.

In the first level we enforce the property that the e_b's, for $b \in L_v$, lie above the a's in S_v. In what follows, we clip all the relevant a's and b's to within σ_v. Actually, we want to have the property that the equator e_b of any such b lie above the upper boundary of any such a. We may replace any e_b by the line containing it (this has no effect on what happens within σ_v), and replace any a by the portion of σ_v lying below the upper boundary of the original a (so we first make a smaller by ignoring its portion outside σ_v, and then make a larger by allowing it to expand downwards within σ_v). We apply a standard duality to the plane (as in [5]), which preserves incidences and the above/below relationship. This duality maps the upper boundary of any $a \in S_v$ to a convex x-monotone curve γ_a, and the (extended) equators e_b of sets in L_v are mapped to points e_b^*. An equator e_b lies above the upper boundary of a if and only if the dual point e_b^* lies above γ_a. Note also that any pair of curves γ_a, $\gamma_{a'}$ intersect each other at most s times, because any such intersection point is the dual of a common tangent to the upper boundaries of a and a', and there can be at most s such common tangents, because ∂a and $\partial a'$ intersect in at most s points.

Thus, in this dual setting, the desired first-level decomposition $\{A'_w \times B'_w\}_w$ of $S_v \times L_v$ has to satisfy the following properties:

(e) For each w, the point e_b^* dual to the equator e_b of any set $b \in B'_w$ lies above γ_a, for every $a \in A'_w$.

(f) For any $a \in S_v$, $b \in L_v$, such that e_b^* lies above γ_a, there is a w such that $a \in A'_w$ and $b \in B'_w$.

Put $m_v = |S_v|$, $n_v = |L_v|$. To achieve this decomposition, we fix some sufficiently large constant parameter ξ, draw a random sample R of ξ sets $a \in S_v$, consider the arrangement $\mathcal{A}_R$ of the corresponding curves γ_a, and apply a vertical decomposition to $\mathcal{A}_R$ that produces $O(\xi^2)$ pseudo-trapezoidal cells. Since each pseudo-trapezoid is determined by at most four curves γ_a, it follows from [4] that with high probability, no pseudo-trapezoid is crossed by more than $\frac{cm_v}{\xi} \log \xi$ curves, for some appropriate constant c. We may assume that our sample R does indeed have this property.

For each pseudo-trapezoid τ, let B_τ denote the subset of L_v consisting of those sets b whose dual points e_b^* lie inside τ, and let A_τ (resp. C_τ) denote the subset of S_v consisting of those sets a whose dual curves γ_a pass below τ (resp. cross τ). Put $k_\tau = |A_\tau|$, $n_\tau = |B_\tau|$, and $m_\tau = |C_\tau|$. We have $\sum_\tau n_\tau = n_v$, $k_\tau \leq m_v$, and $n_\tau \leq \frac{cm_v}{\xi} \log \xi$. By partitioning τ vertically into subcells, if necessary, we may also assume that $n_\tau \leq n_v/\xi^2$ for each τ, while the total number of subcells remains $O(\xi^2)$.

We add to the first-level output collection of complete bipartite graphs all the products $A_\tau \times B_\tau$, and repeat the whole process recursively within each cell τ, with the sets C_τ and B_τ. We stop the recursion when the size of C_τ or of B_τ falls below some specified constant, and then output all appropriate singleton products $\{a\} \times \{b\}$, for $a \in C_\tau$, $b \in B_\tau$.

It is clear from the construction that the resulting decomposition satisfies the required properties (e) and (f). We next estimate its total size $\sum_\tau(|A_\tau| + |B_\tau|)$. At the top level of the recursion we have $\sum_\tau |B_\tau| = n_v$. Since the number of levels of recursion is $O(\log n_v)$ and the points e_b^* are partitioned among the recursive subproblems, it follows that in total we have $\sum_\tau |B_\tau| = O(n_v \log n_v)$. Similarly, at the top level of the recursion we have $\sum_\tau |A_\tau| = O(m_v \xi^2)$, and $\sum_\tau |C_\tau| = O(m_v \xi \log \xi)$ (where both constants of proportionality depend only on s). The maximum depth j of the recursion satisfies $\xi^{2j} \leq n_v$, or $\xi^j \leq n_v^{1/2}$. It follows that, for an appropriate constant c' that depends on s, the overall sum $\sum_\tau |A_\tau|$ is at most proportional to

$$m_v \xi^2 \left(1 + c'\xi \log \xi + \cdots + (c'\xi \log \xi)^{j-1}\right) = O(m_v(c'\xi \log \xi)^j) = O(m_v n_v^{1/2+\varepsilon}),$$

where we can make $\varepsilon > 0$ arbitrarily small by choosing ξ sufficiently large (as a function of ε and s). Hence we have

$$\sum_\tau |A_\tau| = O(m_v n_v^{1/2+\varepsilon}) \quad \text{and} \quad \sum_\tau |B_\tau| = O(n_v \log n_v). \tag{1}$$

Next, fix a pair $A = A_\tau$, $B = B_\tau$ in this decomposition. For each $a \in A$, $b \in B$ the equator e_b lies above the upper boundary of a (within σ_v). This already implies (arguing as above) that any boundary touching ζ between ∂a and ∂b, for any pair $a \in A$, $b \in B$ that lies on the union, must lie on the lower envelope E of the lower boundaries of the sets in B. The complexity of E is at most $\lambda_s(|B|)$, which is nearly linear in $|B|$.

For each $a \in A$, let a_x denote its x-projection, clipped to within σ_v. We construct a secondary segment tree R on the intervals a_x. Each node u of R is associated with a vertical strip $\sigma^{(u)} \subseteq \sigma_v$. In addition to the standard segment-tree list $A^{(u)}$ of x-projections of sets in A that is stored at u, we also store there a list $B^{(u)}$ of the sets in B whose lower boundaries appear in $E \cap \sigma^{(u)}$. We clearly have that $\sum_u |A^{(u)}| = O(|A| \log |A|)$ and that $\sum_{|B^{(u)}| > 1} |B^{(u)}| = O(\lambda_s(|B|) \log |A|)$.

Indeed, the first bound is a standard property of segment trees. The second bound is a consequence of the following observations: (a) The number of breakpoints of E is at most $\lambda_s(|B|)$. (b) Each breakpoint belongs to at most $\log |A|$ strips $\sigma^{(u)}$. (c) For any u, the size of $B^{(u)}$ is upper bounded by 1 plus the number of breakpoints of E in $\sigma^{(u)}$.

Now suppose that there is a pair $a \in A$, $b \in B$, with a touching ζ between the upper part of ∂a and the lower part of ∂b that lies on the union. There is a unique node u of R such that $a \in A^{(u)}$ and $\zeta \in \sigma^{(u)}$. Hence ∂b appears on E within $\sigma^{(u)}$, so $b \in B^{(u)}$. Note also that the line containing e_a passes fully below $b \cap \sigma^{(u)}$ (because the x-projection of a fully contains the projection of $\sigma^{(u)}$).

We now fix a node u for which $|B^{(u)}| > 1$, and apply a symmetric version of the first-level decomposition to $A^{(u)} \times B^{(u)}$, to obtain a collection $\{A'_j \times B'_j\}_j$ of complete bipartite graphs, such that

(g) For each j, we have $A'_j \subseteq A^{(u)}$ and $B'_j \subseteq B^{(u)}$; we also associate the strip $\sigma^{(u)}$ with j.

(h) For each j, each $a \in A'_j$ and each $b \in B'_j$, the equator e_a lies fully below b within $\sigma^{(u)}$.

(i) For any boundary touching ζ as above, there is an index j such that $a \in A'_j$ and $b \in B'_j$ (and ζ lies in the strip of j, that is, in $\sigma^{(u)}$).

(j) $\sum_j (|A'_j| + |B'_j|) = O(|A^{(u)}| \log |A^{(u)}| + |B^{(u)}||A^{(u)}|^{1/2+\varepsilon})$, for any $\varepsilon > 0$.

If $|B^{(u)}| = 1$ we output only one bipartite graph $\tilde{A}^{(u)} \times B^{(u)}$ where $\tilde{A}^{(u)}$ consists of all $a \in A^{(u)}$ such that e_a lies fully below b within $\sigma^{(u)}$.

The grand collection of complete bipartite graphs $A'_j \times B'_j$, gathered over all nodes u of R, and over all first-level pairs $A_\tau \times B_\tau$, is the desired output collection. It clearly satisfies properties (a)–(c). Concerning (d), we first sum the bounds (j) over all $u \in R$, to obtain the bound

$$O(|A| \log^2 |A| + |B||A|^{1/2+\varepsilon}),$$

for a slightly larger but still arbitrarily small $\varepsilon > 0$, where the constant of proportionality depends on ε and on s. Clearly, this bound also subsumes the case $|B^{(u)}| = 1$.

Finally, we sum these bounds over all pairs $A = A_\tau$, $B = B_\tau$, in the first-level decomposition, and make use of (1), to conclude that this sum is at most proportional to

$$\sum_\tau O(|A_\tau| \log^2 |A_\tau| + |B_\tau||A_\tau|^{1/2+\varepsilon}) = O(m_v n_v^{1/2+\varepsilon} + n_v m_v^{1/2+\varepsilon}),$$

again, for a slightly larger but still arbitrarily small $\varepsilon > 0$, where the constant of proportionality depends on ε and on s. Hence (d) is also satisfied. As already noted, this completes the proof of the theorem. $\square$

Remark: An obvious open problem is to close the gap between this upper bound and the lower bound $\Omega(n^{4/3})$ noted earlier.

5 Arbitrary Regions: Proof of Theorem 3

We prove the theorem using the following 'forbidden subgraph' argument. We first transform C as described in Lemma 4, but continue to denote the transformed collection by C. Now we define a graph H whose vertices are the regions in C and whose edges connect pairs of regions that touch at a point on the boundary of the union. We claim that H does not contain a complete bipartite graph $K_{r,g}$ for some sufficiently large constants $r = r(s)$, $g = g(s)$, where s is the maximum number of intersections between any two region boundaries.

Suppose to the contrary that there exist subsets $\mathcal{R}, \mathcal{G}$ of C of sizes r, g, respectively, such that for each $(c, c') \in \mathcal{R} \times \mathcal{G}$, c and c' touch at only one point that lies on the boundary of the union U. With no loss of generality we may assume that $C = \mathcal{R} \cup \mathcal{G}$. We refer to sets of $\mathcal{R}$ (resp. $\mathcal{G}$) as "red" (resp. "green").

Consider the arrangement $\mathcal{A}(\mathcal{G})$ and let $U_\mathcal{G}$ denote the union of $\mathcal{G}$. We claim that any $c \in \mathcal{R}$ is fully contained in the closure of a single *hole* (connected component of the complement) of $U_\mathcal{G}$. Indeed, if this were false, ∂c would have to properly cross some boundary of a green region c', which is impossible by construction.

Consider the collection of holes of $U_\mathcal{G}$ that contain red regions. We call them *interesting green holes*. Since each red region must touch the boundary of every green set at a point that lies on $\partial U_\mathcal{G}$, it follows that all interesting green holes are part of the *zone* in $\mathcal{A}(\mathcal{G})$ of any green boundary, which is the collection of cells in the arrangement met by the boundary. Hence the overall complexity of all these holes is at most $a\lambda_{s+2}(g)$, for some absolute constant a (see [14]). Symmetrically, the overall complexity of interesting red holes is at most $a\lambda_{s+2}(r)$.

We now construct a planar bipartite graph G, whose nodes are the arcs forming the boundaries of interesting green and red holes; each edge of G connects a green arc ζ to a red arc η if these two arcs touch at a regular vertex of the overall union. It is easy to verify that G is indeed planar, and that it has no multiple edges. Since the graph is bipartite, the number of its edges is at most twice the number of its nodes, that is, at most $2a(\lambda_{s+2}(g) + \lambda_{s+2}(r))$. On the other hand, the number of edges of G must be rg, because every green set and every red set touch at some regular vertex on ∂U. We thus obtain: $rg \leq 2a(\lambda_{s+2}(g) + \lambda_{s+2}(r))$, which is impossible if r and g are sufficiently large constants (that depend on s).

Using standard results in extremal graph theory (see [14]), it follows that the number of regular vertices on the boundary of the union of C is $O(n^{2-1/r})$. This completes the proof of the theorem. □

References

[1] B. Aronov and M. Sharir, The common exterior of convex polygons in the plane, *Comput. Geom. Theory Appls.* 8 (1997), 139–149.

[2] M. de Berg, M. van Kreveld, M. Overmars, and O. Schwarzkopf, *Computational Geometry: Algorithms and Applications*, Springer-Verlag, Berlin 1997.

[3] B. Chazelle, H. Edelsbrunner, L. Guibas, and M. Sharir, Algorithms for bichromatic line segment problems and polyhedral terrains, *Algorithmica* 11 (1994), 116–132.

[4] K. Clarkson, New applications of random sampling in computational geometry, *Discrete Comput. Geom.* 2 (1987), 195–222.

[5] H. Edelsbrunner, *Algorithms in Combinatorial Geometry*, Springer-Verlag, Heidelberg 1987.

[6] A. Efrat and M. Sharir, The complexity of the union of fat objects in the plane, *Proc. 13th ACM Symp. on Computational Geometry*, 1997, 104–112.

[7] D. Halperin and M. Sharir, On disjoint concave chains in arrangements of (pseudo)lines, *Inform. Process. Lett.* 40 (1991), 189–192.

[8] D. Halperin and M. Sharir, Corrigendum: On disjoint concave chains in arrangements of (pseudo) lines, *Inform. Process. Lett.* 51 (1994), 53–56.

[9] D. Halperin and M. Sharir, Arrangements and their applications in robotics: Recent developments, *Proc. Workshop on Algorithmic Foundations of Robotics*, K. Goldberg, D. Halperin, J.-C. Latombe, R. Wilson, Editors, A.K. Peters, Boston, MA 1995.

[10] D. Haussler and E. Welzl, Epsilon-nets and simplex range queries, *Discrete Comput. Geom.* 2 (1987), 127–151.

[11] K. Kedem, R. Livne, J. Pach, and M. Sharir, On the union of Jordan regions and collision-free translational motion amidst polygonal obstacles, *Discrete Comput. Geom.* 1 (1986), 59–71.

[12] J. Matoušek, J. Pach, M. Sharir, S. Sifrony, and E. Welzl, Fat triangles determine linearly many holes, *SIAM J. Comput.* 23 (1994), 154–169.

[13] E.E. Moise, *Geometric Topology in Dimension 2 and 3*, Springer-Verlag, New York, 1977.

[14] J. Pach and P.K. Agarwal, *Combinatorial Geometry*, Wiley, New York, 1995.

[15] J. Pach and M. Sharir, On the boundary of the union of planar convex sets, to appear in *Discrete Comput. Geom.*

[16] M. Sharir and P.K. Agarwal, *Davenport Schinzel Sequences and Their Geometric Applications*, Cambridge University Press, New York, 1995.

Distribution-Sensitive Algorithms

Sandeep Sen and Neelima Gupta

Department of Computer Science and Engineering,
I.I.T., New Delhi 110016, India
{ssen , neelima}@cse.iitd.ernet.in

Abstract. We investigate a new paradigm of algorithm design for geometric problems that can be termed *distribution-sensitive*. Our notion of distribution is more combinatorial in nature than spatial. We illustrate this on problems like planar-hulls and 2D-maxima where some of the previously known output-sensitive algorithms are recast in this setting. In a number of cases, the distribution-sensitive analysis yields superior results for the above problems. Moreover these bounds are shown to be tight for a certain class of algorithms.
Our approach owes its spirit to the results known for sorting multisets and we exploit this relationship further to derive fast and efficient parallel algorithms for sorting multisets along with the geometric problems.

1 Introduction

The practice of measuring efficiency of algorithms only with respect to input-size has a major drawback. By restricting ourselves to only one parameter, we are trying to optimize the number of steps assuming worst-case scenario for that parameter only. One of the early examples of this type is that of constructing the convex-hull of point sets. On the plane, the known lower bound [2, 25] was $\Omega(n \log n)$ and a number of algorithms matched this bound. However, it was also known that if the number of hull vertices are small - say $O(1)$, the hull can be constructed in $O(n)$ time. This lacuna was resolved by Kirkpatrick and Seidel [13] who showed that the complexity of this problem is $\Theta(n \log h)$ where h is the number of hull vertices (h is not known initially).

A *multiset* is a a set with repeated elements. It is a classical result from information-theoretic arguments that sorting n elements requires $\Omega(n \log n)$ comparisons in the worst case. However, if there are repeated elements, say the input consists of only 0's and 1's, then such a set can be sorted in linear time easily. It was shown in [19] that the complexity for sorting a multiset is $\Theta((n \log n - \sum_i n_i \log n_i) + n)$ where n_i is the number of elements that have the i-th largest value. The starting point of our work is a simple observation that can be summarized as follows.

Observation 1 *Given a multiset of cardinality n and h distinct values, it can be sorted using $O(n \log h)$ comparisons.*

This follows from the fact that $n \log n - \sum_i n_i \log n_i$ is maximized at $n_i = \frac{n}{h}$. The bound in the previous observation is explicit in several earlier papers investigating the running time of quicksort on multisets ([23]). However, in some sense this is a *loss of information* or a weaker characterization of the complexity of sorting multisets. In a number of cases, especially for very non-uniform distribution of the n_i's, $O(n \log n - \sum_i n_i \log n_i)$ can be much less than $O(n \log h)$. For example, consider all n_i's to be 1 except n_h which is $n - h + 1$ - in this case the running time of $O((n-h+1) \log \frac{n}{n-h+1} + (h-1) \log n)$ is linear when $h \leq \frac{n}{\log n}$. Contrast this with the bound of $O(n \log n)$ from Observation 1.

Turning around this argument, one suspects that there may be a more precise characterization of the complexity of some of the geometric problems in the context of output-size sensitivity. As we shall show in the subsequent sections that this is indeed so and we will design algorithms that take advantage of *distribution* of the input points *without prior knowledge of the distribution*. We will define the notion of *distribution* more precisely in the context of the specific problems. In most cases we will be able to reanalyze the existing algorithms in this light - so we don't claim much originality of the algorithms as such. However, it may be viewed as an alternative framework for analyzing algorithms that can pay rich dividends for other problems.

2 Maximal Vectors in plane

Let T be a set of vectors in R^d. The partial order $\leq_M$ is defined for two vectors $x = (x_1, x_2, \ldots x_d)$ and $y = (y_1, y_2, \ldots y_d)$ as follows : $x \leq_M y$ (x is *dominated by* y) if and only if $x_i \leq y_i$ for all $1 \leq i \leq d$.

Definition A vector $v \in T$ is said to be maximal element if it is not dominated by any other vector $w \in T$.

The problem of maximal vectors is to determine all maximal vectors in a given set of input vectors. Here we will consider the special case when the vectors are two-dimensional. The output looks like a staircase (Figure 3). We will denote the x and y projections of a vector p by $X(p)$ and $Y(p)$ respectively.

An early work [15] on maximal vectors established that the complexity of the maximal vectors (on plane) is $\Theta(n \log n)$ for an input of n vectors. A later development due to Kirkpatrick and Seidel [14] showed that a more precise characterization of the complexity can be obtained by considering the size of the output, namely the number of maximal vectors. Using a technique that they called *marriage before conquest*, they developed an algorithm that had running time $O(n \log h)$ where h is the number of maximal vectors. No prior knowledge of h was required and further they showed that the lower bound is also $\Omega(n \log h)$ in terms of n and h. In the following description we will use the term vector and point interchangeably.

Definition Given T, let $t_1, t_2 \ldots t_h$ denote the maximal vectors in a reverse sorted order along x direction (i.e. as we go up the staircase). Let us denote the set of input vectors that lie between $X(t_i)$ and $X(t_{i+1})$ including t_i by N_i - note that except t_i none of these vectors are maximal. If $|N_i| = n_i$, then we say that the *distribution* of T is characterized by $(n_1, n_2 \ldots n_h)$. Also for all the vectors $v \in N_i$, we will define $slab(v) = i$. Note that $\sum_i n_i = n$.

See Figure 3 for an illustration of the definition.

Algorithm Max1(T)

1. If $|T| \leq 1$ then return T
2. else partition T into two (nearly) equal size sets T_1 and T_2 using the median of T with respect to x coordinate.
(* T_1 has smaller x-coordinate than T_2 *)
3. Determine the point $y_m \in T_2$ that has the largest y coordinate. Let $T_1' = T_1 - \{q|q \in T_1$ and $q \leq_M y_m\}$. $T_2' = T_2 - \{p|p \in T_2$ and $p \leq_M y_m\}$.
(* Discard all points of T that are dominated by y_m *)
4. Return (Max1 (T_2') $\cup$ Max1 (T_1'))

Fig. 1. A divide-and-conquer Algorithm for computing maximal vectors on plane

2.1 A distribution-sensitive algorithm

The algorithm in Figure 1 is a slight modification of the output-sensitive algorithm described by Kirkpatrick and Seidel [14]. The time complexity of the algorithm is dominated by the Steps 2 and 3, which involves median finding and filtering with respect to the closest maximal vector to the right of the median line. If the current interval (in the recursive call) is $\mathcal{I}$ then the above steps take $O(|\mathcal{I}|)$ steps - for concreteness let us assume $c|\mathcal{I}|$ steps for some constant $c > 1$. We will view this as an average of c steps for every point in the interval, i.e. we will assign a cost of c to every point. We will analyze the above algorithm by keeping track of the total cost assigned to each point over the entire algorithm. As the algorithm progresses, the size of the intervals (or equivalently the sub-problem size in the recursive calls) become smaller. With each point p, we can associate an interval $I(p)$ containing the point p. When the algorithm is called recursively on $I(p)$, then the subsequent interval containing p becomes about half the size of $I(p)$ and p is assigned an additional cost c. We will denote the sequence of intervals containing p (as the algorithm proceeds) by $I_0(p), I_1(p) \ldots I_k(p)$ where $I_0(p)$ is the given set T and $I_{j+1}(p)$ is obtained by splitting $I_j(p)$. The total cost accumulated by p is ck. Clearly $k \leq \log n$ but we can get a more accurate estimate on k that will depend on the *distribution* of T.

Lemma 1. *The total cost accumulated by a point $p \in T$ is no more than $c \cdot \log \frac{n}{n_r}$ where $r = slab(p)$.*

Proof This follows from the observation that $|I_k(p)| \geq n_r$ and $|I_{j+1}(p)| \leq |I_j(p)/2|$.

Therefore we obtain the following.

Theorem 1. *The total running time of the algorithm can be bounded by* $O(\sum_i n_i \lceil \log \frac{n}{n_i} \rceil + n)$ *or alternately* $O((n \log n - \sum_i n_i \log n_i) + n)$.

As observed earlier, this is always bounded by $O(n \log h)$. This analysis not only gives a better bound but also sheds more light on the somewhat 'mysterious' bound of $O(n \log h)$ that came out of solving a recurrence.

2.2 A heap-based approach

Munro and Raman [18] described a very simple heap-based multiset sorting algorithm (that is in-place) and achieves optimality up to constant factor of the leading term. The same can be adapted for finding maximal vectors that gives an algorithm which has a somewhat different behaviour than the previous algorithm.

The crux of Munro and Raman's method is the observation that while doing heapsort, all the elements that are equal to the maximum (root) element form a sub-tree rooted at the top of the heap. This makes it easy to locate these elements in time proportional to the number of such elements. For example, if there are n_i such elements equal to the maximum, then using a depth-first-search from the root, we can locate all these elements in $O(n_i)$ steps (recall that a heap is similar to a complete binary tree). Now instead of discarding the root as in normal heapsort, we start deleting the n_i elements in decreasing distance from the root. Recall that if an element is distance d from the bottom of the heap (leaf-level of the tree), then deletion and the subsequent restoration of the heap takes $O(d)$ comparisons. Munro and Raman show that the maximum number of comparisons are required when the n_i elements form a balanced tree which yields the bound $2(n_i \log n - n_i \log n_i)$ comparisons for the n_i elements and $2(n \log n - \sum_i n_i \log n_i) + O(n)$ comparisons overall.

For determining maximal vectors, we build a heap on the x-projections of the vectors. Clearly, the vector t_1 with the maximum x-coordinate is at the top of the heap. Analogous to the arguments of Munro and Raman, we claim that the vectors in N_1 (these are the vectors dominated by t_1 only) form a subtree rooted at the top of the heap. However, unlike sorting, it is not straightforward to detect these vectors. If we somehow knew t_2, then our job would have made simple as we would have eliminated all vectors with x coordinates between $X(t_1)$ and $X(t_2)$. The vector t_2 (and in general t_{i+1}) can be characterized by the following straightforward observation.

Observation 2 $t_{i+1} = \max\{X(p) : Y(p) > Y(t_i)\}$, that is, t_{i+1} has the largest x-coordinate among all vectors that have larger y-coordinate than t_i.

We locate the elements of N_i as follows. We do depth-first-search until we hit a vector that has larger y-coordinate than t_i. This gives us a set of vectors N_i' containing N_i. By observation 2 t_{i+1} is at the root of the heap after restoration.

By using the same analysis as Munro and Raman, this algorithm has running time of $O(n \log n - \sum_i n_i' \log n_i')$, where $n_i' = |N_i'|$. This running time is not in conformity with our definition of *distribution* and cannot be compared directly with the result of Theorem 1. In fact, n_i''s could be different for different permutations of the same input ! This is because the n_i''s are dependent on the initial heap which can be different for different permutations.

Remark This also exposes an inherent difficulty of the distribution-based approach, namely that of a robust definition of the distribution itself. Certain definitions would be suitable for certain approaches as was the case for the divide-and-conquer based approach. However, note that even in the heap-based approach the running time can always be bounded by $O(n \log h)$ as $\sum_i n_i' = n$.

3 Planar Hulls

After Kirkpatrick and Seidel's discovery of the output-sensitive algorithm, there have been some recent simplification by Chan et al. [5] and Wenger[24] and

Bhattacharya and Sen [3]. The latter two are randomized and variations of a folklore method called quickhull.

The definition of *distribution* that we will use for planar hulls is similar in spirit to the two-dimensional maximal vector problem. We will confine ourselves to upper-hulls only - for the lower hull it is symmetric. Let $v_1, v_2 \ldots v_h$ be the hull vertices in anti-clockwise order. We will denote the input points that lie below the edge v_i, v_{i+1} by N_i and let $n_i = |N_i|$. In other words, for all points $p \in N_i$, the vertical visibility edge is $v_i v_{i+1}$. Also for any point $p \in N_i$, we define $slab(p) = i$. The vector $(n_1, n_2 \ldots n_h)$ is the distribution vector for the convex-hull problem. As before $\sum_i n_i = n$.

3.1 Kirkpatrick and Seidel's approach

We will first analyze the original approach of [13] in the context of distribution-sensitive algorithm. Later, we shall show that the simplified algorithms that do not use linear-programming also lead to comparable running times.

Algorithm Hull1(T)

1. If $|T| \leq 1$ then return T
2. else partition T into two (nearly) equal size sets T_1 and T_2 using the median of T with respect to x coordinate.
(* T_1 has smaller x-coordinate than T_2 *)
3. Find the hull-edge (also called the bridge) between T_1 and T_2. This can be done using linear programming with two variables.
4. If $t_1 t_2$ is the bridge (supporting points are t_1 and t_2 in T_1 and T_2 respectively), let $T_1' = T_1 - \{q | q \in T_1$ and q lies below $t_1 t_2\}$. $T_2' = T_2 - \{p | p \in T_2$ and p lies below $t_1 t_2\}$.
(* Discard all points of T that lie below the edge $t_1 t_2$. *)
5. Return (Hull1 $(T_2') \cup$ edge $t_1 t_2 \cup$ Hull1 (T_1')).

Fig. 2. KS algorithm for planar hull

The underlying structure of the algorithm is very similar to that of Algorithm Max1 (Figure 1). Step 3 can be done in $O(|T|)$ steps [17, 6] as are the other steps. The analysis can also be done along the lines of Algorithm Max 1. By accumulating cost with every point that participates in the interval corresponding to the recursive calls, we obtain a bound identical to Lemma 1. Thus we can summarize as follows.

Theorem 2. *The total running time of the algorithm Hull1 can be bounded by* $O(\sum_i n_i \lceil \log \frac{n}{n_i} \rceil + n)$ *or alternately* $O((n \log n - \sum_i n_i \log n_i) + n)$.

The simplified algorithms in [5, 3, 24] are also divide-and-conquer algorithms but they avoid the complicated linear-time linear programming of Step 3. The algorithm of [5] as well as the randomized version [3] detects an output vertex instead of an edge and simultaneously split the problem into roughly equal size sub-problems (each sub-problem no more than 3/4 the original size). Subsequently all the points in the triangle defined by the extreme points and the output-vertex are eliminated. To preserve our definition of the *distribution* vector, we can modify the algorithms slightly as follows. After detecting the output vertex and splitting the problem into roughly equal size subproblems, we can find an incident edge (by a simple angular scan) and then eliminate all points below this edge. Then we can use the previous analysis directly to obtain a bound identical to Theorem 2.

Remark Without this modification, we will run into similar problems as the heap-based maxima algorithm. Namely, the points eliminated from the triangle will depend on the initial permutation and cannot be characterized by the distribution vector. However, the running time will still be bounded by $O(n \log h)$.

4 Lower bounds

In this section, we will address the maximal vector problem *with distribution*, that is the distribution vector must be output in addition to the set of maximal vectors. This is not a very natural restriction as observed in the heap-based approach. The upper-bounds obtained in the previous section are optimal to within a constant factor in the algebraic model where the optimality holds for algorithms that also reports/verifies the *distribution vector*. We will use a technique due to Kapoor and Ramanan [12] to prove a lower bound.

Their technique involves defining a related problem $\mathcal{B}$, that is easily reducible (typically linear-time) to the original problem $\mathcal{A}$ and use Ben-Or's [2] component-based lower-bound in the algebraic model. The problem $\mathcal{B}$ is defined such that it is easier to count the number of connected components in the underlying solution space. Ben-Or's result can be summarized as

Theorem 3. *Let W be a subset of $\mathbb{R}^d$ and let $\#(W)$ represent the number of (disjoint) connected components of W. Then any algebraic decision tree that correctly determines the membership query $x \in W$ has $\Omega(\log \#(W) - d)$ height.*

In our case, we will use a construction due to [12] to define problem $\mathcal{B}$. Let $S_1 = \{p_1, p_2 \ldots, p_m\}$ where $p_i = (i, m - i + 1)$ for $1 \leq i \leq m$. Let L_1 and L_2 be the lines whose equations are $x + y = m + 1$ and $x + y = m$ respectively. All points in S_1 are maximal and they lie on L_1 (see Figure 4).

Definition Dominance Testing with restriction (DTR). Given a set $S_2 = \{q_1, q_2 \ldots q_{n-m}\}$ with $n - m$ points, determine whether they are all dominated by the points in S_1 and lie strictly above the line L_2.

This problem is clearly $O(n)$ time transformable to the problem of determining the maximal vectors[1] of $S_1 \cup S_2$. If the number of maximal elements in $|S_1 \cup S_2| \neq m$, then the answer to the DTR problem is NO. Otherwise, the answer to DTR is YES iff all the points in S_2 lie between L_1 and L_2.

The above transformation can be extended to include the distribution vector. Let us define the problem of Dominance Testing with Restriction and Distribution (DTRD) to be DTR with a distribution vector $(n_1, n_2, \ldots n_m)$ where $\sum_i n_i = n - m$. This problem is $O(n)$ transformable to the maximal vectors

---

[1] Strictly speaking, we should consider the decision version of this problem, i.e., if the output contains m vectors. Since it does not cause any technical difficulty we have used the standard version.

problem with distribution $(n_1, n_2, \ldots n_m)$ (recall that this algorithm reports the distribution) by using a slight modification of the previous argument.

We shall now prove a lower bound for the DTRD problem. The effective problem in DTRD is to determine whether each of the points of S_2 lies in one of the regions T_i and moreover, T_i contains exactly n_i points. Now, each instance of a DTRD problem can be considered to be a point in $\mathbb{R}^{2(n-m)}$. Consider the number of ways of assigning $n - m$ points to the T_i's such that T_i contains n_i points - denote this by K. Each YES instance of DTRD corresponds to such an assignment. Moreover, if S_2' and S_2'' are two distinct assignments, then there are at least two indices l such that $q_l' \in S_2'$ and $q_l'' \in S_2''$ are in different T_i's. Any continuous path from S_2' to S_2'' in $\mathbb{R}^{2(n-m)}$ must necessarily pass through a point that is not a YES instance to the DTRD (either it is not above L_2 or not dominated by a point in S_1). Hence these are in separate connected components of the solution space.

¿From the elementary combinatorics of assigning (non-identical) balls to bins, we know that $K = \frac{(n-m)!}{n_1! \cdot n_2! \ldots n_m!}$. By simplifying using Stirling's approximation, we obtain the following result as a consequence of Theorem 3

Lemma 2. *Any algebraic computation tree that solves the maximal vector problem with distribution vector $(n_1, n_2 \ldots n_m)$ has height at least $\Omega(\sum_i n_i \log\left(\frac{(n-m)}{n_i+1}\right))$.*

An $\Omega(m \log m)$ lower-bound easily follows from the order-types of the output points (for example see [12]). In conjunction with Lemma 2, we obtain the following theorem (for example by separating the cases $m \leq n/2$ and $m > n/2$)

Theorem 4. *Any algebraic computation tree that solves the maximal vector problem with distribution vector $(n_1, n_2 \ldots n_m)$ has height at least $\Omega(\sum_i (n_i + 1) \log\left(\frac{n}{n_i+1}\right))$.*

Remark Our definition of distribution vector is actually $(n_1 + 1, n_2 + 1 \ldots)$ that includes the maximal vectors. So the expressions for lower and upper bounds are identical.

An analogous argument yields a similar bound for planar hulls. See Kapoor and Ramanan [12] for details of the construction used for planar hulls. The obvious weakness of the lower bound is that it is restricted to algorithms that also reports the distribution vector.

5 Optimal distribution-sensitive parallel algorithms

In this section we present a parallel algorithm to sort the multisets. The underlying structure of the algorithm is similar to that of **Max1** and **Hull1**, i.e. the algorithm is recursive and we use the technique of 'marriage before conquest' of Kirkpatrick and Seidel. Subsequently we show that the algorithms **Max1** for planar vector maxima and **Hull1** for planar convex hulls can be directly implemented in parallel in the same bounds as the algorithm **ParMset1** for sorting multisets. Let T be the input set with h distinct elements.

Algorithm ParMset1(T)

1. If the total size of the problem at the ith stage is $\leq n/\log n$ then solve the problem directly using an $(n, \log n)$ algorithm.
2. else
   (a) Find the median x_m of T. (* An approximate median will suffice *)
   (b) Let $T_{left} = \{x \in T | x < x_m\}$ and $T_{right} = \{x \in T | x_m < x\}$. Let $T_1 = T_{left} - \{x | x = x_m\}$ $T_2 = T_{right} - \{x | x = x_m\}$.
   (c) If T_1 is not empty then ParMset1(T_1).
   (d) If T_2 is not empty then ParMset1(T_2).

5.1 Analysis

The following results for *extremal-selection*, *approximate-median* and *approximate compaction* will be used for analyzing the previous algorithm in the CRCW PRAM.

Lemma 3. *There is a CRCW algorithm that finds an element with rank k such that $\frac{n}{3} \leq k \leq \frac{2n}{3}$ deterministically in $O(\log \log n)$ time using $(n/\log \log n)$ processors [11]. Extremal selection can also be done in the same bounds.*

Compacting m non-zero elements in an array of size at most $\lambda \cdot m$ for some constant $\lambda > 1$ is called *approximate compaction* that can be used for load-balancing.

Lemma 4 ([11]). *Approximate compaction can be done deterministically in $O(\log \log n)$ steps using $n/\log \log n$ CRCW processors.*

Claim In Algorithm **ParMset1**, let N_i denote the total size of the sub-problems after the ith stage. If n_t is the number of elements with the t-th largest value,
(i) The total size of the sub-problems can be bound by $n/\log n$ after $O(\log h + \log \log n)$ stages.
(ii) $\sum N_i \leq O(\sum_{t=1}^{h} n_t \log(n/n_t))$.

Proof. In the ith stage, the maximum size of a sub-problem is $< n/c^i$, for some constant $c > 1$. Thus total size of sub-problems $< \frac{n}{c^i} h < n/\log n$ after $O(\log h + \log \log n)$ stages.
Part (ii) Follows along the lines of Lemma 1.

Then the total time spent before the problem is directly solved is $O(\sum_i N_i/p + \log \log n(\log h + \log \log n))$ using Brent's slow-down technique with work redistribution. ¿From Lemma 3 an approximate median can be computed in linear work and $O(N_i/p + \log \log n)$ time. Hence the running time is bounded by $O(\log \log n \cdot (\log h + \log \log n) + \sum_i N_i/p + \log n)$ steps.

Theorem 5. *Multiset sorting problem can be solved deterministically in time $O((\sum_{t=1}^{h} n_t \log(n/n_t))/p)$ with $p \leq n/\log n \log \log n$ processors, where n is the size of the input set and h is the number of distinct elements in it.*

The algorithm **Max1** for planar vector maxima can be directly implemented in parallel in the same bounds as the algorithm **ParMset1**. Using the simplified algorithms in [5, 3, 24] that do not require the linear-programming step, we can also construct planar hulls in the the same time and work bounds.

5.2 Expected bounds using randomization

By substituting the deterministic routines by their faster randomized counterparts we obtain faster algorithms (expected running times). In particular, we make use of constant-time algorithms for extremal selection and approximate-median ([22]) and $O(\log^* n)$ algorithms for approximate-compaction and *semisorting*.

Definition For all $n, m \in N$ and $\lambda \geq 1$, the *m-color semisorting* problem of size n and with slack λ is the following: Given n integers $x_1, \ldots, x_n$ in the range $0, \ldots m$, compute n nonnegative integers $y_1, \ldots, y_n$ (the placement of x_i) such that

(1) All the x_i of the same colour are placed in contiguous locations (not necessarily consecutive).

(2) $\max\{y_j : 1 \leq j \leq n\} = O(\lambda n)$.

Lemma 5 ([1]). *There is a constant $\epsilon > 0$ such that for all given $n, k \in N$, the problems of **semisorting** and **approximate compaction** can be solved on a CRCW PRAM using $O(k)$ time, $O(n \log^{(k)} n)$ processors and $O(n \log^{(k)} n)$ space with probability at least $1 - 2^{-n^{\epsilon}}$. Alternatively, it can be done in $\tilde{O}(t)$ steps, $t \geq \log^* n$ using n/t processors.*

Then the total time spent before the problem is solved directly is $O(\sum N_i/p, \log^* n(\log h + \log \log n))$.

Theorem 6. *Multiset sorting problem can be solved in $O(\sum_{t=1}^{h} n_t \log(n/n_t)/p)$ steps with high probability with $p \leq n/\log^* n \log n$ processors, where n is the size of the input set and h is the number of distinct elements in it.*

Theorem 7. *The planar convex hull and the planar maxima problems can be solved in $O(\sum_{t=1}^{h} n_t \log(n/n_t)/p)$ steps with high probability using $p \leq n/\log^* n \log n$ processors, where n and h are the input and the output sizes respectively.*

6 Optimal multiset sorting

In this section, we present a very simple and fast output-sensitive algorithm for sorting multisets. The optimality is achieved with respect to the number of distinct elements, h which we will refer to as *slabs*. This is the best possible if all the distinct elements have the same cardinality, i.e., $\Theta(n/h)$. The algorithm is randomized and is described below.

Algorithm Mset2(T)

1. Pick up a random sample R of size $\sqrt{n}$ and sort it.
2. Delete the duplicate elements. Denote the resulting set by R^*. This partitions the input set into $O(h)$ partitions.
3. Delete an element of the input set T in slab t if an element from slab t has been picked up in the sample . Denote the resulting set by T^*.
4. Sort T^*. Delete the duplicate elements and denote the resulting set by T^{**}.
5. Merge R^* and T^{**}. Denote the resulting set by S.
6. Label each element of T by its rank in S and apply $h-$color semisorting ([1]).

6.1 Analysis

The following result due to Hagerup[Hag92] is useful in our analysis.

Fact 1 *For every fixed $\epsilon > 0$, the 1's in an array A can be compacted, in constant time, into an array of size $k^{1+\epsilon}$ using $O(n)$ processors, where n is the size of the array A and k is the number of 1's in it.*

Assume that we have a linear number of processors. Assume $h = O(n^\delta)$ for an appropriate $\delta > 0$, for otherwise the problem can be solved in $O(\log n) = O(\log h)$ time.

A sample of size $\sqrt{n}$ can be sorted in constant time with n processors. The sample is compressed, that is duplicate elements are deleted, by an application of compaction. Using Fact 1 the sorted sample can be compacted into an array of size $O(h^{1+\epsilon})$ for every fixed $\epsilon > 0$ in constant time. Therefore, the size of R^* is $O(h^{1+\epsilon})$ and Step 3 can be done in $O(\log h)$ time. Call an element of T in slab t *active* if no element from slab t was picked up in the sample.

Claim. The number of active elements in T is $O(h\sqrt{n}\log^2 n)$ with high probability.

Proof. The probability that none of the elements in slab t is picked up in the sample is $(1 - 1/\sqrt{n})^{n_t}$ where n_t is the number of elements in slab t. Probability that none of the elements in slab t is picked up and $n_t \geq \sqrt{n}\log^2 n$ is $\leq 1/n^2$, giving us the required result.

Thus for every $\epsilon > 0$, active elements of T can be compacted into an array of size $O((h\sqrt{n}\log^2 n)^{1+\epsilon})$ in constant time. Thus, for $h = O(n^\delta)$, $\delta > 0$, size of $T^* = O(n^{\epsilon'})$ for some $\epsilon' > 0$ with high probability and hence T^* can be sorted and compacted into T^{**} in constant time with high probability. Merging can also be done in $O(\log h)$ time. Semisorting takes $O(\log^* n)$ time. Hence, we have the following

Theorem 8. *The multiset sorting problem can be solved in $O(\log h + \log^* n)$ time with high probability with n processors, where n is the size of the input and h is the number of distinct elements in it.*

Remark Given the $\Omega(\log^* n)$ lower bound for approximate compaction [16], $O(\log h + \log^* n)$ is an optimal time bound for Multisorting.

A closely related problem called *Renaming* involves determining the rank of each element, where the rank of an element is determined by the number of *distinct* elements less than the element. Farach and Muthukrishnan [7] described an $O(\log h)$ time $O(n \log h)$ work randomized algorithm for this problem which implies an algorithm for multisorting with $O(\log^* n + \log h)$ running time.

With $p = n/\log^* n$ processors and redistributing processors evenly our algorithm runs in $O(\log^* n \log h)$ time with high probability. Hence we have the following

Theorem 9. *The multiset sorting problem can be solved in $O(\log^* n \log h)$ time with high probability with $n/\log^* n$ processors, where n is the size of the input and h is the number of distinct elements in it.*

Concluding Remarks

There is a natural extension of the notion of distribution to three dimensions but the present techniques fall short of giving any improved results. Moreover, a weakness of our definition of distribution is that it is related to certain algorithms. To make the results more appealing, we have to dissociate the distribution from the algorithms completely.

References

1. H Bast and T Hagerup. Fast parallel space allocation,estimation and integer sorting. *Technical Report, MPI-I-93-123*, June 1993.
2. M. Ben-Or Lower bounds for algebraic computation trees. *Proc. of the 15th ACM STOC*, pp. 80–86, 1983.
3. B. Bhattacharya and S. Sen. On a simple, practical, optimal, output-sensitive randomized planar convex-hull algorithm. *Journal of Algorithms*, 25, pp. 177–193, 1997.
4. T.M. Chan. Output-Sensitive Results on convex hulls, extreme points and related problems. ACM Symp. on Comput. Geom., 1995.
5. Timothy.M.Y. Chan, Jack Snoeyink, Chee-Keng Yap. Output-Sensitive Construction of Polytopes in Four Dimensions and Clipped Voronoi Diagrams in Three. *Proc. 6th ACM-SIAM Sympos. Discrete Algorithms* 1995, pp 282-291.
6. M. Dyer Linear time algorithms for two and three variable linear programs. *SIAM Journal on Computing*, 13(1), pp. 31–45, 1984.
7. Farach and Muthukrishnan. Optimal parallel randomized renaming. *Information Processing Letters*, pp. 12–15, Jan 1997.
8. R L Graham. An efficient algorithm for determining the convex hull of a finite planar set. *Information Proc. Lett.*, 1:132–133, 1972.
9. N. Gupta and S. Sen. Optimal, output-sensitive algorithms for constructing planar hulls in parallel. *Computational Geometry: Theory and Applications*, 8, 151–166, 1997.
10. T. Hagerup. Fast deterministic processor allocation. *Proc. of the 4th ACM Symposium on Discrete Algorithms* 1993, pp. 1–10.
11. T. Goldberg and U. Zwick. Optimal Deterministic Approximate Parallel Prefix Sums and Their Applications. In *Proc. Israel Symp. on Theory and Computing Systems (ISTCS'95)*, (1995), pp. 220-228.
12. S. Kapoor and P. Ramanan. Lower bounds for maximal and convex layer problems. *Algorithmica*, pages 447–459, 1989.
13. D G Kirkpatrick and R Seidel. The ultimate planar convex hull algorithm. *SIAM Jl. of Comput.*, 15(1):287–299, Feb. 1986.
14. D G Kirkpatrick and R Seidel. Output-Size sensitive algorithms for finding maximal vectors. *Proc. of ACM Symp. on Computational Geometry* 1985, pp. 89–96.
15. H.T. Kung, F. Luccio and F.P. Preparata. On finding the maxima of a set of vectors. *Journal of the ACM*, 22, pp. 469–476, 1975.
16. P. D. MacKenzie. Load balancing requires $\Omega(\log *n)$ expected time. *Symp. on Discrete Algorithms*, pp. 94–99, 1992.
17. N. Megiddo Linear time algorithms for linear programming in R^3 and related problems. *SIAM Journal on Computing* 12(4), pp. 759–776, 1983.
18. I.J. Munro and V. Raman. Sorting multisets and vectors In-place. *Proc. of the 2nd WADS*, LNCS 519, pp. 473–479, 1991.

19. I. Munro and P.M. Spira Sorting and Searching in Multisets. *SIAM journal on Computing*, 5(1), pp. 1–8, 1976.
20. F P Preparata and S J Hong. Convex hulls of finite sets of points in two and three dimensions. *Comm. ACM*, 20:87–93, 1977.
21. F P Preparata and M I Shamos. *Computational Geometry : An Introduction.* Springer-verlag, New York, 1985.
22. S. Rajasekaran and S. Sen. *Random sampling Techniques and parallel algorithm design.* J.H. Reif editor. Morgan, Kaufman Publishers, 1993.
23. L.M. Wenger. Sorting a linked list with equal keys. *Information Processing Letters*, 15(5), pp. 205–208, 1982.
24. R. Wenger. Randomized Quick Hull. *to appear in* Algorithmica.
25. A.C. Yao A lower bound to finding convex hulls. *Journal of the ACM*, 28, pp. 780–787, 1981.

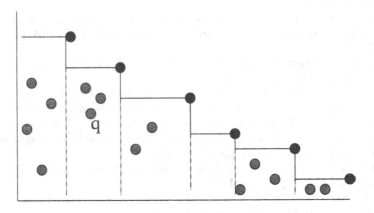

Fig. 3. Distribution vectors for 2-D Maximal vectors is $\{3, 4, 1, 3, 4, 5\}$. $h = 6$ and $\text{slab}(q) = 5$.

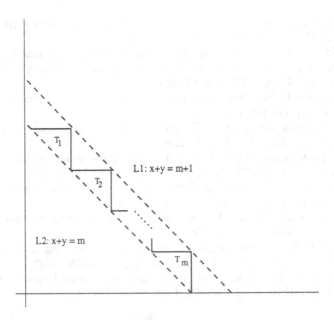

Fig. 4.

Author Index

Aleksandrov L.   11
Aronov B.   322
Aspvall B.   47, 222

Barrett C.   234
de Berg M.   83
Berman P.   255
Bhatia R.   23
Brodal G. S.   107, 158

Chen D.Z.   169
Coulston C.   255

Dahlhaus E.   210
De Marco G.   193

Efrat A.   322

Gargano L.   193
Goldberg A. V.   1
Graf T.   59
Grossi R.   181
Guha S.   23
Gupta N.   335

Halldórsson M. M.   47
Halperin D.   322
Hirsch E. A.   246
Hoffmann F.   71
Håstad J.   205

Icking C.   71

Jacob R.   234
Jansen K.   35

Kaplan H.   119
Katajainen J.   107
Katz M. J.   83, 95
Kedem K.   95
Khuller S.   23
Klein R.   71
Kriegel K.   71

Lanthier M.   11

Mahajan M.   276
Maheshwari A.   11
Manne F.   47
Marathe M.   234

Okasaki C.   119
Overmars M.   83

Pandu Rangan C.   143
Pietracaprina A.   181
Pinotti M. C.   158
Proskurowski A.   222
Pucci G.   181

Rajasekar K.   143
Raman V.   131
Ramnath S.   131

Sack J.-R.   11
Segal M.   95
Seiden S. S.   264
Sen S.   335
Sharir M.   322
Sibeyn J. F.   288
Sleumer N.   300
Sridhar R.   143
van der Stappen A. F.   83
Sussmann Y. J.   23

Tarjan R. E.   119
Telle J. A.   222

Vaccaro U.   193
Veezhinathan K.   59
Vinay V   276
Vleugels J.   83

Will H. M.   310

Xu J.   169

Zuckerman D.   155

Lecture Notes in Computer Science

For information about Vols. 1–1359

please contact your bookseller or Springer-Verlag

Vol. 1360: D. Wang (Ed.), Automated Deduction in Geometry. Proceedings, 1996. VII, 235 pages. 1998. (Subseries LNAI).

Vol. 1361: B. Christianson, B. Crispo, M. Lomas, M. Roe (Eds.), Security Protocols. Proceedings, 1997. VIII, 217 pages. 1998.

Vol. 1362: D.K. Panda, C.B. Stunkel (Eds.), Network-Based Parallel Computing. Proceedings, 1998. X, 247 pages. 1998.

Vol. 1363: J.-K. Hao, E. Lutton, E. Ronald, M. Schoenauer, D. Snyers (Eds.), Artificial Evolution. XI, 349 pages. 1998.

Vol. 1364: W. Conen, G. Neumann (Eds.), Coordination Technology for Collaborative Applications. VIII, 282 pages. 1998.

Vol. 1365: M.P. Singh, A. Rao, M.J. Wooldridge (Eds.), Intelligent Agents IV. Proceedings, 1997. XII, 351 pages. 1998. (Subseries LNAI).

Vol. 1366: Z. Li, P.-C. Yew, S. Chatterjee, C.-H. Huang, P. Sadayappan, D. Sehr (Eds.), Languages and Compilers for Parallel Computing. Proceedings, 1997. XII, 428 pages. 1998.

Vol. 1367: E.W. Mayr, H.J. Prömel, A. Steger (Eds.), Lectures on Proof Verification and Approximation Algorithms. XII, 344 pages. 1998.

Vol. 1368: Y. Masunaga, T. Katayama, M. Tsukamoto (Eds.), Worldwide Computing and Its Applications — WWCA'98. Proceedings, 1998. XIV, 473 pages. 1998.

Vol. 1370: N.A. Streitz, S. Konomi, H.-J. Burkhardt (Eds.), Cooperative Buildings. Proceedings, 1998. XI, 267 pages. 1998.

Vol. 1371: I. Wachsmuth, M. Fröhlich (Eds.), Gesture and Sign Language in Human-Computer Interaction. Proceedings, 1997. XI, 309 pages. 1998. (Subseries LNAI).

Vol. 1372: S. Vaudenay (Ed.), Fast Software Encryption. Proceedings, 1998. VIII, 297 pages. 1998.

Vol. 1373: M. Morvan, C. Meinel, D. Krob (Eds.), STACS 98. Proceedings, 1998. XV, 630 pages. 1998.

Vol. 1374: H. Bunt, R.-J. Beun, T. Borghuis (Eds.), Multimodal Human-Computer Communication. VIII, 345 pages. 1998. (Subseries LNAI).

Vol. 1375: R. D. Hersch, J. André, H. Brown (Eds.), Electronic Publishing, Artistic Imaging, and Digital Typography. Proceedings, 1998. XIII, 575 pages. 1998.

Vol. 1376: F. Parisi Presicce (Ed.), Recent Trends in Algebraic Development Techniques. Proceedings, 1997. VIII, 435 pages. 1998.

Vol. 1377: H.-J. Schek, F. Saltor, I. Ramos, G. Alonso (Eds.), Advances in Database Technology – EDBT'98. Proceedings, 1998. XII, 515 pages. 1998.

Vol. 1378: M. Nivat (Ed.), Foundations of Software Science and Computation Structures. Proceedings, 1998. X, 289 pages. 1998.

Vol. 1379: T. Nipkow (Ed.), Rewriting Techniques and Applications. Proceedings, 1998. X, 343 pages. 1998.

Vol. 1380: C.L. Lucchesi, A.V. Moura (Eds.), LATIN'98: Theoretical Informatics. Proceedings, 1998. XI, 391 pages. 1998.

Vol. 1381: C. Hankin (Ed.), Programming Languages and Systems. Proceedings, 1998. X, 283 pages. 1998.

Vol. 1382: E. Astesiano (Ed.), Fundamental Approaches to Software Engineering. Proceedings, 1998. XII, 331 pages. 1998.

Vol. 1383: K. Koskimies (Ed.), Compiler Construction. Proceedings, 1998. X, 309 pages. 1998.

Vol. 1384: B. Steffen (Ed.), Tools and Algorithms for the Construction and Analysis of Systems. Proceedings, 1998. XIII, 457 pages. 1998.

Vol. 1385: T. Margaria, B. Steffen, R. Rückert, J. Posegga (Eds.), Services and Visualization. Proceedings, 1997/1998. XII, 323 pages. 1998.

Vol. 1386: T.A. Henzinger, S. Sastry (Eds.), Hybrid Systems: Computation and Control. Proceedings, 1998. VIII, 417 pages. 1998.

Vol. 1387: C. Lee Giles, M. Gori (Eds.), Adaptive Processing of Sequences and Data Structures. Proceedings, 1997. XII, 434 pages. 1998. (Subseries LNAI).

Vol. 1388: J. Rolim (Ed.), Parallel and Distributed Processing. Proceedings, 1998. XVII, 1168 pages. 1998.

Vol. 1389: K. Tombre, A.K. Chhabra (Eds.), Graphics Recognition. Proceedings, 1997. XII, 421 pages. 1998.

Vol. 1390: C. Scheideler, Universal Routing Strategies for Interconnection Networks. XVII, 234 pages. 1998.

Vol. 1391: W. Banzhaf, R. Poli, M. Schoenauer, T.C. Fogarty (Eds.), Genetic Programming. Proceedings, 1998. X, 232 pages. 1998.

Vol. 1392: A. Barth, M. Breu, A. Endres, A. de Kemp (Eds.), Digital Libraries in Computer Science: The MeDoc Approach. VIII, 239 pages. 1998.

Vol. 1393: D. Bert (Ed.), B'98: Recent Advances in the Development and Use of the B Method. Proceedings, 1998. VIII, 313 pages. 1998.

Vol. 1394: X. Wu. R. Kotagiri, K.B. Korb (Eds.), Research and Development in Knowledge Discovery and Data Mining. Proceedings, 1998. XVI, 424 pages. 1998. (Subseries LNAI).

Vol. 1395: H. Kitano (Ed.), RoboCup-97: Robot Soccer World Cup I. XIV, 520 pages. 1998. (Subseries LNAI).

Vol. 1396: E. Okamoto, G. Davida, M. Mambo (Eds.), Information Security. Proceedings, 1997. XII, 357 pages. 1998.

Vol. 1397: H. de Swart (Ed.), Automated Reasoning with Analytic Tableaux and Related Methods. Proceedings, 1998. X, 325 pages. 1998. (Subseries LNAI).

Vol. 1398: C. Nédellec, C. Rouveirol (Eds.), Machine Learning: ECML-98. Proceedings, 1998. XII, 420 pages. 1998. (Subseries LNAI).

Vol. 1399: O. Etzion, S. Jajodia, S. Sripada (Eds.), Temporal Databases: Research and Practice. X, 429 pages. 1998.

Vol. 1400: M. Lenz, B. Bartsch-Spörl, H.-D. Burkhard, S. Wess (Eds.), Case-Based Reasoning Technology. XVIII, 405 pages. 1998. (Subseries LNAI).

Vol. 1401: P. Sloot, M. Bubak, B. Hertzberger (Eds.), High-Performance Computing and Networking. Proceedings, 1998. XX, 1309 pages. 1998.

Vol. 1402: W. Lamersdorf, M. Merz (Eds.), Trends in Distributed Systems for Electronic Commerce. Proceedings, 1998. XII, 255 pages. 1998.

Vol. 1403: K. Nyberg (Ed.), Advances in Cryptology – EUROCRYPT '98. Proceedings, 1998. X, 607 pages. 1998.

Vol. 1404: C. Freksa, C. Habel. K.F. Wender (Eds.), Spatial Cognition. VIII, 491 pages. 1998. (Subseries LNAI).

Vol. 1405: S.M. Embury, N.J. Fiddian, W.A. Gray, A.C. Jones (Eds.), Advances in Databases. Proceedings, 1998. XII, 183 pages. 1998.

Vol. 1406: H. Burkhardt, B. Neumann (Eds.), Computer Vision – ECCV'98. Vol. I. Proceedings, 1998. XVI, 927 pages. 1998.

Vol. 1407: H. Burkhardt, B. Neumann (Eds.), Computer Vision – ECCV'98. Vol. II. Proceedings, 1998. XVI, 881 pages. 1998.

Vol. 1409: T. Schaub, The Automation of Reasoning with Incomplete Information. XI, 159 pages. 1998. (Subseries LNAI).

Vol. 1411: L. Asplund (Ed.), Reliable Software Technologies – Ada-Europe. Proceedings, 1998. XI, 297 pages. 1998.

Vol. 1412: R.E. Bixby, E.A. Boyd, R.Z. Ríos-Mercado (Eds.), Integer Programming and Combinatorial Optimization. Proceedings, 1998. IX, 437 pages. 1998.

Vol. 1413: B. Pernici, C. Thanos (Eds.), Advanced Information Systems Engineering. Proceedings, 1998. X, 423 pages. 1998.

Vol. 1414: M. Nielsen, W. Thomas (Eds.), Computer Science Logic. Selected Papers, 1997. VIII, 511 pages. 1998.

Vol. 1415: J. Mira, A.P. del Pobil, M.Ali (Eds.), Methodology and Tools in Knowledge-Based Systems. Vol. I. Proceedings, 1998. XXIV, 887 pages. 1998. (Subseries LNAI).

Vol. 1416: A.P. del Pobil, J. Mira, M.Ali (Eds.), Tasks and Methods in Applied Artificial Intelligence. Vol.II. Proceedings, 1998. XXIII, 943 pages. 1998. (Subseries LNAI).

Vol. 1417: S. Yalamanchili, J. Duato (Eds.), Parallel Computer Routing and Communication. Proceedings, 1997. XII, 309 pages. 1998.

Vol. 1418: R. Mercer, E. Neufeld (Eds.), Advances in Artificial Intelligence. Proceedings, 1998. XII, 467 pages. 1998. (Subseries LNAI).

Vol. 1420: J. Desel, M. Silva (Eds.), Application and Theory of Petri Nets 1998. Proceedings, 1998. VIII, 385 pages. 1998.

Vol. 1421: C. Kirchner, H. Kirchner (Eds.), Automated Deduction – CADE-15. Proceedings, 1998. XIV, 443 pages. 1998. (Subseries LNAI).

Vol. 1422: J. Jeuring (Ed.), Mathematics of Program Construction. Proceedings, 1998. X, 383 pages. 1998.

Vol. 1423: J.P. Buhler (Ed.), Algorithmic Number Theory. Proceedings, 1998. X, 640 pages. 1998.

Vol. 1424: L. Polkowski, A. Skowron (Eds.), Rough Sets and Current Trends in Computing. Proceedings, 1998. XIII, 626 pages. 1998. (Subseries LNAI).

Vol. 1425: D. Hutchison, R. Schäfer (Eds.), Multimedia Applications, Services and Techniques – ECMAST'98. Proceedings, 1998. XVI, 532 pages. 1998.

Vol. 1427: A.J. Hu, M.Y. Vardi (Eds.), Computer Aided Verification. Proceedings, 1998. IX, 552 pages. 1998.

Vol. 1430: S. Trigila, A. Mullery, M. Campolargo, H. Vanderstraeten, M. Mampaey (Eds.), Intelligence in Services and Networks: Technology for Ubiquitous Telecom Services. Proceedings, 1998. XII, 550 pages. 1998.

Vol. 1431: H. Imai, Y. Zheng (Eds.), Public Key Cryptography. Proceedings, 1998. XI, 263 pages. 1998.

Vol. 1432: S. Arnborg, L. Ivansson (Eds.), Algorithm Theory – SWAT '98. Proceedings, 1998. IX, 347 pages. 1998.

Vol. 1433: V. Honovar, G. Slutzki (Eds.), Grammatical Inference. Proceedings, 1998. X, 271 pages. 1998. (Subseries LNAI).

Vol. 1435: M. Klusch, G. Weiß (Eds.), Cooperative Information Agents II. Proceedings, 1998. IX, 307 pages. 1998. (Subseries LNAI).

Vol. 1436: D. Wood, S. Yu (Eds.), Automata Implementation. Proceedings, 1997. VIII, 253 pages. 1998.

Vol. 1437: S. Albayrak, F.J. Garijo (Eds.), Intelligent Agents for Telecommunication Applications. Proceedings, 1998. XII, 251 pages. 1998. (Subseries LNAI).

Vol. 1438: C. Boyd, E. Dawson (Eds.), Information Security and Privacy. Proceedings, 1998. XI, 423 pages. 1998.

Vol. 1439: B. Magnusson (Ed.), Software Configuration Management. Proceedings, 1998. X, 207 pages. 1998.

Vol. 1441: M. Pagnucco, W. Wobcke, C. Zhang (Eds.), Agents and Multi-Agent Systems. Proceedings, 1997. XII, 241 pages. 1998. (Subseries LNAI).

Vol. 1444: K. Jansen, J. Rolim (Eds.), Approximation Algorithms for Combinatorial Optimization. Proceedings, 1998. VIII, 201 pages. 1998.

Vol. 1445: E. Jul (Ed.), ECOOP'98 – Object-Oriented Programming. Proceedings, 1998. XII, 635 pages. 1998.

Vol. 1446: D. Page (Ed.), Inductive Logic Programming. Proceedings, 1998. VIII, 301 pages. 1998. (Subseries LNAI).

Vol. 1448: M. Farach-Colton (Ed.), Combinatorial Pattern Matching. Proceedings, 1998. VIII, 251 pages. 1998.